Macroscopic Quantum Coherence and Quantum Computing

Macroscopic Quantum Coherence and Quantum Computing

Edited by

Dmitri V. Averin
State University of New York at Stony Brook
Stony Brook, New York

Berardo Ruggiero
Instituto di Cibernetica
Naples, Italy

and

Paolo Silvestrini
Instituto di Cibernetica
Naples, Italy

Produced under the auspices of the
Istituto Italiano Studi Filosofici

Kluwer Academic / Plenum Publishers
New York, Boston, Dordrecht, London, Moscow

Proceedings of the International Workshop on Macroscopic Quantum Coherence and Computing-MQC^2 Istituto Italiano per gli Studi Filosofici, held June 14–17, 2000, in Naples, Italy

ISBN 0-306-46565-5

©2001 Kluwer Academic / Plenum Publishers, New York
233 Spring Street, New York, N.Y. 10013

http://www.wkap.nl/

10 9 8 7 6 5 4 3 2 1

A C.I.P. record for this book is available from the Library of Congress.

Printed in the United States of America

PREFACE

This volume is an outgrowth of the Second International Workshop on Macroscopic Quantum Coherence and Computing held in Napoli, Italy, in June 2000. This workshop gathered a number of experts from the major Universities and Research Institutions of several countries. The choice of the location, which recognizes the role and the traditions of Naples in this field, guaranteed the participants a stimulating atmosphere.

The aim of the workshop has been to report on the recent theoretical and experimental results on the macroscopic quantum coherence of macroscopic systems. Particular attention was devoted to Josephson devices. The correlation with other atomic and molecular systems, exhibiting a macroscopic quantum behaviour, was also discussed. The seminars provided both historical overview and recent theoretical ground on the topic, as well as information on new experimental results relative to the quantum computing area.

The first workshop on this topic, held in Napoli in 1998, has been ennobled by important reports on observations of Macroscopic Quantum Coherence in mesoscopic systems. The current workshop proposed, among many stimulating results, the first observations of Macroscopic Quantum Coherence between macroscopically distinct fluxoid states in rf SQUIDs, 20 years after the Leggett's proposal to experimentally test the quantum behavior of macroscopic systems. Reports on observations of quantum behaviour in molecular and magnetic systems, small Josephson devices, quantum dots have also been particularly stimulating in view of the realization of several possible q-bits.

The present volume, far from being exhaustive, represents an interesting update of the subject and hopefully, a further stimulation. We hope that this will be a useful tool in promoting new experiments.

In conclusion, we wish to thank the Istituto Italiano per gli Studi Filosofici, the Istituto di Cibernetica del Consiglio Nazionale delle Ricerche, the State University of New York at Stony Brook, the Istituto Nazionale di Fisica Nucleare, the Istituto Nazionale di Fisica della Materia, the Dipartimento Scienze Fisiche dell' Università di Napoli "Federico II", and the Gruppo Nazionale di Fisica della Materia del CNR for their support. This initiative is within the scope of the MQC2 Association, Naples, Italy for international interchanges on "Macroscopic Quantum Coherence and Computing."

We are indebted to Valentina Corato for scientific assistance, and to E. Degrazia and A. M. Mazzarella for their valuable assistance in all the tasks connected to the organization of the Workshop.

We are also grateful to E. Esposito, C. Granata, L. Keller, A. Monaco, and S. Piscitelli for hints and help during the organization of the Workshop.

Thanks are due to L. De Felice and V. Sindoni for the organization of the social event.

Paolo Silvestrini

CONTENTS

MACROSCOPIC QUANTUM COHERENCE AND DECOHERENCE IN SQUIDS

A. J. Leggett

Department of Physics, University of Illinois at Urbana-Champaign, 1110 West Green Street, Urbana, IL 61801-3080, USA; E-mail: tony@cromwell.physics.uiuc.edu

Abstract   I consider the general problem of decoherence in Josephson systems, with particular reference to the question of how far we can reliably infer the degree of such decoherence, in an experiment of the MQC type, either by a priori order-of-magnitude arguments or from other types of experiment which may be easier at the initial stages of the program, such as measurements of dissipation in the quasiclassical regime or of tunnelling rates out of a metastablestate. Potential sources of decoherence which will be considered include normal electrons, phonons, nuclear spins, the radiation field and the "passing truck." It is concluded that the effects of all but the last can be reaosnably estimated, and that there are good prospects (confirmed by recent experiments) of reducing them to manageable levels. The implications for the use of suchsystems for practical quantum computation are briefly considered.

The goal of preparing a (reasonably) macroscopic object in a quantum-mechanical superposition of states which are (reasonably) macroscopically distinct, in such a way that it can be demonstrated that it must indeed have been in such a superposition rather than in a classical mixture ("MQC") is one which has been consciously pursued for the last twenty years. Of course, what exactly is to count as "reasonably" macroscopically distinct, etc., is a legitimate matter of debate, but I believe more of us have a relatively well-defined intuitive notion of what we mean by the phrase. Thus, the elegant series of experiments in quantum optics carried out at the ENS (1) where the relevant superposition is of states of a cavity which differ in the behavior of ~\$10 photons, while a beautiful demonstration (inter alia) of the phenomenon of decoherence, would probably not qualify under the present heading, and one is forced to look at the messier and less well-characterized systems typical of condensed-matter physics. In fact, experiments directly or indirectly related to the question of interest have been carried out on magnetic systems (2), mesoscopic grains (3), dislocations (4) and other systems. However, the area in which this program has been most systematically pursued is that of electronic devices incorporating the Josephson effect (SQUIDS),[1] and very recently a milestone in this area has been passed with a couple of experiments (5,

---

[1]For brevity I will use the term "SQUID" to refer generically to a system of this type, even though it is sometimes not technically correct.

6) which give very strong evidence that MQC has indeed been obtained in such systems. In this paper I will consider specifically the case of SQUIDS, though many of my remarks apply equally to the other systems mentioned above.

What distinguishes SQUIDS from these other candidate systems for MQC? Two features in particular. First, in distinction to most of the latter, the "ensembles" we are dealing with in the experiments correspond to a <u>single</u> physical system subjected to repeated trials under the same conditions; by contrast, in an experiment to look for MQC in (e.g.) biomolecules, we typically measure the summed response of a large number of independent and equivalent systems; while it is certainly possible to look for the effects of (e.g.) quantum tunneling out of a metastable well at the single-grain or single-molecular level (see, e.g., (7)), to the best of my knowledge evidence for MQC as such has not been claimed at this level.[1] Although this "time-ensemble" aspect is of no special relevance so long as one is interested only in verifying that the predictions of quantum mechanics continue to work at the relevant level of "macroscopicness," it becomes crucial both in the context of quantum computing and for tests (8) of alternative classes of "macrorealistic" theories which compete with quantum mechanics at this level.

The second feature which distinguishes the SQUID system is simply that the physics of the relevant degree of freedom (in this case the trapped flux or circulating current) is believed to be extremely well understood at the classical level, to the extent that all the parameters relevant to MQC can be independently measured in experiments which stay within the (quasi-) classical regime; see below. By contrast, in most other candidate systems there are a number of parameters and/or effects which can affect the MQC behavior quantitatively, and which have to be guessed at since independent measurement is not usually possible.

In the remainder of this paper I review briefly the question of the extent to which one can use independently obtained information to predict quantitatively the MQC behavior of SQUID systems, with special reference to effects which are likely to lead to decoherence. This is a topic on which I and others have written extensively in the past (see in particular ref. (9)), so in most cases I will simply refer to this literature for the quantitative details of the argument. For the purposes of illustration I shall consider the very simplest system relevant to the observation of MQC, namely a single ring closed with a single Josephson junction ("of SQUID") and subject to an external flux Φ_x which for present purposes I treat as a classical control parameter.

For such a system the interacting dynamical variable is the trapped flux Φ (or equivalently the circulating current I); the "macroscopically distinct" states whose quantum superposition we seek to produce correspond to values of Φ which differ by some substantial fraction of the flux quantum $\varphi_o \equiv h/2e$. The very simplest model of the dynamics of Φ treats the latter as completely analogous to the position variable x of a particle moving in one dimension; the role of particle mass is played by the effective capacitance C of the junction (for a detailed

---

[1]In this respect, however, experiments on mesoscopic (Coulomb-blockade) grains (3) are similar to those on SQUIDs.

discussion of the meaning of this concept, see ref. (9)), and the "potential" U(Φ) in which it moves is a sum of a self-inductance term of the form $(\Phi - \Phi_x)^2/2L$ (L = self-inductance of ring) and a coupling term of the form $-I_c\varphi_0/2\pi\Phi/\varphi_0$) due to the Josephson junction. For the observation of MQC, we need to adjust the parameters L, I_c and Φ_x so that U(Φ) has two nearly degenerate minima. This simple model, when supplemented by phenomenological terms describing dissipation (cf. below), appears to describe rather well the dynamics of the flux in the "classical" regime, that is the regime where we expect the effects of treating Φ as a quantum-mechanical operator to be unimportant; see e.g. ref. (10).

To extend the description to the quantum regime, the standard procedure is to impose a canonical commutation relation between the dynamical "coordinate" Φ and its conjugate momentum $p_\Phi \equiv \partial\mathcal{L}/\partial\dot{\Phi}$ (where $\mathcal{L}$ is the Lagrangian). It is not entirely obvious a priori that this procedure is valid; for a justification, see ref. (9), section 3. If one accepts it, one has in place of Newton's equation a Schrödinger equation for the quantum-mechanical probability amplitude $\Psi(0\Phi)$ for the system to possess a given value of Φ, and can begin discussing, at least schematically, superpositions of the type envisaged in the MQC experiment.

So far, the discussion has proceeded by analogy with the description of a single isolated one-particle system. However, an important property which distinguishes the "mascroscopic" coordinates characteristic of condensed-matter systems from the simple one-particle coordinates whose motion is described in textbooks of quantum mechanics (and to a very considerable degree also from the systems which are the subject of quantum optics) is that the former are inevitably coupled strongly to other degrees of freedom. More specifically, the idea of writing down a closed Hamiltonian (or Lagrangian) for the flux Φ in terms only of Φ and its time derivative, without reference to any of the myriad other variables characterizing the physical system, is clearly wholly unviable. This is already seen in experiments in the classical regime, which demonstrate unambiguously that the motion of the flux involves substantial dissipation. Thus it is necessary to build into the description from the start some account of the interaction of the flux with all the other relevant degrees of freedom (in accordance with the terminology which has become standard in the literature, I shall refer to the latter generically as the "environment"). From general notions concerning the effects of system-environment interaction (see e.g. ref. (11), one would expect that in the quantum regime one important effect of the latter would be to "decohere" (reduce to classical mixtures) superpositions of states corresponding to appreciably different values of Φ, and thereby to tend to wash out the phenomenon of MQC. It is therefore very important to develop methods to estimate the degree of decoherence ahead of time.

Let's first briefly list and comment on some important components of the "environment." Crudely speaking, any other physical degree of freedom which "notices" the value of the flux will tend, prima facie, to induce decoherence. Since the dynamics of the (isolated) flux is associated entirely with the superconducting fraction of the electrons (Cooper pairs), one obvious ingredient in the environment is the unpaired (normal) electrons. Perhaps surprisingly, in the conditions of a real-life experiment this turns out to be a negligible source of decoherence, for the simple reason that if we consider, say a 1 cm³ block of Nb($T_c \cong$ 9K) in thermal equilibrium at 10 mK, the average number (not fraction!) of normal electrons is much less than one! However, this statement refers to the bulk metal, and it is much less obvious that there are no normal electrons in the metallic "shorts" which may shunt the junction (see below). Other possible sources of decoherence intrinsic to the system include the ionic motion (phonons) and the nuclear spins. The most obvious "external" source of decoherence is the blackbody radiation field (which for everyday objects at room temperature is probably the overwhelmingly dominant mechanism of decoherence, see ref. (11)). However, in real life, as experimentalists know to their cost, sources of noise which one would tend to think of as purely classical, such as the 50 Hz electromagnetic background, are ubiquitous, and it turns out that in the context of the MQC experiment and related areas they are at least as efficient in destroying superpositions[1] as intrinsically quantum-mechanical sources of noise; I refer to this as the "passing truck" problem.

To estimate a crude order of magnitude for the effectiveness or not of those various sources of decoherence, we can use the following rough-and-ready procedure: Suppose that the "MQC rate" (that is, the tunneling splitting between the even- and odd-parity states of the iolated system in the double-well potential) is $\hbar\omega_o$. Consider now that part of the energy difference ΔE between the two states of the superposition which derives from the coupling to the relevant part of the environments (As an example, in the case of nuclear spins ΔE would be of the order of $(\Delta\Phi M/A)$ where A is the area of the SQUID ring. M is the relevant nuclear spin magnetization and $\Delta\Phi \sim \sim_o$ is the difference in flux between the superposed states). If the root-mean-square fluctuations of Δ in a bandwidth $\sim\omega_o$ are large compared to $\hbar\omega_o$ then this ingredient of the environment will provide an effective decoherence mechanism in the context of MQC; if on the other hand the fluctuations are $\ll \hbar\omega_o$, then it will be essentially irrelevant. Using this criterion it is easily seen that for typical SQUID parameters neither the ionic motion nor the radiation field is an effective source of decoherence. This also turns out to be true, though somewhat less obvious, for the nuclear spins (in this case it is necessary to take into account both the frequency spectrum of the nuclear spin dynamics and the fact that the "relevant" nuclear spins are only those lying within a London penetration depth of the surface of the ring, since the rest feel no magnetic field). At first sight these results may be surprising, since they contradict a long-held dogma in the quantum measurement literature to the effect that by mere virtue of possessing very closely spaced energy levels a macroscopic body must be overwhelmingly prone to decoherence; for the refutation of this claim, see ref. (12).

Unfortunately, the intrinsic mechanism of decoherence which is likely to be dominant in real life, namely the effect of conduction in the normal-metal shorts shunting the junction (or, in a more complicated geometry in the external leads) cannot readily be estimated a priori by the above technique. It is serendipitous that

---

[1]Strictly speaking, the "passing truck", if described as a classical noise source, does not exactly destroy the superposition but injects a random relative phase into its components, which in the context of predictions of physical quantities comes to the same thing.

in this case (and, in fact, also in the case of most of the mechanisms discussed above) we have an alternative way of estimating the decoherence (and more generally the effects of the environment on quantum-mechanical processes) which relies not on a priori estimates of the coupling strength but on the experimentally observed strength of dissipation (and other properties) in the quasi-classical regime. Specifically, it can be shown that provided the system-plus environment complex satisfies a few very general and plausible conditions, a knowledge of the classical (in general dissipative) equation of motion of the flux uniquely determines its quantum-mechanical behavior. This conclusion is established, for the case of a general linear impedance mechanism, in ref. (13); while it has obviously not been possible up to now to test it in the context of MQC, it has been tested (14) in the context of the related problem of quantum tunneling out of a metastable well ("MQT"), and has correctly predicted the suppression of tunneling by dissipation without any fitted parameters. Thus there seems no obvious reason to doubt the predictions for the case of MQC, with however the caveat that since the frequency regime relevant to the latter phenomenon is different by several orders of magnitude from that for MQT, one cannot necessarily simply take on values of the dissipation parameters (etc.) measured in the latter context but may have to redetermine them from scratch. In the simple case in which the classical dissipation in the frequency regime relevant to MQC has a "simple ohmic" form corresponding to a constant resistance R, the upshot is that the decoherence in MQC is proportional to the dimensionless parameter $\alpha \equiv (\Delta\Phi)^2/2\pi R\hbar$, in such a way that the clear observation of simplest form of MQC would require say $\alpha \lesssim 10^{-2}$. For $\Delta\Phi \sim \varphi_o$ this condition is equivalent to $R \gtrsim 100$ kΩ, which is much less than values obtained (in the MQT regime) in existing experiments. Thus there appears no obvious a priori obstacle to the observation of MQC in SQUIDS, as has indeed been verified in recent experiments.(5, 6).

In the light of the above considerations, my guess is that if a serious program to explore the use of SQUIDS or related devices as elements of a quantum computer gets under way, the main long-term difficulties will come not so much from intrinsic decoherence mechanisms as from the "passing-truck" problem and (a somewhat related point) the external electronics necessary to perform the sophisticated manipulations involved in such a computer. Whether and how this kind of difficulty can be overcome is likely to be very much a matter of trial and error, and not something to which the theorist as such is likely to be able to contribute much fundamental input.

ACKNOWLEDGMENT. This research was supported by the MacArthur Chair endowed by the John D. and Catherine T. MacArthur Foundation at the University of Illinois.

REFERENCES

1. M. Brune, E. Hagley, J. Dreyer, X. Maitre, A. Maali, C. Wunderlich, J-M. Raimond and S. Haroche, Phys. Rev. Letters 77, 4887 (1996).

2. S. Gider, D. D. Awschalom, T. Douglas and M. Chaparala, Science 268, 77 (1995).
3. Y. Nakamura, Y. A. Pashkin and J. S. Tsai, Nature 398, 786 (1999).
4. A. Hikata and C. Elbaum, Phys. Rev. Letters 54, 2418 (1985).
5. J. R. Friedman, V. Patel, W. Chen, S. K. Tolpygo and J. E. Lukens, cond-mat/0004293.
6. H. Mooij et al., talk presented at APS March Meeting, Minneapolis, March 2000.
7. W. Wernsdorfer, E. B. Orozco, K. Hasselbach, A. Benoit, D. Mailly, O. Kubo, H. Nakano and B. Barbara, Phys. Rev. Letters 79, 4014 (1997).
8. A. J. Leggett and A. Garg, Phys. Rev. Letters 54, 857 (1985).
9. A. J. Leggett, in *Chance and Matter* (Les Houches Session XLVI), ed. J. Souletie, J. Vannimenus and R. Stora, North-Holland, Amsterdam 1987.
10. A. Barone and G. Paterno, Physics and Applications of the Josephson Effect, Wiley, New York 1982.
11. E. Joos and H. D. Zeh, Z. Phys. B 59, 223 (1985).
12. A. J. Leggett, in *Applications of Statistical and Field Theory Methods to Condensed Matter*, eds. D. Baeriswyl, A. R. Bishop and J. Carmelo, NATO ASI series B, Physics vol. 218, Plenum, New York 1990.
13. A. J. Leggett, Phys. Rev. B 30, 1208 (1984).
14. J. M. Martinis, M. H. Devoret and J. Clarke, Phys. Rev. B 35, 4682 (1987).

MACROSCOPIC QUANTUM COHERENCE IN AN RF-SQUID

Jonathan R. Friedman,* Vijay Patel, W. Chen, S. K. Tolpygo and J. E. Lukens
Department of Physics and Astronomy, The State University of New York at Stony Brook, Stony Brook, NY 11794-3800 USA

Abstract   We present experimental evidence for a coherent superposition of macroscopically distinct flux states in an rf-SQUID. When the external flux Φ_x applied to the SQUID is near 1/2 of a flux quantum Φ_0, the SQUID has two nearly degenerate configurations: the zero- and one-fluxoid states, corresponding to a few microamperes of current flowing clockwise or counterclockwise, respectively. The system is modeled as a particle in a double-well potential where each well represents a distinct fluxoid state (0 or 1) and the barrier between the wells can be controlled *in situ*. For low damping and a sufficiently high barrier, the system has a set of quantized energy levels localized in each well. The relative energies of these levels can be varied with Φ_x. External microwaves are used to pump the system from the well-localized ground state of one well into one of a pair of excited states nearer the top of the barrier. We spectroscopically map out the energy of these levels in the neighborhood of their degeneracy point by varying Φ_x as well as the barrier height. We find a splitting between the two states at this point, when both states are below the classical energy barrier, indicating that the system attains a coherent superposition of flux basis states that are macroscopically distinct in that their mean fluxes differ by more than $1/4\ \Phi_0$ and their currents differ by several microamperes.

Can macroscopic objects be put into a quantum superposition of macroscopically distinct states? This question lies at the heart of Schrödinger's so-called cat paradox, a thought experiment in which a cat is put in a quantum superposition of alive and dead states [1]. The cat in Schrödinger's paradox is "macroscopic" in two senses: 1) it is a large, complex object comprising many microscopic degrees of freedom and 2) it is (supposedly) in a superposition of states (alive and dead) that are distinct in the "every-day world" in which quantum effects are

---

*Corresponding author. Electronic mail: jonathan.friedman@sunysb.edu

usually suppressed. In the early 1980s Leggett and co-workers [2, 3, 4] proposed that under suitable conditions such a macroscopic object could behave quantum mechanically provided that it was sufficiently decoupled from its environment. The canonical example of such a macrosocpic system has been the supercon-ducting quantum interference device (SQUID), which can be macroscopic in both of the above senses: 1) it may have many (billions of) Cooper pairs in the superconducting condensate, all acting in tandem and 2) its flux states differ in mean flux by a substantial fraction of a flux quantum Φ_0 and differ in current by several μA. Much progress has been made in demonstrating the macroscopic quantum behaviour of SQUIDs [5, 6] as well as other superconducting systems [7, 8, 9], nanoscale magnets [10, 11, 12], laser-cooled trapped ions [13], pho-tons in a microwave cavity [14] and C_{60} molecules [15]. However, heretofore there has been no experimental demonstration of a quantum superposition of truly macroscopically distinct states. Here we present experimental evidence that a SQUID can be put into a superposition of two macroscopically distinct states, a phenomenon known as Macroscopic Quantum Coherence (MQC).

The simplest SQUID (the rf-SQUID) is a superconducting loop of inductance L broken by a Josephson tunnel junction with capacitance C and critical current I_c. In equilibrium, a dissipationless supercurrent can flow around this loop, driven by the difference between the flux Φ that threads the loops and the external flux Φ_x applied to the loop. The dynamics of the SQUID can be described in terms of the variable Φ and are analogous to those of a particle of "mass" C (and kinetic energy $1/2C\dot{\Phi}^2$) moving in a one-dimensional potential (Figure 1a) given by the sum of the magnetic energy of the loop and the Josephson coupling energy of the junction:

$$U = U_0 \left[\tfrac{1}{2} \left(\frac{2\pi(\Phi - \Phi_x)}{\Phi_0} \right)^2 - \beta_L \cos(2\pi\Phi/\Phi_0) \right], \qquad (1.1)$$

where $U_0 \equiv \Phi_0^2/4\pi^2 L$ and $\beta_L \equiv 2\pi L I_c/\Phi_0$. For the parameters used in our experiment, this is a double-well potential separated by a barrier with a height that depends on I_c. When $\Phi_x = \Phi_0/2$ the potential is symmetric. Any change in Φ_x then tilts the potential, as shown in Figure 1a.

Each well can be labelled by the quantum number f, which defines the "flux-oid" state of the SQUID.[1] In Figure 1a, the left (right) well corresponds to f = 0 (1) in which a static current (>1 μA for our system) flows around the loop in such a way as to tend to cancel (augment) Φ_x. Classically, a transition between the f = 0 and f = 1 wells involves passage over the top of the barrier. Quantum mechanically, however, the system can tunnel through the barrier. For weak damping, the system has quantized energy levels that, well below the barrier, are localized in each well. At various values of Φ_x, levels in opposite wells will align, giving rise to resonant tunnelling between the wells [5]. During this interwell transition Φ changes by some fraction of Φ_0, or - stated in different terms - the magnetic moment of the system changes by a macroscopic amount, in our case over $10^{10} \mu_B$. Until now, however, there has been no evidence that the tunnelling process between these macroscopically distinct states could be coherent, that is, that the SQUID could be put into a coherent superposition of two flux states in different wells.

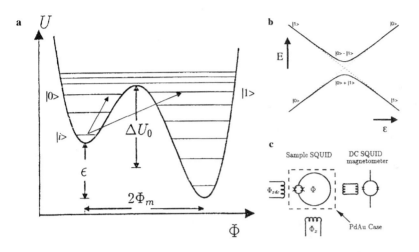

Figure 1.1   a) SQUID potential. The left well corresponds to the zero-fluxoid state of the SQUID and the right well to the one-fluxoid state. Energy levels are localized in each well. Both the tilt ε and energy barrier at zero tilt ΔU_0 can be varied *insitu* in the experiments. The process of photon-induced interwell transitions is illustrated by the arrows, where the system is excited out of the initial state $|i\rangle$ and into one of two excited states $|0\rangle$ or $|1\rangle$. b) Schematic anticrossing. When the two states $|0\rangle$ and $|1\rangle$ would become degenerate in the absence of coherence, the degeneracy is lifted and the states of the system are the symmetric and antisymmetric superpositions of the flux-basis states: $\frac{1}{\sqrt{2}}(|0\rangle + |1\rangle)$ and $\frac{1}{\sqrt{2}}(|0\rangle - |1\rangle)$. c) Experimental set-up. Our SQUID contains a "tuneable junction", a small dc-SQUID. A flux Φ_{xdc} applied to this small loop tunes the barrier height ΔU_0. Another flux Φ_x tunes the tilt ε of the potential. A separate dc-SQUID acts as a magnetometer, measuring the flux state of the sample. The sample SQUID used in our experiments is characterised by the following three energies: the charging energy $E_c \equiv e^2/2C = 9.0mK$, the inductive energy $E_L \equiv \Phi_0^2/2L = 645K$ and a tuneable Josephson coupling energy $E_J \equiv (I_c\Phi_0/2\pi)\cos(\pi\Phi_{xdc}/\Phi_0) = 76K \cos(\pi\Phi_{xdc}/\Phi_0)$. The plasma frequency ω_J (the frequency of small oscillations in the bottom of a well) associated with these parameters is 1.5-1.8×10^{11} rad/s, depending on the value of Φ_{xdc}. The fact that $E_c \ll E_L, E_J$ confirms that flux is the proper basis to describe the SQUID's dynamics.

Such a superposition would manifest itself in an anticrossing, as illustrated in Figure 1b, where the energy-level diagram of two levels of different flux-oid states (labelled $|0\rangle$ and $|1\rangle$) is shown in the neighbourhood in which they would become degenerate without coherent interaction (dashed lines). Coherent tunnelling lifts the degeneracy (solid lines) so that at the degeneracy point the energy eigenstates are $\frac{1}{\sqrt{2}}(|0\rangle + |1\rangle)$ and $\frac{1}{\sqrt{2}}(|0\rangle - |1\rangle)$, the symmetric and anti-symmetric superpositions. The energy difference ΔE between the two states is approximately given by $\Delta E = \sqrt{\varepsilon^2 + \Delta^2}$, where Δ is known as the tunnel splitting. The goal of the present work is to demonstrate the existence of such a splitting and, thereby, the coherent superposition of macroscopically distinct flux states. A necessary condition for resolving this splitting is that the experimental linewidth of the states be smaller than Δ [16]. The SQUID is extremely sensitive to external noise and dissipation (including that due to the measurement of Φ), both of which broaden the linewidth. Thus, the experimental challenges to observing MQC are severe. The measurement apparatus must be weakly coupled to the system to preserve coherence while the signal strength must be sufficiently large to resolve the closely spaced levels. In addition, the system must be well shielded from external noise. These challenges have frustrated previous attempts [5, 6] to observe coherence in SQUIDs.

The SQUID used in these experiments is made up of two Nb/AlOx/Nb tunnel junctions in parallel, as shown in Figure 1c; this essentially acts as a tuneable junction in which I_c can be adjusted with an applied flux Φ_{xdc}: $I_c = I_{c0} \cos(\pi \Phi_{xdc}/\Phi_0)$, where I_{c0} is the critical current when $\Phi_{xdc} = 0$. Thus, with Φ_x we control the tilt ε of the potential in Figure 1a, while with Φ_{xdc} we control ΔU_0, the height of the energy barrier at $\varepsilon = 0$. The flux state of our sample is measured by a separate dc-SQUID magnetometer inductively coupled to the sample. The sample is encased in a PdAu shield that screens it from unwanted radiation; a coaxial cable entering the shield allows the application of controlled external microwaves. The set-up is carefully filtered and shielded, as described elsewhere, [5, 6] and cooled to ≈ 40 mK in a dilution refrigerator.

In our experiments, we probe the anticrossing of two excited levels in the potential by using microwaves to produce photon-assisted tunnelling. Figure 1a depicts this process for the case where the levels $|0\rangle$ and $|1\rangle$ are each localized in opposite wells. The system is initially prepared in the lowest state in the left well (labelled $|i\rangle$) with the barrier high enough that the rate for tunnelling out of $|i\rangle$ is negligible on the time scale of the measurement. Microwave radiation is then applied. When the energy difference between the initial state and an excited state matches the radiation frequency, the system has an appreciable probability of being excited into this state and subsequently decaying into the right well. This transition between wells results in a change in flux that can be detected by the magnetometer. Figure 2 shows the photon-assisted process when the excited levels are coherent. From the SQUID's parameters (see below), it is straightforward to numerically diagonalize the SQUID Hamiltonian and solve for the energy levels of the system in the zero-damping (completely coherent) limit. In the figure, calculated levels (for $\Delta U_0 = 9.117K$) are plotted as a function of Φ_x (thin solid lines). The calculated top of the barrier is indicated by the thick solid line. The dot-dashed line represents level $|i\rangle$ shifted upward by the energy of the microwaves. At values of Φ_x for which this line intersects one of the excited levels (indicated by the arrows), the system can absorb a photon and make an interwell transition. When the barrier is reduced (to $\Delta U_0 = 8.956K$), the excited levels and top of the barrier move to lower energy relative to $|i\rangle$ (dotted lines in the figure) and photon absorption occurs at different values of Φ_x. For a fixed frequency,[2] we can map out the anticrossing by progressively reducing the barrier and thus moving the levels through the dashed photon line.

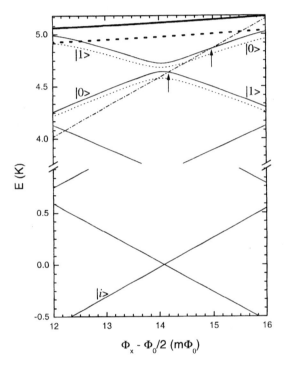

Figure 1.2 Calculated energy levels and the photon-assisted tunnelling process. The thin solid lines represent the calculated SQUID energy levels as a function of Φ_x for a barrier height of $\Delta U_0 = 9.117K$. The thick solid line is the calculated top of the energy barrier. The 96.0 GHz (4.61 K) applied microwaves boost the system out of the initial state $|i\rangle$, bringing it virtually to the dot-dashed line. At certain values of Φ_x for which this line intersects one of the excited states (indicated by the arrows), a photon is absorbed and the system has a large probability of making an interwell transition. When ΔU_0 is reduced (to 8.956 K), the levels and top of the barrier move down relative to $|i\rangle$ (dotted lines), changing the values of Φ_x at which photon absorption occurs. All energies are calculated relative to the mean energy of $|i\rangle$ and the lowest state of the right well; for clarity, the zero of energy is shifted to the point where the levels in lower part of the figure cross.

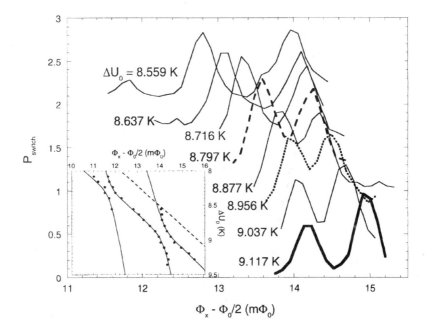

Figure 1.3   Experimental data. The main figure shows, as a function of Φ_x, the probability of making an interwell transition when a millisecond pulse of 96-GHz microwave radiation is applied. For clarity, each curve is shifted vertically by 0.3 relative to the previous one. As the energy barrier is reduced, the two observed peaks move closer together and then separate: the signature of an anticrossing. The inset shows the position of the observed peaks in the $\Delta U_0 - \Phi_x$ plane. Also shown is the calculated locus of points at which the virtual photon level (dot-dashed line in Figure 2) intersects an excited level (solid lines) or the top of the classical energy barrier (dashed line).

We use pulsed microwaves to excite the system to the upper levels. Before each pulse, the system is prepared in state $|i\rangle$ and the values of ΔU_0 and ε are set. Millisecond pulses of 96-GHz microwave radiation at a fixed power are applied and the probability of making a transition is measured. The experiment is repeated for various values of ε and ΔU_0. Data from these measurements are shown in Figure 3, where the probability of making a photon-assisted interwell transition is plotted as a function of Φ_x. Each curve, for a given ΔU_0, is shifted vertically for clarity. Two sets of peaks are clearly seen. As ΔU_0 is decreased, these peaks move closer together and then separate without crossing. For ΔU_0 = 9.117 K (thick solid curve), the right peak roughly corresponds to level $|0\rangle$, which is localized in the same well as $|i\rangle$ (compare with Figure 2). The relative amplitude of the two peaks is due to the asymmetry of the potential (for this data, $|0\rangle$ is the fourth excited level in the left well while $|1\rangle$ is the tenth excited level in the right well) and is in accordance with recent calculations [16]. When ΔU_0 is decreased to 8.956 K (dotted curve), the peaks move closer together and the asymmetry disappears. The two states are now in the coherent regime and correspond approximately to the symmetric and anti-symmetric superpositions of the $|0\rangle$ and $|1\rangle$ states. As the barrier is decreased further (8.797 K - dashed curve), the peaks move apart again and the asymmetry reappears, now with the left peak being larger and corresponding to $|0\rangle$. The two levels have thus passed through the anticrossing, changing roles without actually intersecting. The inset shows the positions of the peaks in the main figure (as well as other peaks) in the $\Delta U_0 - \Phi_x$ plane; two anticrossings are clearly seen. The solid (dashed) lines in the inset represent the locus of points when the calculated energy levels (top of the barrier) are 96 GHz above the state $|i\rangle$. All of our data lie to the left of the dashed line and, therefore, correspond to levels that are below the top of the barrier. Hence, the flux-basis states $|0\rangle$ and $|1\rangle$ are macroscopically distinct with mean fluxes that, we calculate, differ by about $1/4 \, \Phi_0$. Each observed anticrossing thus represents the coherent superposition of macroscopically distinct states.

For one of the anticrossings, Figure 4 shows the energy of the levels $|0\rangle$ and $|1\rangle$ as a function of ε relative to their mean energy, $E_{mean} (\Delta U_0, \Phi_x)$, calculated for each experimental point using the parameters listed below. This makes the data manifestly similar to Figure 1b. At the middle of the anticrossing, the two levels have a tunnel splitting Δ of $\approx 0.1 K$ in energy while the upper level is still $\approx 0.15 K$ below the top of the classical energy barrier.

There are three parameters used for the calculations presented in Figures 3 and 4: L, $Z \equiv \sqrt{L/C}$ and β_L, all of which can be independently determined from measurements of classical phenomena or incoherent resonant tunnelling in the absence of radiation. From these independent measurements, we find $L = 240 \pm 15$ pH, $Z = 48.0 \pm 0.1 \Omega$ and $\beta_L = 2.33 \pm 0.01$. The values used in the calculation that yielded the best agreement with the data are $L = 238$ pH, $Z = 48 \, \Omega$ and $\beta_L = 2.35$, all in good agreement with the independently determined values.

In summary, we have spectroscopically observed an anticrosing – the signature of a quantum superpostion – in two excited states of an rf-SQUID. Thus, we have provided the first evidence for MQC in a SQUID. Our SQUID shows macroscopic quantum behaviour in both of the senses given above: 1) The quantum dynamics of the SQUID is determined by the flux through the loop, a collective coordinate representing the motion of $\approx 10^9$ Cooper pairs acting in tandem. Since the experimental temperature is ≈ 500 times smaller than the superconducting energy gap, almost all microscopic degrees of freedom are frozen out and only the collective flux coordinate retains any dynamical relevance. 2) The two flux-basis states that we find to be superposed are macroscopically distinct. We calculate that for the anticrossings measured, the states

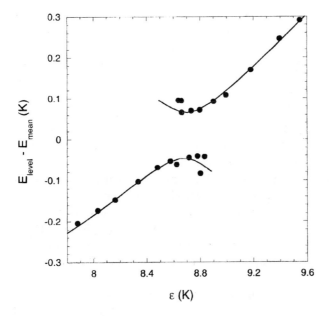

Figure 1.4 Energy of the measured peaks relative to the calculated mean of the two levels as a function of ε. At the midpoint of the figure, the measured tunnel splitting Δ between the two states in this anticrossing is $\approx 0.1K$ and the upper level is $\approx 0.15K$ below the top of the classical energy barrier. Calculated energy levels are indicated by the lines.

$|0\rangle$ and $|1\rangle$ differ in flux by more than $1/4\ \Phi_0$ and differ in current by 2 - 3 microamperes. Given the SQUID's geometry, this corresponds to a local magnetic moment of $\approx 10^{10}\mu_B$, a truly macroscopic moment.

Acknowledgments

We thank D. Averin and S. Han for many useful conversations, J. Männik, R. Rouse and A. Lipski for technical advice and/or assistance and M. P. Sarachik for the loan of some equipment. This work was supported by the US Army Research Office and the US National Science Foundation. All figures and some text reprinted by permission from Nature **406**, 43 (2000) copyright 2000 Macmillan Magazines Ltd.

Notes

1. As required by quantum mechanics, in the dissipationless state the phase of the macroscopic superconducting wave function must vary continuously around the loop, increasing by an integer f times 2π when one winds once around the loop.

2. A fixed frequency is used to ensure that the microwave power coupled to the sample remains constant for all resonances.

References

[1] E. Schrödinger, Die gegenwärtige situation in der quantenmechanik, *Naturwissenschaften*, **23**, 807–812, 823–828, 844–849 (1935).

[2] A.O. Caldeira and A.J. Leggett, Influence of dissipation on quantum tunneling in macroscopic systems, *Phys. Rev. Lett.*, **46**, 211–214 (1981).

[3] A.J. Leggett, S. Chakravarty, A.T. Dorsey, M.P.A. Fisher, A. Garg, and W. Zwerger, Dynamics of the dissipative 2-state system, *Rev. Mod. Phys.*, **59**, 1–85 (1987).

[4] U. Weiss, H. Grabert, and S. Linkwitz, Influence of friction and temperature on coherent quantum tunneling, *J. Low Temp. Phys.*, **68**, 213–244 (1987).

[5] R. Rouse, S. Han, and J.E. Lukens, Observation of resonant tunneling between macroscopically distinct quantum levels, *Phys. Rev. Lett.*, **75**, 1614–1617 (1995).

[6] R. Rouse, S. Han, and J.E. Lukens, in *Phenomenology of Unification from Present to Future*, G.D. Palazzi, C. Cosmelli, and L. Zanello, eds., 207–224 (World Scientific, Singapore, 1998).

[7] J. Clarke, A.N. Cleland, M.H. Devoret, D. Esteve and J.M. Martinis, Quantum mechanics of a macroscopic variable: the phase difference of a Josephson junction, *Science*, **239**, 992–997 (1988).

[8] P. Silvestrini, V.G. Palmieri, B. Ruggiero and M. Russo, Observation of energy level quantization in underdamped Josephson junctions above the classical-quantum regime crossover temperature, *Phys. Rev. Lett.*, **79**, 3046–3049 (1997).

[9] Y. Nakamura, Y.A. Pashkin and J.S. Tsai, Coherent control of macroscopic quantum states in a single-Cooper-pair box, *Nature*, **398**, 786–788 (1999).

[10] J.R. Friedman, M.P. Sarachik, J. Tejada and R. Ziolo Macroscopic measurement of resonant magnetization tunneling in high-spin molecules, *Phys. Rev. Lett.*, **76**, 3830–3833 (1996).

[11] E. del Barco, N. Vernier, J.M. Hernández, J. Tejada, E.M. Chudnovsky, E. Molins, and G. Bellessa, Quantum coherence in Fe_8 molecular nanomagnets, *Europhys. Lett.*, **47**, 722–728 (1999).

[12] W. Wernsdorfer, B. Bonet Orozco, K. Hasselbach, A. Benoit, D. Mailly, O. Kubo, H. Nakano, and B. Barbara, Macroscopic quantum tunneling of magnetization of single ferrimagnetic nanoparticles of barium ferrite, *Phys. Rev. Lett.*, **79**, 4014–4017 (1997).

[13] C. Monroe, D.M. Meekhof, B.E. King and D.J. Wineland, A "Schrödinger cat" superposition state of an atom, *Science*, **272**, 1131–1136 (1996).

[14] M. Brune, E. Hagley, J. Dreyer, X. Maitre, A. Maali, C. Wunderlich, J.M. Raimond and S. Haroche, Observing the progressive decoherence of the "meter" in a quantum measurement, *Phys. Rev. Lett.*, **77**, 4887–4890 (1996).

[15] M. Arndt, O. Nairz, J. Vos-Andreae, C. Keller, G. van der Zouw and A. Zeilinger, Wave-particle duality of C_{60} molecules, *Nature*, **401**, 680-682 (1999).

[16] D. Averin, J.R. Friedman and J.E. Lukens, Macroscopic resonant tunneling of magnetic flux, cond-mat/0005081 (2000).

QUANTUM-STATE INTERFERENCE IN A COOPER-PAIR BOX

Y. Nakamura
NEC Fundamental Research Laboratories
Tsukuba, Ibaraki 305-8501, Japan
yasunobu@frl.cl.nec.co.jp

Yu. A. Pashkin
CREST, Japan Science and Technology Corporation (JST)
Kawaguchi, Saitama 332-0012, Japan
pashkin@frl.cl.nec.co.jp

J. S. Tsai
NEC Fundamental Research Laboratories
Tsukuba, Ibaraki 305-8501, Japan
tsai@frl.cl.nec.co.jp

Keywords: Cooper-pair box, two-level system, interference, decoherence

Abstract Responses of an artificial two-level system in a Cooper-pair box to two sequential gate-voltage pulses have been investigated. We observed interference fringes due to the quantum phase accumulation during the delay time between the two pulses. The observed short decay time suggests that the effect of charge fluctuations on the dephasing is significant.

1.   INTRODUCTION

As a result of the development of quantum information theory, coherent control of a quantum system has attracted much more attention recently. Many kinds of two-level systems have been extensively studied as a candidate for a qubit, a basic unit of the quantum information [1].

A Cooper-pair box is one possible solid-state implementation of a qubit [2, 3, 4]. It has been experimentally confirmed that coherent su-

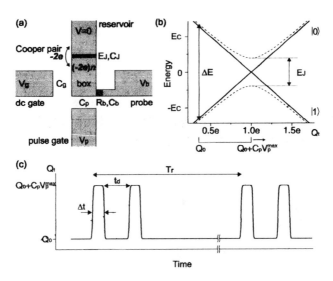

Figure 1 (a) Schematics of a Cooper-pair box. (b) Energy diagram of a Cooper-pair box as a function of Q_t, the total gate-induced charge in the box. The two-level system prepared in the ground state at Q_0 is brought by a sharp gate-voltage pulse to $Q_0 + C_p V_p^{max}$ (not necessarily equals $1e$) for the quantum-state control. In the experiment, we sweep Q_0 with a fixed pulse height V_p^{max}, and measure the dc current I through the probe junction for each Q_0. (c) Schematic shape of the two-pulse array.

perposition between two charge states exists [5, 6] as well as that the control of the superposed state is possible by using a high-speed gate-voltage pulse [7]. However, the mechanism and the time scale of the decoherence, which will probably be the most serious problem in the implementation of the qubit, have not been clarified yet. Although coherent oscillations of the quantum states have been observed up to a few nanoseconds [7], this short time scale is obviously not enough for quantum computing nor comparable to the theoretically expected decoherence time.

In this paper, we describe our experiment on quantum-state control by applying two sequential gate-voltage pulses on a Cooper-pair box. In a previous experiment with a single pulse [7], we have already observed quantum-state oscillations. However, the result did not necessarily mean that the phase coherence exists even *after* (not during) the pulse. In the present two-pulse experiment, on the other hand, we see the interference due to the phase accumulation between two non-degenerate states during the delay time, similarly to the Ramsey-fringe experiment in atom optics. Therefore, we can confirm the existence of the phase coherence after the first pulse. We also see the effect of decoherence at the off-resonant point, and this might be different from that at the resonance. Furthermore, at the off-resonant point, we can measure the energy-relaxation rate from the upper eigenstates by using our charge-sensitive measurement.

2.   EXPERIMENT

A Cooper-pair box has a small superconducting 'box' electrode coupled to an electrically-grounded reservoir electrode via a Josephson junction (Fig. 1(a)). One or more gate electrodes are also capacitively coupled to the box to control the electrostatic potential in the box by the gate voltage. The change of n can be restricted to one because of the charging effect, where n is the excess number of Cooper pairs in the box. The two charge states, that is, superconducting ground states with a fixed number of Cooper pairs in the box, for example, $|n = 0\rangle$ and $|1\rangle$, are coherently coupled by single-Cooper-pair tunneling due to the Josephson effect, and they can be used as a coherent two-level system. To measure the quantum state (see below), we also attach an additional voltage-biased 'probe' electrode to the box via a highly resistive tunnel junction. Cooper-pair tunneling at the probe junction can be ignored because of the large resistance R_b.

Figure 1(b) shows an energy diagram of the two-level system. Using the energy difference $\delta E(Q_t) = 4E_C(Q_t/e - 1)$ between the charge states (solid lines) and the Josephson energy E_J, we can write the 'spin' Hamiltonian of the Cooper-pair box as $H = \frac{1}{2}\delta E(Q_t)\sigma_z - \frac{1}{2}E_J\sigma_x$, where σ_z and σ_x are Pauli matrices, E_C is the single-electron charging energy of the box, $Q_t \equiv Q_0 + C_p V_p(t)$ is the total gate-induced charge in the box, and $Q_0 \equiv C_g V_g + C_b V_b$ is the static gate-induced charge. The energy difference $\Delta E(Q_t)$ between the eigenstates (dashed curves) equals $(\delta E(Q_t)^2 + E_J^2)^{1/2}$. In the quantum-state control with a single pulse [7], $\delta E(Q_t)$ is changed abruptly from $\delta E(Q_0)$ to $\delta E(Q_0 + C_p V_p^{\mathrm{max}})$ for a short pulse length Δt. This works, in the spin language, as a spin-rotating operation around a magnetic-field vector $(-E_J, 0, \delta E(Q_0 + C_p V_p^{\mathrm{max}}))$ during Δt. After the pulse, the final state will be a superposition of the two charge states that depends on Δt. The population (square of the coefficient in the wave function) of $|1\rangle$ is measured by using the quasiparticle tunneling through the probe junction. $|1\rangle$ is unstable against the quasiparticle tunneling and emits two electrons to relax into $|0\rangle$, while $|0\rangle$ is stable and does nothing. By taking an average over many pulse operations and subsequent relaxations of $|1\rangle$ repeated rapidly (but slowly compared to the relaxation of $|1\rangle$), we can accumulate the tunneling quasiparticle and measure dc probe-junction current I which is

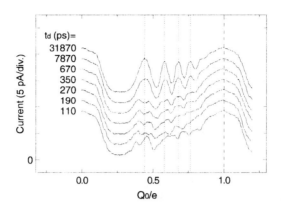

Figure 2 Current through the probe junction vs. Q_0, the static gate-induced charge in the box, for various t_d. The dotted lines are guides for the positions of current peaks in the topmost curve. Each curve is shifted by 2 pA for clarity.

proportional to the population of $|1\rangle$. Although the probe junction is always connected to the box, it does not disturb the coherence until the quasiparticle tunneling takes place. We therefore need a large R_b for the small tunneling rate.

In the two-pulse experiment, the array of twin pulses shown schematically in Fig. 1(c) were applied on a Cooper-pair box. For example, starting from $|0\rangle$ and if the first pulse works as a $(-\frac{\pi}{2})_x$-pulse, the quantum state right after the pulse is $\frac{1}{\sqrt{2}}(|0\rangle - i|1\rangle)$. During the delay time t_d at $Q_t = Q_0$ the quantum state acquires a phase factor and becomes $\frac{1}{\sqrt{2}}(|0\rangle - ie^{i\varphi}|1\rangle)$, where $\varphi = \Delta E(Q_0)t_d/\hbar$. The final state after the second $(-\frac{\pi}{2})_x$-pulse is $\frac{1}{2}\{(1 - e^{i\varphi})|0\rangle - i(1 + e^{i\varphi})|1\rangle\}$, in which the population of $|1\rangle$ $(= \cos^2(\varphi/2))$ and the measured current depend on the acquired phase φ. The pulse length Δt and the pulse-pattern repetition time T_r were fixed to be 80 ps and 64 ns, respectively. The delay time t_d was controlled by the combination of a discrete delay with an 80-ps step by the pulse-pattern generator and a continuous delay (0–100 ps) by a PC-controlled mechanical delay line.

The Cooper-pair box was fabricated by electron-beam lithography and shadow evaporation of Al films. The sample parameters were as follows: $E_J = 34 \pm 2$ μeV; $E_C = 122 \pm 3$ μeV; the superconducting-gap energy in the electrodes $\Delta = 235 \pm 10$ μeV; $R_b = 42.6$ MΩ. The probe electrode was voltage-biased at $V_b = 645$ μV. The details of the experimental setup and the measurement scheme have been described elsewhere [8].

3. RESULTS AND DISCUSSIONS

Figure 2 shows $I - Q_0$ curves for various t_d. When t_d is small, complicated curves due to the phase coherent effect are observed. We discuss this later in more detail. When t_d is longest, the curve is almost equivalent to that of a single-pulse experiment with $T_r = 32$ ns, where after each pulse the quantum state relaxes to the ground state almost completely before the next pulse arrives. The peak positions indicated by the dotted lines correspond to Q_0 where the single-pulse operation works nearly like a π-pulse that flips the ground state to the upper state which results in the current maximum. On the other hand, in the medium range of t_d, a dip, instead of a peak, is observed at those Q_0 values. This means that the second pulse comes before the relaxation of the upper state created by the first pulse and flips the state back to the ground state which results in the current minimum. Similarly, if we neglect the phase-coherent effect between the two pulses, the twin peaks accompanying the dip can be attributed to a $\frac{\pi}{2}$-pulse and a $\frac{3\pi}{2}$-pulse. The time scale of the change from the dip to the peak and that of the saturation of the peak height are consistent with the quasiparticle tunneling rate through the probe junction. This fact suggests that the energy relaxation of the two-level system was dominated by our measurement and that no faster relaxation processes existed.

When t_d is shorter, the phase coherence gives a significant effect on the observed current. Figure 3(a) shows $I - Q_0$ curves for $110 < t_d < 230$ ps. We observed periodic change of the curves as a function of t_d. For three different values of Q_0, labeled A, B and C, $I - t_d$ curves are plotted in Fig. 3(b). The oscillation period depends on Q_0 and agrees, within our experimental accuracy, with $h/\Delta E(Q_0)$, indicated by the length of the horizontal bar in Fig. 3(b). Therefore, the oscillations are attributed to the interference fringe due to the phase accumulation during the delay time between the two pulses, that is, $\varphi = \Delta E(Q_0)t_d/\hbar$. The complicated features in Fig. 3(a) could be a result of the convolution of the quantum-state transformation by the pulses and the phase evolution during the delay time, though they are not yet analyzed completely.

Such oscillations corresponding to point A decay quite rapidly as shown in Fig. 3(c). The decay rate is not significantly different for points B and C, and it is higher than that of coherent oscillations observed in a single-pulse control at the resonance; for the same set of parameters, the coherent oscillations were observed up to about 5 ns. Anyway, these time scales are much shorter than the expected decoherence time (~ 100 ns for this particular sample) due to the dissipative electromagnetic environment around the Cooper-pair box [4]. They are still shorter than

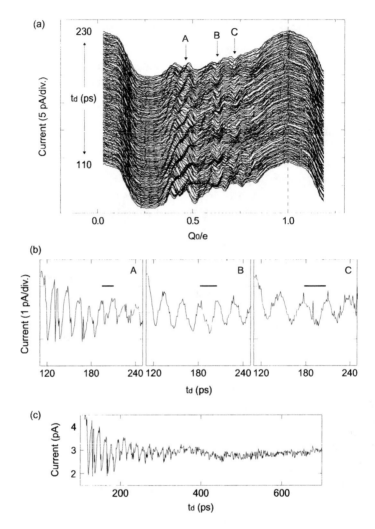

Figure 3 (a) Current through the probe junction vs. Q_0 for various t_d from 110 to 230 ps with a 1-ps step. Each curve is shifted by 0.2 pA for clarity. (b) Cross-sections of (a) at three points labeled A, B, and C. At each point, current shows oscillations as a function of t_d. The length of the horizontal bar indicates the expected oscillation period $h/\Delta E(Q_0)$. (c) Decay of the oscillation in a longer time scale at point A.

the limitation of the coherence caused by the quantum-state measurement. The measurement, which is based on the quasiparticle tunneling, destroys the coherence at the same rate as the tunneling $(\sim (9 \text{ ns})^{-1})$.

One possible explanation of the short decay time is the dephasing due to background charge fluctuations. It has recently been suggested in a single-electron pump experiment that the spectrum of $1/f$ background charge fluctuations extends to the 100-GHz range [9]. Even if the energy-relaxation rate due to the fluctuations were small, the fluctuations in the wide frequency range could work as an inhomogeneous background for the ensemble of pulse operations in the time domain. Hence, they could be a reason of 'inhomogeneous dephasing' caused by the difference in the phase evolution speed in each pulse operation, and could decrease the observed decay time. Note that we usually average the signals from 10^5–10^6 pulse operations in 20 ms. Therefore, even the low-frequency component of the $1/f$ noise could result in the dephasing. By simply extrapolating the $1/f$ noise observed at low frequencies to the higher frequency, we estimated the dephasing time of our smaple to be ~ 1 ns. Though such estimation is not very precise, the estimation is not in contradiction with the observed time scale. We mention that other fluctuation like voltage noise from the measurement instruments can also contribute to the decoherence. On the other hand, the quantum-state evolution at the resonance could be more robust than the evolution at the off-resonant point because of the less sensitivity of ΔE to the charge fluctuations. The fact that the single-pulse experiment showed a longer decay time than the two-pulse experiment is consistent with this explanation. A technique similar to the spin-echo experiment in NMR will be able to remove the low-frequency contribution and increase the lifetime of the observed oscillations.

4.   SUMMARY

In the quantum-state-control experiment with two sequential gate-voltage pulses applied on a Cooper-pair box, we observed interference fringes due to the quantum phase accumulation during the delay time between the two pulses. The observed short decay time of the interference suggests that the effect of charge fluctuations on the dephasing is significant. Such a decoherence factor could pose a serious problem to the implementation of solid-state qubits related to the charge degree of freedom.

Acknowledgments

We thank Daniel Esteve for useful discussions on the decoherence. This work was supported by the Core Research for Evolutional Science and Technology (CREST) of the Japan Science and Technology Corporation (JST).

References

[1] See for instance: A. Ekert and R. Jozsa, *Rev. Mod. Phys.* **68**, 733 (1996); *Introduction to Quantum Computation and Information*, edited by H.-K. Lo, T. Spiller, and S. Popescu (World Scientific, Singapore, 1998); articles in *Proc. Roy. Soc. London A* **454** 257–486 (1998).

[2] A. Shnirman, G. Schön, and Z. Hermon, *Phys. Rev. Lett.* **79**, 2371 (1997).

[3] D. V. Averin, *Solid State Commun.* **105**, 659 (1998).

[4] Yu. Makhlin, G. Schön, and A. Shnirman, *Nature* **398**, 305 (1999).

[5] V. Bouchiat, D. Vion, P. Joyez, D. Esteve, and M. H. Devoret, *Phys. Scr.* **T76**, 165 (1998).

[6] Y. Nakamura, C. D. Chen and J. S. Tsai, *Phys. Rev. Lett.* **78**, 2328 (1997).

[7] Y. Nakamura, Yu. A. Pashkin, and J. S. Tsai, *Nature* **398**, 786 (1999); *Physica B* **280**, 405 (2000).

[8] Y. Nakamura and J. S. Tsai, *J. Low Temp. Phys.* **118**, 765 (2000).

[9] M. Covington, M. W. Keller, R. L. Kautz, and J. M. Martinis, *Phys. Rev. Lett.* **84**, 5192 (2000).

MACROSCOPIC QUANTUM SUPERPOSITION IN A THREE-JOSEPHSON-JUNCTION LOOP

Caspar H. van der Wal[1], A. C. J. ter Haar[1], F. K. Wilhelm[1],
R. N. Schouten[1], C. J. P. M. Harmans[1], T. P. Orlando[2],
Seth Lloyd[3], J. E. Mooij[1,2]

[1] Department of Applied Physics and

Delft Institute for Micro Electronics and Submicron Technology (DIMES)

Delft University of Technology, P. O. Box 5046, 2600 GA Delft, the Netherlands

[2] Department of Electrical Engineering and Computer Science and

[3] Department of Mechanical Engineering, MIT, Cambridge, MA 02139, USA

Corresponding author, e-mail: caspar@qt.tn.tudelft.nl

Keywords: Macroscopic quantum mechanics, Josephson effect, SQUIDs, microwave spectroscopy

Abstract   We present microwave-spectroscopy experiments on two quantum levels of a superconducting loop with three Josephson junctions. The level separation between the ground state and first excited state shows an anti-crossing where two classical persistent-current states with opposite polarity are degenerate. This is evidence for symmetric and anti-symmetric quantum superpositions of two macroscopic states; the classical states have persistent currents of 0.5 μA and correspond to the center-of-mass motion of millions of Cooper pairs. A study of the thermal occupancies of the two quantum levels shows that the loop is at low temperatures in a non-equilibrium state.

1.   INTRODUCTION

A Josephson supercurrent is a macroscopic degree of freedom in the sense that it corresponds to the center-of-mass motion of a condensate with a very large number of Cooper pairs [1]. Even though the Josephson effect itself (with classical current and voltage variables) is often called a macroscopic quantum phenomena, Anderson [1], Leggett [2] and Likharev [3] discussed that a quantum superposition of Josephson currents would be a "true" [3] manifestation of quantum mechanics at a

25

macroscopic scale. A simple system in which such a superposition can be studied is a superconducting loop containing one or more Josephson tunnel junctions, where an external magnetic field is used to induce a persistent current in the loop. When the enclosed magnetic flux is close to half a superconducting flux quantum Φ_0, the loop may have multiple stable persistent-current states. The weak coupling of the Josephson junctions then allows for transitions between the states. At very low temperatures, the persistent-current states are very well decoupled from environmental degrees of freedom; excitations of individual charge carriers around the center of mass of the Cooper-pair condensate are prohibited by the superconducting gap. As a result, the transitions between the states can be a quantum coherent process, and superpositions of the macroscopic persistent-current states should be possible (loss of quantum coherence results from coupling to an environment with many degrees of freedom [4]). Josephson junction loops therefore rank among the best systems for experimental tests of the validity of quantum mechanics for systems containing a macroscopic number of particles [2, 5]. The potential for quantum coherent dynamics has stimulated research aimed at applying Josephson junction loops as basic building blocks for quantum computation (qubits) [6, 7, 8, 9].

We report in this chapter on microwave-spectroscopy experiments that demonstrate quantum superpositions of two macroscopic persistent-current states in a small loop with three Josephson junctions (Fig. 1, this is the qubit system discussed in [8, 9]). At an applied magnetic flux of $\frac{1}{2}\Phi_0$ this system behaves as a particle in a double-well potential, where the classical states in each well correspond to persistent currents of opposite sign (Fig. 1c). The two classical states are coupled via quantum tunneling through the barrier between the wells, and the loop is a macroscopic quantum two-level system. The energy levels vary with the applied flux as shown. While classically the levels should cross at $\frac{1}{2}\Phi_0$, quantum tunneling leads to an avoided crossing with symmetric and anti-symmetric superpositions of the two macroscopic persistent-current states. An inductively coupled DC-SQUID magnetometer was used to measure the flux generated by the loop's persistent current, while at the same time low-amplitude microwaves were applied to induce transitions between the levels (Fig. 2). We observed narrow resonance lines at magnetic field values where the level separation ΔE was resonant with the microwave frequency. The level separation shows the expected anti-crossing at $\frac{1}{2}\Phi_0$ (Fig. 3), which is interpreted as evidence for macroscopic superposition states [10, 11]. A study of the thermal broadening of the transition between the two states at $\frac{1}{2}\Phi_0$ shows that the loop is at low temperatures in a non-equilibrium state (Fig. 4).

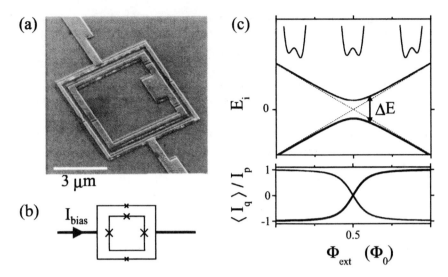

Figure 1 SEM-image (a) and schematic (b) of the small superconducting loop with three Josephson junctions (denoted by the crosses). The loop is inductively coupled to an underdamped DC-SQUID which is positioned around the loop. (c) Energy levels and persistent currents of the loop as a function of applied flux Φ_{ext}. The insets of the top plot show the double-well potential that is formed by the loop's total Josephson energy, plotted for a Φ_{ext}-value below $\frac{1}{2}\Phi_0$ (left), at $\frac{1}{2}\Phi_0$ (middle), and above $\frac{1}{2}\Phi_0$ (right). The horizontal axis for these potentials is a Josephson phase coordinate. The loop's two classical persistent-current states are degenerate at $\Phi_{ext} = \frac{1}{2}\Phi_0$ (dashed lines). The quantum levels (solid lines) show level repulsion at this point, and are separated in energy by ΔE. The bottom plot shows the quantum mechanical expectation value $\langle I_q \rangle = -\partial E_i / \partial \Phi_{ext}$ of the persistent current in the loop, for the ground state (black) and the excited state (grey), plotted in units of I_p.

Note that we have a scheme in which the meter (the DC-SQUID) is performing a measurement on a single quantum system. We should therefore expect that the measuring process is limiting the coherence of our system. While the system is pumped by the microwaves, the SQUID is actively measuring the flux produced by the persistent currents of the two states. Detecting the quantum levels of the loop is still possible since the meter is only weakly coupled to the loop. The flux signal needs to be built up by averaging over many repeated measurements on the same system, such that effectively an ensemble average is determined (time-ensemble). We measure the level separation, i. e. energy, rather than flux, since we perform spectroscopy; we observe a shift in averaged flux when the microwaves are resonant with the level separation. In our experiment we also chose to work with an extremely underdamped DC-SQUID with unshunted junctions to minimize damping of the quantum system via the inductive coupling to the SQUID.

A recent paper by Friedman *et al.* [12] reports on similar results obtained from spectroscopy on excited states in a loop with a single junction (RF-SQUID). Previous experiments on RF-SQUIDs have demonstrated resonant tunneling between discrete quantum states in two wells [13, 14] and microwave-induced transitions between the wells [15]. Other observations that have been related to macroscopic superposition states are tunnel splittings observed with magnetic molecular clusters [16] and quantum interference of C_{60} molecules [17]. In quantum dots [18] and superconducting circuits where charge effects dominate over the Josephson effect [19, 20, 21] superpositions of charge states have been observed, as well as quantum coherent charge oscillations [22].

A quantum description of our system was reported in Refs. [8, 9]. It is a low-inductance loop intersected by three extremely underdamped Josephson junctions (Fig. 1), which are characterized by their Josephson coupling E_J and charging energy $E_C = e^2/2C$. Here C is the junction capacitance and e the electron charge. The critical current of a junction is $I_{C0} = \frac{2e}{\hbar}E_J$, where $\hbar = \frac{h}{2\pi}$ is Planck's reduced constant. One of the junctions in the loop has E_J and C smaller by a factor $\beta \approx 0.8$. At an applied flux Φ_{ext} close to $\frac{1}{2}\Phi_0$ the total Josephson energy forms a double well potential. The classical states at the bottom of each well have persistent currents of opposite sign, with a magnitude I_p very close to I_{C0} of the weakest junction, and with energies $E = \pm I_p(\Phi_{ext} - \frac{1}{2}\Phi_0)$ (dashed lines in Fig. 1c). We assume here Φ_{ext} to be the total flux in the loop (the small self-generated flux due to the persistent currents leads to a constant lowering of the energies, but the crossing remains at $\frac{1}{2}\Phi_0$). The system can be pictured as a particle with a mass proportional to C in the Josephson potential; the electrostatic energy is the particle's kinetic energy. The charging effects are conjugate to the Josephson effect. For low-capacitance junctions (small mass) quantum tunneling of the particle through the barrier gives a tunnel coupling t between the persistent-current states. In the presence of quantum tunneling and for E_J/E_C-values between 10 and 100, the system should have two low-energy quantum levels E_0 and E_1, which can be described using a simple quantum two-level picture [8, 9], $E_{0(1)} = -(+)\sqrt{t^2 + \left(I_p(\Phi_{ext} - \frac{1}{2}\Phi_0)\right)^2}$. The loop's level separation $\Delta E = E_1 - E_0$ is then

$$\Delta E = \sqrt{(2t)^2 + \left(2I_p(\Phi_{ext} - \frac{1}{2}\Phi_0)\right)^2}. \qquad (1.1)$$

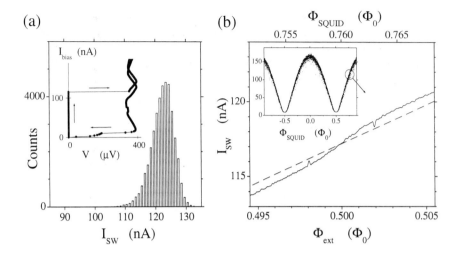

Figure 2 (a) Current-voltage characteristic (inset) and switching-current histogram of the underdamped DC-SQUID. The I_{bias}-level where the SQUID switches from the supercurrent branch to a finite voltage state –the switching current I_{SW}– is a measure for the flux in the loop of the DC-SQUID. The histogram in the main plot shows that the variance in I_{SW} is much larger than the flux signal of the inner loop's persistent current, which gives a shift in I_{SW} of about 1 nA. (b) The inset shows the modulation of I_{SW} versus the flux Φ_{SQUID} applied to the DC-SQUID loop (data not averaged, one point per switching event). The main figure shows the averaged level of I_{SW} (solid line) near $\Phi_{\text{SQUID}} = 0.76\,\Phi_0$. At this point the flux in the inner loop $\Phi_{\text{ext}} \approx \frac{1}{2}\Phi_0$. The rounded step at $\Phi_{\text{ext}} = \frac{1}{2}\Phi_0$ indicates the change of sign in the persistent current of the loop's ground state. In the presence of continuous-wave microwaves (here 5.895 GHz) a peak and a dip appear in the signal, symmetrically around $\frac{1}{2}\Phi_0$. The background signal of the DC-SQUID that results from flux directly applied to its loop (dashed line) is subtracted from the data presented in Figs. 3a and 4a.

2.   EXPERIMENTAL REALIZATION

The system was realized by microfabricating an aluminum micrometer-sized loop with unshunted Josephson junctions (Fig. 1a). The sample consisted of a 5 x 5 μm^2 aluminum loop with aluminum-oxide tunnel junctions, microfabricated with e-beam lithography and shadow-evaporation techniques on a SiO$_2$ substrate. The electrodes of the loop were 450 nm wide and 80 nm thick. The DC-SQUID magnetometer was fabricated in the same layer around the inner loop, with a 7 x 7 μm^2 loop and smaller Josephson junctions that were as underdamped as the junctions of the inner loop. The DC-SQUID had an on-chip superconducting shunt capacitance of 2 pF and superconducting leads in four-point configuration. The sample was mounted in a dilution refrigerator, inside a microwave-tight copper measurement box, magnetically shielded by two mu-metal and one superconducting shield. All spectroscopy measurements were taken with the temperature stabilized at 30 ± 0.05 mK. Microwaves were applied to the sample by a coaxial line, which was shorted at the end by a small loop of 5 mm diameter. This loop was positioned parallel to the sample plane at about 1 mm distance. Switching currents were measured with dedicated electronics, with repetition rates up to 9 kHz and bias currents ramped at typically 1 μA/ms (further details of the fabrication and experimental techniques can be found in Ref. [23]). Loop parameters estimated from test junctions fabricated on the same chip and electron-microscope inspection of the measured device give $I_p = 470 \pm 50$ nA, $\beta = 0.82 \pm 0.1$, $C = 2.6 \pm 0.4$ fF for the largest junctions in the loop, giving $E_J/E_C = 38 \pm 8$. Due to the exponential dependence of the tunnel coupling t on the mass (C) and the size of the tunnel barrier, these parameters allow for a value for t/h between 0.2 and 5 GHz. The parameters of the DC-SQUID junctions were $I_{C0} = 109 \pm 5$ nA and $C = 0.6 \pm 0.1$ fF. The self inductance of the inner loop and the DC-SQUID loop were numerically estimated to be 11 ± 1 pH and 16 ± 1 pH respectively, and the mutual inductance between the loop and the SQUID was 7 ± 1 pH.

The flux in the DC-SQUID was measured by ramping a bias current through the DC-SQUID and recording the current level I_{SW} where the SQUID switches from the supercurrent branch to a finite voltage (Fig. 2a). Traces of the loop's flux signal were recorded by continuously repeating switching-current measurements while slowly sweeping the flux Φ_{ext} (Fig. 2b). The measured flux signal from the inner loop will be presented as $\tilde{I}_{SW}$, which is directly deduced from the raw switching-current data, as described in the following three points:

1) Because the variance in I_{SW} was much larger than the signature from the loop's flux (Fig. 2a) we applied low-pass FFT-filtering in Φ_{ext}-space (over 10^7 switching events for the highest trace, and $2 \cdot 10^8$ events for the lowest trace in Fig. 3a).

2) By applying Φ_{ext} we also apply flux directly to the DC-SQUID. The resulting background signal (dashed line in Fig. 2b) was subtracted.

3) Applying microwaves and changing the sample temperature influenced the switching current levels significantly. To make the flux signal of all data sets comparable we scaled all data sets to $I_{SW} = 100$ nA at $\Phi_{ext} = \frac{1}{2}\Phi_0$. Data taken in the presence of microwaves could only be obtained at specific frequencies where I_{SW} was not strongly suppressed by the microwaves. At temperatures above 300 mK drift in the I_{SW}-level due to thermal instabilities of the refrigerator obscured the signal.

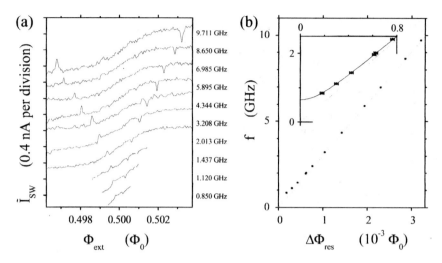

Figure 3 (a) Resonance lines in traces of the scaled switching current $\tilde{I}_{SW}$ versus Φ_{ext}, measured at different microwave frequencies f (labels on the right). (b) Half the distance in Φ_{ext} between the resonant peak and dip $\Delta\Phi_{res}$ at different microwave frequencies f. Peak and dip positions are determined from traces as in Fig. 3a. The inset zooms in on the low frequency data points. The grey line is a linear fit through the high frequency data and zero. The black line is a fit of (1).

3.   RESULTS

Figs. 2b and 3a show the flux signal of the inner loop, measured in the presence of low-amplitude continuous-wave microwaves at different frequencies f. The rounded step in each trace at $\frac{1}{2}\Phi_0$ is due to the change in direction of the persistent current of the loop's ground state (see also Fig. 1c). Symmetrically around $\Phi_{ext} = \frac{1}{2}\Phi_0$ each trace shows a peak and a dip, which were absent when no microwaves were applied. The positions of the peaks and dips in Φ_{ext} depend on microwave frequency but not on amplitude. The peaks and dips result from microwave-induced transitions to the state with a persistent current of opposite sign. These occur when the level separation is resonant with the microwave frequency, $\Delta E = hf$.

In Fig. 3b half the distance in Φ_{ext} between the resonant peak and dip $\Delta\Phi_{res}$ is plotted for all the frequencies f. The relation between ΔE and Φ_{ext} is linear for the high-frequency data. This gives $I_p = 484 \pm 2$ nA, in good agreement with the predicted value. At lower frequencies $\Delta\Phi_{res}$ significantly deviates from this linear relation, demonstrating the

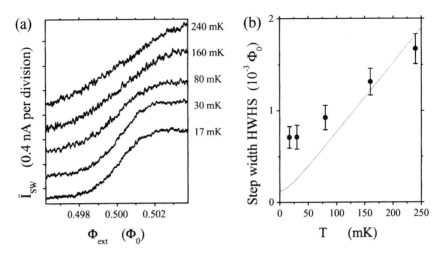

Figure 4 (a) $\tilde{I}_{SW}$ versus Φ_{ext}, measured at different temperatures T (labels on the right). No microwaves were applied. The step in $\tilde{I}_{SW}$ broadens with temperature. (b) The width of the step as a function of temperature. The half-width-half-step (HWHS) is defined as the distance in Φ_{ext} from $\frac{1}{2}\Phi_0$ to the point where the amplitude of the step is half completed. The solid line is the calculated HWHS for thermally mixed levels, using (1) and the I_p and t-value from the spectroscopy results, with a saturating width on the scale of t at low temperatures.

presence of a finite tunnel splitting at $\Phi_{ext} = \frac{1}{2}\Phi_0$. A fit to Eq. (1) yields $t/h = 0.33 \pm 0.03$ GHz, in agreement with the estimate from fabrication parameters. The level separation very close to $\frac{1}{2}\Phi_0$ could not be measured directly since at this point the expectation value for the persistent current is zero for both the ground state and the excited state (Fig. 1c). Nevertheless, the narrow resonance lines allow for an accurate mapping of the level separation near $\frac{1}{2}\Phi_0$, and the observed tunnel splitting gives clear evidence for quantum superpositions of the persistent-current states. The large uncertainty in the predicted t-value does not allow for a quantitative analysis of a possible suppression of t due to a coupling between our two-level system and a bosonic environment [24] or a spin-bath environment [25, 26]. However, the fact that we see a finite tunnel splitting indicates that the damping of our quantum system by environmental degrees of freedom is weak. The dimensionless dissipation parameter α introduced by Leggett *et al.* [24] must be $\alpha < 1$.

The width of the rounded steps in the measured flux in Figs. 2b and 3a is much broader than expected from quantum rounding on the scale of the value of t that was found with spectroscopy (see also Fig. 1c). The temperature dependence of the step width presented in Fig. 4 confirms that the step width HWHS (defined in the caption of Fig. 4b) is too wide at low temperatures T. At temperatures above 100 mK the step width is in agreement with the thermally averaged expectation value for the persistent current $< I_{\text{th}} >= I_{\text{p}} \tanh\left(\frac{\Delta E}{2k_{\text{B}}T}\right)$ (k_{B} is Boltzmann's constant), where we use the level separation ΔE and I_{p} found with spectroscopy. However, when lowering the temperature the observed step width saturates at an effective temperature of about 100 mK. We checked that the effective temperature for the SQUID's switching events did not saturate at the lowest temperatures. The high effective temperature of the loop is a result of the loop being in a non-equilibrium state. Cooling the sample longer after the dissipative switching events did not make the step narrower. The step width at $T = 30$ mK was measured with 100 μs and 50 ms dead time between switching events, but no significant differences were found. This indicates that the out-of-equilibrium population of the excited state is caused by the measurement process with the SQUID or other weakly coupled external processes, in combination with a long time scale for cooling the system to equilibrium (as can be expected since it is very well isolated from the environment). Note that the observed line width and the level separation near $\frac{1}{2}\Phi_0$ are small compared to the effective temperature of 100 mK. Silvestrini *et al.* [27] showed that this can be the case in a Josephson junction system when the transitions between the levels occur much faster than the thermal mixing time, a phenomena that is also well known from e. g. room-temperature NMR on liquids.

4.   CONCLUSION

We have presented clear evidence that a quantum superposition of two macroscopic persistent currents can occur in a small Josephson junction loop. Even though the measuring DC-SQUID is contributing significantly to the decoherence of our system (see also Ref. [11]), it was possible to detect the superposition states since the SQUID was only weakly coupled to the loop. The present results demonstrate the potential of three-junction persistent-current loops for research on macroscopic quantum coherence and quantum computation.

Acknowledgments

We thank J. B. Majer, A. C. Wallast, L. Tian, D. S. Crankshaw, J. Schmidt, A. Wallraff and L. Levitov for help and stimulating discussions. This work was financially supported by the Dutch Foundation for Fundamental Research on Matter

(FOM), the European TMR research network on superconducting nanocircuits (SUP-NAN), the USA Army Research Office (grant DAAG55-98-1-0369) and the NEDO joint research program (NTDP-98).

References

[1] P. W. Anderson, in *Lectures on the Many-Body Problem*, E. R. Ca-ianiello, Ed. (Academic Press, New York, 1964), vol. 2, pp. 113-135.

[2] A. J. Leggett, *Prog. Theor. Phys. Suppl.* **69**, 80 (1980).

[3] K. K. Likharev, *Sov. Phys. Usp.* **26**, 87 (1983).

[4] W. H. Zurek, *Phys. Today* **44**, 36 (October 1991).

[5] A. J. Leggett, A. Garg, *Phys. Rev. Lett.* **54**, 857 (1985).

[6] M. F. Bocko *et al.*, *IEEE Trans. Appl. Supercond.* **7**, 3638 (1997).

[7] L. B. Ioffe *et al.*, *Nature* **398**, 679 (1999).

[8] J. E. Mooij *et al.*, *Science* **285**, 1036 (1999).

[9] T. P. Orlando *et al.*, *Phys. Rev. B.* **60**, 15398 (1999).

[10] We acknowledge that the results presented here do not exclude alternative theories for quantum mechanics (e. g. macro-realistic theories). This would require a type of experiment as proposed by Leggett *et al.* [5].

[11] C. H. van der Wal *et al.*, submitted to *Science*.

[12] J. R. Friedman *et al.*, *Nature* **406**, 43 (2000).

[13] R. Rouse, S. Han, J. E. Lukens, *Phys. Rev. Lett.* **75**, 1614 (1995).

[14] C. Cosmelli *et al.*, *Phys. Rev. Lett.* **82**, 5357 (1999).

[15] S. Han, R. Rouse, J. E. Lukens, *Phys. Rev. Lett.* **84**, 1300 (2000).

[16] W. Wernsdorfer, R. Sessoli, *Science* **284**, 133 (1999).

[17] M. Arndt *et al.*, *Nature* **401**, 680 (1999).

[18] T. H. Oosterkamp *et al.*, *Nature* **395**, 873 (1998).

[19] Y. Nakamura *et al.*, *Phys. Rev. Lett.* **79**, 2328 (1997).

[20] V. Bouchiat *et al.*, *Phys. Scr.* **T76**, 165 (1998).

[21] D. J. Flees, S. Han, J. E. Lukens, *J. Supercond.* **12**, 813 (1999).

[22] Y. Nakamura, Yu. A. Pashkin, J. S. Tsai, *Nature* **398**, 786 (1999).

[23] C. H. van der Wal, J. E. Mooij, *J. Supercond.* **12**, 807 (1999).

[24] A. J. Leggett *et al.*, *Rev. Mod. Phys.* **59**, 1 (1987).

[25] N. Prokof'ev, P. Stamp, *Rep. Prog. Phys.* **63**, 669 (2000).

[26] L. Tian *et al.*, to be published; cond-mat/9910062.

[27] P. Silvestrini *et al.*, *Phys. Rev. Lett.* **79**, 3046 (1997).

BOSE-EINSTEIN CONDENSATION WITH ATTRACTIVE INTERACTION FATE OF A FALSE VACUUM

Masahito Ueda[1] and Hiroki Saito[2]
[1]*Department of Physics, Tokyo Institute of Technology, Meguro-ku, Tokyo 152-8551, Japan*
[2]*Institute of Physics, University of Tokyo, Tokyo 153-8902, Japan*

Abstract   Among the species that have Bose-condensed, lithium 7 and rubidium 85 form unique condensates in that atoms interact attractively and that the energy barrier that stabilizes the gaseous Bose-Einstein condensates originates from the zero-point motion of atoms. This paper reviews some unique features of these new states of matter that are distinguished from those of other species with repulsive interactions. These include the density instability, macroscopic quantum tunneling, intermittent implosion, shell-structure formation, and partial quantization of circulation.

1. METASTABILITY AND DENSITY INSTABILITY OF THE CONDENSATE

Atoms with attractive interactions are believed not to undergo Bose-Einstein condensation in the uniform 3D system. It is therefore appropriate to begin by discussing why attractive Bose-Einstein condensates (BECs) have been realized in trapped systems [1, 2].

A trapped BEC has three characteristic energies, namely, the kinetic, potential and interaction energies. The kinetic energy arises from the uncertainty principle, and is proportional to R^{-2} where R is the linear dimension of BEC. The trapping potential is usually harmonic, and therefore the potential energy is proportional to R^2. The mean-field interaction energy is proportional to $N_0 a R^{-3}$, where N_0 and a denote the number of condensate atoms and the s-wave scattering length, respectively. The interaction energy and hence a are negative for attractive systems. The sum of these three energies $F = N_0(AR^{-2} + BR^2 - CN_0|a|R^{-3})$, where A, B, C are positive constants, gives the total energy of the system which has a metastable minimum if N_0 is smaller than a certain critical value N_c. It is at this local minimum where a metastable BEC is believed to be formed. If N_0 is larger than N_c, there is no metastable minimum and the condensate is bound to collapse to the $R = 0$ state, which is a metallic phase. It is clear from the above argument that the zero-point energy due to confinement serves as a kinetic obstacle against collapse, allowing the metastable BEC to be formed.

For the case of attractive interactions, the lowest-energy collective mode is the monopole mode in which the condensate alternatively contracts and expands isotropically.The restoring force against the contraction is provided by the zero-point kinetic pressure, and the monopole mode becomes softer as N_0 approaches its critical value N_c [3]. Near the critical point it is shown that [4]

$$\omega_M = 160^{\frac{1}{4}} \left(1 - \frac{N_0}{N_c}\right)^{\frac{1}{4}}. \tag{1}$$

This formula agrees very well with the result obtained by the numerical diag-onalizaion of Bogoliubov theory if we take the numerically obtained precise value for N_c [5]. The softening of the collective mode signals the density instability ofthe condensate, and its spectacular consequence is macroscopic quantum tunneling.

2. Macroscopic Quantum Tunneling

Macroscopic quantum tunneling (MQT) is triggered by zero-point density fluctuations.Applying the instanton technique to the Gross-Pitaevskii energy functional yields the MQT rate Γ as

$$\Gamma = Ae^{-S/\hbar}, \tag{2}$$

where the attempt frequency A and the tunneling exponent S are given by [4]

$$A = 11.8N_0^{\frac{1}{2}}\left(1 - \frac{N_0}{N_c}\right)^{\frac{7}{8}}, \quad S/\hbar = 4.58n_0\left(1 - \frac{N_0}{N_c}\right)^{\frac{5}{4}}. \tag{3}$$

Remarkably, the tunneling exponent is proportional to the five-fourth power of $1 - N_0/N_c$, in contrast to the unit power found in previous literature [6]. Because of this extra power of 1/4, and a relatively large attempt frequency, MQT becomes important faster than previouslythought as N_0 approaches N_c. For example, when the number of condensate atoms is 99.5% ofits critical value, the decay rate is about 2 per seccond, and we may therefore expect to observe MQT in such immediate vicinity of the criticality.

Macroscopic quantum tunneling described here is different from the usual one in that the energy barrier to be overcome is generated dynamically. One might therefore wonder if such a barrier really exists [7]. A recent numerical study [8] based on the Hamiltonian that includes a full binary contact interaction shows the existence of sucha dynamically generated energy barrier, and that the dependence on N_0 of the MQT rate that is evaluated from the decay of the overlap integral between the initial wave function and the time-evolved one qualitatively agrees with the rate found in Ref. [4] except in the region $1 - N/0/N_c \leq 10^{-3}$, where the MQT rate is found to be substantially enhanced by quantum-field fluctuations.

3. Fate of BEC for $N_0 > N_c$

What happens when the number of atoms is larger than its critical value?A variational study suggests the collapse of the entire system, and we have no BEC.

The truth of this conclusion, however, is not so obvious because the density of BEC at its periphery is too low to collapse. Moreover, as the peak density becomes high, inelastic collisions become important. To find the fate of BEC for $N_0 > N_c$, these effects must be taken into account.

The time evolution of BEC in this case may be examined by expressing thetransition amplitude of the condensate wave function from an initial state to a final one in terms of path integrals [9]. A Feynman path describes a possible history of the condensate wave function, and the most probable path is obtained by requiring that the Euclidean action be minimized. The resultant equation is nothing but the Gross-Pitaevskii equation but in imaginary time. It is first-order in imaginary time and not second-order; it therefore describes nonlinear diffusion rather than wave propagation. Unlike the relativistic problem,the motion does not obey Newtonian mechanics in an inverted potential. We therefore have no bounces [10] or no instantons [11].

The solution of the imaginary-time Gross-Pitaevskii equation reveals a remarkable feature of the collapsing dynamics. We now have a certain radius. Inside it, which we shall refer to as "black hole," the wave function grows in time. Outside it, the wave function attenuates. The radius of the black hole is determined by the strength of interaction and fits nicely the square root of the logarithmic curve [9]; the "black hole" opens when N_0 exceeds N_c and closes when N_0 goes below N_c.

Inside the "black hole" the density becomes very high, so the atomshave an appreciable probability of undergoing inelastic collisions and draining out of the trap. In Rice experiments [1], there are abundant above-condensate atoms which will replenish the lost atoms. We may therefore expect collapse-and-growth cycles of BEC [12]. Macroscopic quantum tunneling and thermal-assisted tunneling also induce the collapse of BEC, and the ensuing refilling of the condensate atoms from the thermal gas again leads to the collapse-and-growth cycles of BEC. Kagan *et al.* [13] showed that the loss due to three-bodyrecombination alone gives rise to similar oscillations. And indeed, recent experiments [14] investigating the spread in the number of condensate atoms have suggested the occurrence of dynamic collapse-and-growth cyles of BEC, but have neither favored nor excluded any one of these possibilities. It is likely that all of these contribute to the dynamics of this system, which is quite complex and certainly merits further experimentaland theoretical study.

4. Intermittent Implosion and Shell-Structure Formation

Once the collapsing process of the condensate starts, the peak density grows very high, and we must therefore include the atomic loss due to the two-body dipolar loss and three-body recombination loss. Based on a generalized Gross-Pitaevskii equation that includes these losses, we found that at an initial stage of the collapse the atomic density grows very slowly and suddenly the rapid implosion breaks out. Remarkably, the implosion occurs not once but several times intermittently [15], and at each implosion several tens of atoms are lost from the condensate. We have also confirmed that the atomic loss occurs within a very small spatial region (about one hundredth of the trap dimension) around the center of BEC.

The physics of the intermittent implosion may be understood as follows. At extremely low temperature and for N_0 slightly above N_c, the attractive interaction barely dominates the kinetic energy. The condensate therefore has a negative pressure, and BEC shrinks towards the central region.Since the kinetic-energy density and the interaction-energy density are proportional to the atomic density and its square, the inelastic collision decreases the interaction energy twice as much as the kinetic energy [13]. The atomic loss therefore makes the kinetic energy dominate over the interaction energy, thereby turning the sign of pressure to positive. This causes the remaining atoms to expand after the inelastic collisions take place. However, the inward flow outside the region of implosion replenishes the lost atoms, turning the sign of pressure again to negative, which causes the subsequent implosion.

We next consider the situation in which initially a large BEC with repulsive interaction is prepared and then suddenly the sign of interaction is switched to attractive. Such experiments have recently been realized at JILA [2] using the Feshbach resonance [16]. When the initial number of atoms is sufficiently large, the initial wavefunction takes the Thomas-Fermi form [17]. After the change in the sign of a to negative, the wave function begins to shrink and gives rise to a ripple, which grows to be a series of pulses due to a self-focusing effect associated with the attractive interaction. The column density integrated along the z axis, which can directly be measured by the absorption imaging and the phase-contrast imaging, exhibits concentric circles. This indicates the formation of the shell structure in the atomic density [15].

5. Superfluidity

Superfluidity represents a complex of phenomena, including persistent current, vortices, and nonclassical rotational inertia [18], and it is of interest to investigate whether these phenomena occur in BEC systems with attractive interaction.

Suppose that we rotate a torus that contains neutral atoms at frequency ω which is smaller than a certain critical value, and measure themoment of rotational inertia I. Experiments on liquid helium 4 show that above the transition temperature, I equals its classical value. However, below the transition temperature, it becomes smaller than its classical value, and goes to zero as temperature goes to zero.

Now the system of neutral particles in a rotating frame of reference is equivalent to that of charged particles in a magnetic field. Therefore, the Hess-Fairbank effect—disappearance of the angular momentum of the system as it is cooled down to absolute zero (with its container kept rotating slowly)—is an analogue of the Meissner effect in superconductivity, and it may therefore be regarded as a hallmark of superfluidity.

For the case of repulsive interaction, the circulation is quantized as in the case of liquid helium 4. What's new with the case of attractive interaction? To understand this, let us consider the delta-function-type binary interaction. For a genuine condensate in which all particles occupy the same quantum state, we have only the Hartree energy. If the condensate is fragmented or hybridized into two states, we have not only the Hartree but also the Fock exchange energy. Therefore, for the case of repulsive interaction, hybridization costs an extensive Fock exchange energy, and therefore a genuine condensate is energetically favorable [19]. However, for the case of attractive interaction, hybridization decreases the interaction energy, but it also increases the kinetic energy. Thus, we may expect an interesting competition between the interaction energy and the kinetic energy.

Accordingly, the circulation shows a gradual transition, in contrast with the sharp transition for the case of repulsive interaction [20]. Also, the plateaus of quantized circulation appear only when the mean-field interaction energy per particle is smaller than the single-particle energy-level spacing, with the lengths of the plateaus reduced due to hybridization of the condensate over different angular-momentum states. Only when experiments are performed in a reduced plateau, we may observe the Hess-Fairbank effect.

Wilkin *et al.* [21] considered a situation in which BEC with very weak attractive interaction is confined in a rotating two-dimensional isotropic parabola and found that AM is carried entirely by the center of mass, in sharp contrast to the case of repulsive interaction where AM is carried by vortices. The underlying physics is the attractive interaction, which causes the particles to move together. Finally, we comment on persistent current. For the case of repulsive interaction, the Fock exchange interaction serves as the energy barrier that stabilizes persistent current. However, for the case of attractive interaction, the exchange energy is negative, so that persistent current does not exist. In fact, a simple calculation shows that the Landau criterion for superflow cannot be met for the case of attractive interaction [22].

ACKNOWLEDGMENTS. This work was supported by a grant for Core Research for Evolutional Science and Technology (CREST) of the Japan Science and Technology Corporation (JST), by a Grant-in-Aid for Scientific Research (Grant No. 11216204) by the Ministry of Education, Science, Sports,and Culture of Japan, and by the Toray Science Foundation. H.S. acknowledges support by the Japan Society for the Promotion of Science.

References

1. C. C. Bradley, C. A. Sackett, J. J. Tollett, and R. G. Hulet, Phys. Rev. Lett. **75**, 1687 (1995); **79**, 1170(E) (1997); C. C. Bradley, C. A. Sackett, and R. G. Hulet, *ibid.* **78**, 895 (1997).
2. S. L. Cornish, N. R. Claussen, J. L. Roberts, E. A. Cornell, and C. E. Wieman, cond-mat/0004290.
3. S. Stringari, Phys. Rev. Lett. **77**, 2360 (1996).
4. M. Ueda and A. J. Leggett, Phys. Rev. Lett. **80**, 1576 (1998).
5. H. Saito and M. Ueda, unpublished.
6. E. V. Shuryak, Phys. Rev. A **54**, 3151 (1996).
7. X.-B. Wang, L. Chang, B.-L. Gu, Phys. Rev. Lett. **81**, 1342 (1998); M. Ueda, A. J. Leggett, Phys. Rev. Lett. **80**, 1576 (1998)

8. H. Saito and Ueda, unpublished.
9. M. Ueda and K. Huang, Phys. Rev. A **60**, 3317 (1999).
10. S. Coleman, Phys. Rev. D**15**, 2929 (1977).
11. G. 'tHooft, Phys. Rev. Lett. **37**, 8 (1976).
12. C. A. Sackett, C. C. Bradley, M. Welling, and R. G. Hulet, Appl. Phys. B **65**, 433 (1997); C. A. Sackett, H. T. C. Stoof, and R. G. Hulet, Phys. Rev. Lett. **80**, 2031 (1998).
13. Yu. Kagan, A. E. Muryshev, and G. V. Shlyapnikov, Phys. Rev. Lett. **81**, 933 (1998).
14. C. A. Sackett, J. M. Gerton, M. Welling, and R. G. Hulet, Phys. Rev. Lett. **82**, 876 (1999).
15. H. Saito and M. Ueda, cond-mat/0002393.
16. S. Inouye, M. R. Andrews, J. Stenger, H.-J. Miesner, D. M. Stamper-Kurn,and W. Ketterle, Nature **392**, 151 (1998).
17. G. Baym and C. J. Pethick, Phys. Rev. Lett. **76**, 6 (1996).
18. A. J. Leggett, Physica Fennica **8**, 125 (1973).
19. P. Nozières and D. Saint James, J. Physique **43**, 1133 (1982).
20. M. Ueda and A. J. Leggett, Phys. Rev. Lett. **83**, 1489 (1999).
21. N. K. Wilkin, J. M. F. Gunn, and R. A. Smith, Phys. Rev. Lett. **80**, 2265 (1998).
22. D. S. Rokhsar, cond-mat/9709212.

MACROSCOPIC QUANTUM PHENOMENA IN ATOMIC BOSE-EINSTEIN CONDENSATES

Fernando Sols and Sigmund Kohler

Dpto. Física Teórica de la Materia Condensada, C-V
Universidad Autonoma de Madrid, E-28043 Madrid, Spain

Keywords: Bose systems, quantum interference, quantum fluctuations, Josephson effect, macroscopic quantum phenomena.

Abstract   We review some aspects of the physics of macroscopic quantum phenomena in Bose-Einstein condensates. Depending on the relative strength of the interaction and Josephson coupling energies, a double condensate may exhibit three different regimes: Rabi, Josephson, and Fock. In a macroscopic quantum interference experiment, the relative phase of two independent condensates may become well-defined when atoms emitted from them are forced to interfere. This process competes with the ballistic randomization of the phase due to interactions. A steady state is reached when the phase becomes defined by the detections at the same rate at which it is randomized. The resulting quantum phase uncertainty sets a fundamental limit to the achievable phase resolution.

1.   INTRODUCTION: THE PENDULUM HAMILTONIAN

Since its theoretical prediction in 1962 [1], the Josephson effect has played a major role in the physics and technology of superconductors [2]. The physics of the Josephson effect manifests itself when a weak link is created (*e.g.* by tunneling or through a point contact) between two systems which have undergone some type of gauge symmetry breaking. In fact, the Josephson effect has also been observed in superfluids [3], and, because of the profound analogies, its observation in the recently achieved [4, 5, 6, 7] Bose condensed atomic gases is generally expected. A system whose gauge symmetry has been spontaneously broken can be described by an order parameter that behaves in many respects like a macroscopic wave function $\Psi(\mathbf{r})$. In the simplest cases, the order parameter reduces to a complex scalar, $\Psi = \sqrt{\rho}e^{i\phi}$, where $\rho(\mathbf{r})$ is the "superfluid density" and $\phi(\mathbf{r})$ is the phase. For Bose-Einstein condensates of dilute alkali gases, $\Psi(\mathbf{r})$ is the wave function of the macroscopically occupied one-atom state.

Josephson predicted [1] that, between two weakly connected superconductors of phases φ_1 and φ_2, a non-dissipative particle (Cooper pair) current flows between them whose value is $I(\varphi) = I_c \sin\varphi$, where I_c is the critical current and

$$\varphi = \varphi_2 - \varphi_1 \equiv \int_1^2 d\mathbf{r} \cdot \nabla\phi(\mathbf{r}) \tag{1.1}$$

is the relative phase. He also predicted that, in the presence of a nonzero chemical potential difference $\mu = \mu_2 - \mu_1$, the relative phase rotates as $\dot{\varphi} = -\mu/\hbar$. The Josephson relations can be obtained from very general considerations [8] and apply to any pair of weakly linked systems that can be described by macroscopic wave functions [9]. They can be obtained as the equations of motion of the "pendulum Hamiltonian"

$$H(\varphi, N) = E_J(1 - \cos\varphi) + \frac{1}{2}E_c n^2, \tag{1.2}$$

where $E_J = \hbar I_c$ is the Josephson coupling energy, $n = (N_2 - N_1)/2$ is the number of transferred particles, and $E_c \equiv \partial\mu/\partial n$ is the capacitive energy due to interactions. In the absence of external constraints, $\mu = E_c n$. For trapped BEC's, E_c can be obtained to a good accuracy from the Thomas-Fermi (TF) calculation [10] of the chemical potential.

When both E_c and $k_B T$ are $\ll E_J$, the Josephson pendulum Hamiltonian (1.2) can be approximated as a harmonic oscillator whose frequency

$$\omega_{JP} = \frac{1}{\hbar}\sqrt{E_J E_c} \tag{1.3}$$

is called the Josephson plasma frequency.

1.1.   INTERNAL JOSEPHSON EFFECT

It is possible to couple different hyperfine states of an atom (say $|F, m\rangle$ and $|F', m'\rangle$) through laser light. In particular, by applying a laser pulse, one may force an atom initially in state $|F, m\rangle$ to evolve in a controllable way towards another state $|F', m'\rangle$. If such a pulse is applied to a macroscopic condensate, then it is the whole ensemble of atoms that evolves coherently. Thus it is possible to prepare a condensate in which each atom is in the same coherent superposition of the two states. Such an experiment has been realized by Hall et $al.$ [11] with the $|1, -1\rangle$ and

$|2, 1\rangle$ states of ^{87}Rb (hereafter, $|A\rangle$ and $|B\rangle$). A natural variant of this experiment consists in using laser light which also induces the opposite rotation (*e.g.* linearly polarized light), in such a way that the two states are coherently connected in both directions, very much like Cooper pairs in a Josephson junction can tunnel from left to right and vice versa.

The analysis of the spatial (external) Josephson effect is based on the knowledge of external bistable potential [12]. A similar study of the internal Josephson effect would not be practical. Fortunately, all is really needed is the value of the matrix element connecting states $|A\rangle$ and $|B\rangle$, which can be known by other means. Then, it is most convenient to employ a two-site description of the Josephson link. We write [13, 14]

$$H = -\frac{\hbar\omega_R}{2}\left(a^\dagger b + b^\dagger a\right) + \frac{E_c}{4}\left[(a^\dagger a)^2 + (b^\dagger b)^2\right], \qquad (1.4)$$

where $a^\dagger$ ($b^\dagger$) creates one atom in state $|A\rangle$ ($|B\rangle$), E_c accounts for the interactions, and ω_R is the Rabi frequency. For simplicity, we assume that the two atomic energies, as well as the intraspecies interactions, are equal, and neglect the interspecies interaction. Particle number conservation requires $a^\dagger a + b^\dagger b = N \equiv 2N_0$.

To establish the connection with the pendulum Hamiltonian, we note that particle number eigenstates admit a phase representation $\Phi_n(\varphi) = (2\pi)^{-1/2}\exp(-in\varphi)$, and thus we may write for (1.2) $\langle n+1|H|n\rangle = -E_J/2$. A similar analysis for the two-site Hamiltonian (1.4) yields

$$\langle n+1|H|n\rangle = -\frac{\hbar\omega_R}{2}\sqrt{(N_A + 1)N_B}. \qquad (1.5)$$

Noting that $N_{A,B} = N_0 \pm n$ we arrive at

$$E_J(n) = \hbar\omega_R\sqrt{N_0(N_0 + 1) - n(n + 1)}. \qquad (1.6)$$

In the limit $1 \ll n \ll N_0$, eq. (1.6) becomes $E_J = N_0\hbar\omega_R$, which is a clear manifestation of the phenomenon of Bosonic amplification.

1.2. RABI, JOSEPHSON, AND FOCK REGIMES

A clear limitation of the standard pendulum Hamiltonian (1.2) is that, in the non-interacting limit $E_c \to 0$, the dynamics is suppressed; in particular, the Josephson plasma frequency for small collective oscillations vanishes. This contradicts our physical notion that noninteracting atoms should indeed exhibit some dynamics. For instance, under the effect of Hamiltonian (1.4), an atom initially prepared in state $|A\rangle$ will undergo

Rabi oscillations between states $|A\rangle$ and $|B\rangle$ with frequency ω_R. This effect is clearly beyond the scope of a rigid (with n-independent E_J) pendulum model. However, it should be describable by (1.4), which has a well-defined non-interacting limit. Thus, it seems of interest to develop a unified description which views collective Josephson behavior and single atom Rabi oscillations as particular cases of a more general dynamics. To that end, (1.6) becomes in the semiclassical limit $E_J(N) = \hbar\omega_R(N_0^2 - n^2)^{1/2}$, resulting in the non-rigid pendulum Hamiltonian,

$$H = -N_0\hbar\omega_R\sqrt{1 - n^2/N_0^2}\cos\varphi + \frac{1}{2}E_c n^2, \qquad (1.7)$$

which has been analyzed by Smerzi et al. [15]. In the limit $\varphi \ll 1$ and $n \ll N_0$, (1.7) also acquires a harmonic form but with a natural oscillation frequency

$$\omega^2 = \omega_{JP}^2 + \omega_R^2 = \frac{N_0 E_c}{\hbar}\omega_R + \omega_R^2. \qquad (1.8)$$

It is thus clear that, for a given interaction strength, the double BEC system can be driven continuously from the Josephson to the Rabi regime by varying ω_R, something feasible with current laser technology. Noting that $E_c \sim \mu_0/N_0$, the Josephson limit corresponds to $\mu_0 \gg \omega_R$.

Equation (1.7) describes in a unified way Josephson and Rabi ($E_c = 0$) dynamics. This is analogous to how the Gross-Pitaevskii equation of a many boson system reduces for zero interaction to the Schrödinger equations. The crossover between collective Josephson and individual Rabi dynamics cannot be studied in superconductors and superfluids, because there interactions are never completely negligible. It is a nice feature of Bose-Einstein condensation that it will allow us to study the crossover between these two qualitatively different dynamical regimes in an elegant fashion.

In the limit of very strong interactions, quantum fluctuations of the phase dominate, the Josephson coupling becomes a small perturbation, and the relative particle number is approximately a good quantum number. This is the Fock regime, where the particle number eigenstates (Fock states) are stationary states.

In summary, the magnitude of the "charging" energy E_c distinguishes three regimes:

(i) $E_c \ll \hbar\omega_R/N_0$ (Rabi)
(ii) $\hbar\omega_R/N_0 \ll E_c \ll N_0\hbar\omega_R$ (Josephson)
(iii) $N_0\hbar\omega_R \ll E_c$ (Fock)

2. PHASE MEASUREMENT VERSUS INTERACTION

Recent macroscopic interference experiments [16, 17] have revived interest on the question of whether two independent Bose condensates can have a well-defined relative phase [18, 19]. Guided by the experimental result that a precise phase was indeed observed, the ensuing theoretical work [20, 21, 22, 23] has noted that, even if the initial relative phase is random, as happens for pairs of condensates prepared in a Fock state, the phase becomes progressively well-defined as atoms are emitted from the compound system and recorded in a phase-sensitive detector. The resulting view accords well with the standard quantum measurement picture according to which the phase is created by the very act of measurement; if one tries to measure the phase one indeed observes a definite phase. Here we study the role of interactions, and show that they modify the above picture considerably. While the final conclusion remains true that a well-defined phase is expected in a wide range of experiments, we argue that this is possible only because the interactions are sufficiently weak, and calculate explicitly how weak they must be in a specific experimental setup. The central idea is that, due to interactions, phase is not a constant of motion. An experiment will be effective in measuring the phase only if the measurement is sufficiently intense to overcome the phase internal dynamics opposing definition.

We consider two independent Bose condensates, each confined in a harmonic trap. A Hamiltonian for the mean field description of the interaction is easily derived from a Taylor expansion of the energy of each condensate [12]. We have N_A (N_B) atoms in condensate A (B), $N = N_A + N_B$ and $n = N_A - N_B$. For $|n| \ll N$,

$$H = 2E(N/2) + E_c n^2/2 \qquad (1.9)$$

describes the coherent dynamics of the total system. $E_c = E''(N/2)/2$ determines to lowest order in n the self-interaction within each condensate, and can be derived from the ground state properties of an interacting Bose gas [24]. In a quantized version, the atom numbers N_a and N_b are given by the operators $a^\dagger a$ and $b^\dagger b$.

Following Ref. [20], we consider an interference experiment in which the condensate atoms are emitted from the traps and guided to a 50-50 atom beam splitter [25] and the emerging beams are directed to two detectors (Fig. 1). This experiment has the attractive feature that, being realistic, it lends itself to a fundamental theoretical analysis, since it captures the essence of phase measurement in its most simple form. The action of the two detectors $D_\pm$ on the system is described by the operators $C_\pm = \sqrt{\gamma/2}\,(a \pm b)$, where γ is the outcoupling rate per atom, and

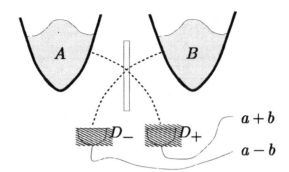

Figure 1 Setup of the interference experiment. The condensates A and B are initially independent. They are outcoupled from the traps with equal rate γN and brought to interference at the beam splitter.

the expectation values $I_\pm(t) = \langle \psi(t)|C_\pm^\dagger C_\pm|\psi(t)\rangle$ give the corresponding detection rates [26].

2.1.   STATIONARY VARIANCES

A convenient representation is provided by the phase states

$$|\phi\rangle_N = \frac{1}{\sqrt{2^N N!}} \left(a^\dagger e^{i\phi} + b^\dagger e^{-i\phi}\right)^N |0,0\rangle. \qquad (1.10)$$

If the wave function of a state $|\psi\rangle$ in the phase representation $\psi(\phi) = {}_N\langle\phi|\psi\rangle$ exhibits a sharp peak at an average value $\bar{\phi}$ of the relative phase, it is easy to see that the detection rates read $I_\pm = \gamma N[1 \pm \cos(2\bar{\phi})]/2$. The number difference operator $n = a^\dagger a - b^\dagger b$ acts on the phase states as $n|\phi\rangle_N = -i(\partial/\partial\phi)|\phi\rangle_N$. Thus n and ϕ are conjugate quantities, and, for a sharp phase distribution, the inequality $\Delta n\,\Delta\phi \geq 1/2$ applies. The Heisenberg equations of motion $\partial\phi/\partial t = -E_c n/\hbar$, $\partial n/\partial t = 0$, lead to a time-dependent interference pattern $I_\pm(t) \propto 1 \pm \cos(2E_c \bar{n}t/\hbar)$.

A necessary feature of phase-sensitive detection is that it does not give information on which condensate the recorded atom comes from. Each detected atom may come with equal probability from either condensate A or B. As a result, the uncertainty in the relative atom number after k detections grows like $\Delta n = \sqrt{k}$. In the absence of interactions [20], a minimal wave packet is formed at a sufficiently advanced stage of the detection process, so that $\Delta\phi = 1/2\sqrt{k} \ll 1$. Let us assume that the same result holds true in the presence of interactions. Then we can conclude that the increase $\Delta n^2 \to \Delta n^2 + 1$ in the number uncertainty occurring in each additional detection, is accompanied by a corresponding decrease in the phase uncertainty,

$$\Delta\phi \rightarrow \frac{1}{2\sqrt{\Delta n^2 + 1}} \approx \Delta\phi - 2\Delta\phi^3. \tag{1.11}$$

On the other hand, interactions generate phase dynamics. The mean value of the phase shows a drift, $\bar{\phi}(t) = \bar{\phi}(0) - E_c \bar{n} t/\hbar$, while its variance undergoes ballistic spreading [19, 27], $\Delta\phi(t) = \Delta\phi(0) + E_c \Delta n\, t/\hbar$. Thus between detections the phase uncertainty grows at a rate proportional to Δn, which in turn grows with the number of detections. Since γN is the total detection rate, the mean increase of phase variance between detections is

$$\Delta\phi \rightarrow \Delta\phi + \frac{E_c}{2\hbar\gamma N \Delta\phi}. \tag{1.12}$$

In the early stages of the measurement process Δn is small and the effect of interactions is negligible; the variation in (1.12) is much smaller than that in (1.11). However, as k grows, the two effects become comparable and cancel each other. Then a steady state is formed in which *the phase becomes defined by the detections at the same rate at which it is randomized by the interactions*. The stationary values of the variances are

$$\Delta\phi_s = \left(\frac{\kappa}{4}\right)^{1/4}, \quad \Delta n_s = \left(\frac{1}{4\kappa}\right)^{1/4}, \quad \kappa \equiv \frac{E_c}{\hbar\gamma N}. \tag{1.13}$$

In the absence of interactions, $\Delta\phi_s = 0$, and the phase can become arbitrarily well-defined. In contrast to that, Eq. (1.13) indicates that *interactions cause an intrinsic limitation in the achievable phase resolution*. A steady state with a well-defined phase $\Delta\phi \ll 2\pi$ can be established only if interactions are sufficiently weak, $\kappa \ll 1$, which marks the range of validity for Eqs. (1.11)–(1.13). For a 3d harmonic trap in the Thomas-Fermi limit, $E_c \sim N^{-3/5}$ [12, 24, 10] and thus the effective interaction parameter, $\kappa \sim N^{-8/5}$, increases during the detection process. However, we shall focus on time scales where N is practically constant.

2.2.   STOCHASTIC TIME EVOLUTION

For a more quantitative analysis we have performed a numerical simulation of the time evolution in a single experiment based on the quantum jump method [26]. Starting from an inital Fock state $|N_A, N_B\rangle$, the time evolution over a sufficiently small time interval Δt can be simulated as follows. If an atom is detected at $D_\pm$, the wave function changes as

$$|\psi(t)\rangle \rightarrow C_\pm|\psi(t)\rangle. \tag{1.14}$$

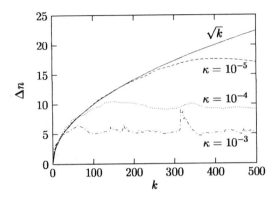

Figure 2 Time evolution of the number uncertainty Δn in a typical run of the simulation for different effective interaction strengths $\kappa \equiv E_c/\hbar\gamma N$. The full line depicts the behavior in absence of interaction. Initially, the total atom number is $N = 10^6$ and the number difference $n_0 = 200$.

These detections occur with probabilities $p_\pm(t, \Delta t) = I_\pm(t)\Delta t \ll 1$. If no detection takes place, the system propagates coherently from t to $t + \Delta t$. For details of the simulation we refer the reader to Ref. [28].

Figure 2 shows how the number variance evolves in a typical experimental run. Starting the stochastic time evolution with the initial state $|N_A, N_B\rangle$, the variance Δn grows with the square root of the number of detected atoms as in the interaction free case [20]. After $k^* = \Delta n_s^2$ atoms have been detected, *i.e.* after a time $t^* = (\hbar/4N\gamma E_c)^{1/2}$, the interaction becomes effective, and the variances saturate at the values (1.13). The average over many simulation runs confirms the analytical estimates (1.13) for $\kappa < 0.1$ [28]. For larger values of κ, the phase variance is no longer well-defined since it becomes of order unity.

3.   SUMMARY AND CONCLUSIONS

We have discussed some macroscopic quantum phenomena that can be observed in Bose condensed systems with emphasis on those that are likely to be observed only (or more easily) in these novel systems. It is possible to explore the crossover between collective Josephson behavior and independent boson Rabi dynamics. The Fock regime corresponding to effective separation, of course, can also be realized.

A study of the influence of self-interaction on the phase showed that in a macroscopic interference experiment between separate Bose condensates the phase resolution is limited by the ratio between interaction strength and detection rate. The limit arises when the phase randomizes due to interactions at the same rate at which it becomes defined by the successive partial measurements performed by each atom detection. We have derived an analytical expression for the phase resolution which has been confirmed by a Monte Carlo simulation. It will be most desirable to perform experiments that explore the fading of the interference pattern with increasing interaction strength.

Acknowledgment

This work has been supported by the EU TMR Programme under Contract No. FMRX-CT96-0042 and by the Dirección General de Investigación Científica y Técnica under Grant No. PB96-0080-C02.

References

[1] B. D. Josephson, Phys. Lett., **1** (1962) 251.

[2] A. Barone and B. Paternò, *Physics and applications of the Josephson effect* (John Wiley & Sons, New York) 1982.

[3] O. Avenel and E. Varoquaux, Phys. Rev. Lett., **55** (1985) 2704.

[4] M. H. Anderson, J. R. Ensher, M. R. Matthews, C. E. Wieman and E. A. Cornell, Science, **269** (1995) 198.

[5] C. C. Bradley, C. A. SacketT, J. J. Tollett, and R. G. Hulet, Phys. Rev. Lett., **75** (1995) 1786.

[6] K. B. Davis *et al.* Phys. Rev. Lett., **75** (1995) 3969.

[7] M. O. Mewes *et al.*, Phys. Rev. Lett., **77** (1996) 416; *ibid.*, 988.

[8] F. Sols, in *Bose-Einstein Condensation in Atomic Gases*, Proc. Int. School of Physics "Enrico Fermi", edited by M. Inguscio, S. Stringari, and C. E. Wieman (IOS Press, Amsterdam, 1999), p. 453.

[9] R. P. Feynman, R. B. Leighton, and M. Sands, *The Feynman Lectures on Physics* (Addison-Wesley Pub. Co., Reading) 1963.

[10] G. Baym and C. Pethick, Phys. Rev. Lett. **76**, 6 (1996).

[11] D. S. Hall, M. R. Matthews, C. E. Wieman, and E. A. Cornell, Phys. Rev. Lett., **81**, 1543 (1998); *ibid.*, 4532 (Erratum).

[12] I. Zapata, F. Sols, and A. J. Leggett, Phys. Rev. A **57**, R28 (1998).

[13] G. J. Milburn, J. Corney, E. M. Wright, and D. F. Walls, Phys. Rev. A, **55** (1997) 4318.

[14] P. Villain *et al.*, J. Mod. Opt., **44** (1997) 1775.

[15] A. Smerzi, S. Fantoni, S. Giovanazzi, and S. R. Shenoy, Phys. Rev. Lett., **79** (1997) 4950.

[16] M. R. Andrews *et al.*, Science **275**, 637 (1997).

[17] B. P. Anderson and M. A. Kasevich, Science **282**, 1686 (1998).

[18] P. W. Anderson, *Basic Notions of Condensed Matter Physics* (Benjamin, Menlo Park, 1984).

[19] A. J. Leggett and F. Sols, Found. Phys. **21**, 353 (1991).

[20] Y. Castin and J. Dalibard, Phys. Rev. A **55**, 4330 (1997).

[21] J. Javanainen and S. M. Yoo, Phys. Rev. Lett. **76**, 161 (1996).

[22] J. I. Cirac, C. W. Gardiner, M. Naraschewski, and P. Zoller, Phys. Rev. A **54**, R3714 (1996).

[23] M. Naraschewski, H. Wallis, A. Schenzle, J. I. Cirac, and P. Zoller, Phys. Rev. A **54**, 2185 (1996).

[24] F. Dalfovo, S. Giorgini, L. P. Pitaevskii, and S. Stringari, Rev. Mod. Phys. **71**, 463 (1999).

[25] C. S. Adams, M. Sigel, and J. Mlynek, Phys. Rep. **240**, 143 (1994).

[26] K. Mølmer and Y. Castin, Quantum Semiclass. Opt. **8**, 49 (1996), and references therein.

[27] F. Sols, Physica B **194-196**, 1389 (1994).

[28] S. Kohler and F. Sols, cond-mat/9912252.

PHASE-COHERENT ELECTRONIC TRANSPORT IN A MULTI-WALL CARBON NANOTUBE

Nam Kim[1*], Jinhee Kim[1], Jong Wan Park[1], Kyung-Hwa Yoo[1], Jeong-O Lee[2], Kicheon Kang[2], Hyun-Woo Lee[3], and Ju-Jin Kim[2]

[1]*Korea Research Institute of Standards and Science, Taejon 305-600, Korea*
[2]*Department of Physics, Chonbuk National University, Chonju 561-756, Korea*
[3]*Department of Physics, Korea Institute of Advanced Science, Seoul 130-012, Korea*
*Present address; Dept. of Physics, University of Jyväskylä, FIN-40351 Jyväskylä, Finland.

Key words: carbon-nanotube, phase-coherent transport, mesoscopic systems

Abstract: Non-local electric transport phenomena were observed for a multi-wall carbon nanotube. The magnetic field dependence of non-local resistance was out of phase with respect to the conventional four-probe resistance, which could be explained in terms of the Landauer-Büttiker formula. Our observations indicate that the phase coherence length of multi-wall carbon nanotube exceeds the whole sample length of 3.8 μm up to the measured temperature 18 K.

1.   INTRODUCTION

Carbon nanotube (CNT) is considered to be a model system for the investigation of electric transport in low-dimensional system where quantum mechanical nature of electric conduction becomes apparent [1]. Despite about a decade of extensive studies, some issues on the electronic transport properties of carbon nanotube (CNT) is still on debates [1]. Especially, the magnitudes of the electron mean free path and the phase coherence length in a CNT is at the center of the debates. The experimental result by Frank *et al.* [2] revealed that the electron transport in CNT is ballistic even at room temperature indicating that the electron mean free path exceeds the sample size. Other experimental results [3,4], on the other hand, suggested that electrons move diffusively rather than ballistically in a CNT, where the phase coherence length was estimated to be much shorter than the CNT length.

To investigate whether the electrons move phase coherently over the whole CNT length or not, we have measured the magnetoresistance (MR) and the differential conductance characteristics for a multi-wall CNT both in the conventional four-probe and the nonlocal configurations. Pronounced variation of non-local resistance as well as four-probe resistance was observed with varying magnetic field. The non-local MR we have observed indicates that electrons move phase coherently over the whole length of the CNT, 3.8 μm, up to our measured temperature of 18 K. The characteristic field dependence of the non-local resistance was explained in terms of the Landauer-Büttiker formula [5].

2.   EXPERIMENTS

The MWCNT synthesized by arc discharge method were dispersed ultrasonically in chloroform for about half an hour. A droplet of dispersed solution containing CNT was dropped on the Si substrate with 500 nm-thick thermally-grown SiO_2 layer and each nanotube was positioned by using optical microscope and scanning electron microscope (SEM).

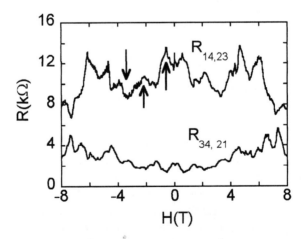

Figure 1. The magnetic field dependence of the four-probe resistance $R_{14,23}$ and the non-local resistance $R_{34,21}$ at T = 200 mK. The bias current was 5 nA. The three arrows denoted correspond to the magnetic field intensity of -0.6 T, -2.0 T, and -3.2 T where the four-probe MR shows global maximum, local maximum, and local minimum, respectively.

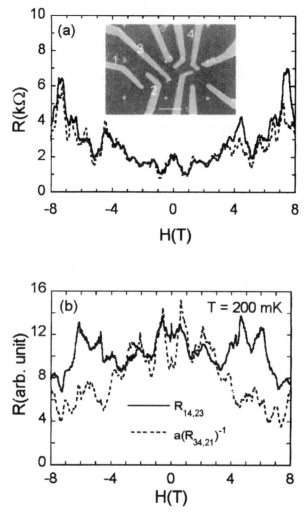

Figure 2. (a) The non-local MR for two different current-voltage configurations at 200 mK. Solid and dotted lines are $R_{34,21}$ and $R_{21,34}$, respectively. Inset shows the SEM photograph of the CNT measured and the electric leads labeled. The length of white-bar corresponds to 1 μm. (b) Comparison of the four-probe MR, $R_{14,23}$ (solid line) and the inverse of the non-local MR, $R_{34,21}^{-1}$ multiplied by a constant a = 20 (dotted line).

The patterns for electrical leads were generated using *e*-beam lithography technique onto the selected CNT and then 20 nm of Ti and 50 nm of Au were deposited successively on the contact area by thermal evaporation [6]. Shown in the inset of Figure 2(a) is the SEM photograph of the CNT with the electric leads labeled. The diameter of the selected CNT was about 25 - 30 nm. To form low-ohmic contacts between the CNT and the Ti/Au electrodes [7], we have performed rapid thermal annealing at 800 °C for 30 s. The contact resistances were below 3.5 kΩ in the whole temperature range measured.

We have employed two different configurations for the MR measurement. One is conventional four-probe configuration $R_{14,23}$: applying current with the leads 1 and 4 and picking up voltage drop between the leads 2 and 3. The other is non-local configuration $R_{34,21}$: injecting current through the leads 3 and 4 and measuring voltage between the leads 2 and 1. Figure 1 shows the four-probe MR, $R_{14,23}(H)$ and the non-local MR, $R_{34,21}(H)$ at $T =$ 200 mK with the bias current 5 nA. The magnetic field was applied along the direction of the tube axis. The four-probe MR curve exhibits a small oscillation imposed on a large variation, which looks very similar to those observed by Bachtold *et al* [3]. The periods of small oscillation and large variation are about 0.25 T and 5.5 T, respectively.

3.   RESULTS AND DISCUSSION

For the non-local MR, two characteristic features should be noticed. First, the non-local MR is almost symmetric with the magnetic field reversal, $R_{21,34}(H) \approx R_{21,34}(-H)$, and remains unchanged under the exchange of current and voltage leads, that is, $R_{21,34}(H) \approx R_{34,21}(H)$ [see Figure 2(a)]. Considering the distance between the leads 3 and 4 is about twice the distance between the leads 1 and 2, one can conclude that the characteristic length scale for this system is not the sample length but the electron phase coherence length which can be greater than the sample length. If the phase coherence length is greater than the sample size, the Büttiker's reciprocity relation [5], $R_{34,21}(H)$ $= R_{21,34}(-H)$, combined with the experimental fact $R_{21,34}(H) \approx R_{34,21}(H)$, gives the relation $R_{21,34}(H) \approx R_{21,34}(-H)$. This means that the non-local MR should be symmetric with the magnetic field, which was exactly the case as shown in the Figure 2(a). Actually, the non-local MR was not perfectly symmetric with the magnetic field but the asymmetric component was very small.

The other characteristic feature of the non-local MR becomes apparent if we compare the four-probe MR and the non-local MR. The non-local MR was out of phase with respect to the four-probe MR: the local minima of non-local MR correspond to the local maxima of four-probe MR and vice versa. To further clarify this point, we replotted in Figure 2(b) the four-probe MR and the reciprocal of the non-local MR multiplied by a constant. As shown in the figure, $R_{14,23}$ and $R_{34,21}^{-1}$ are almost in phase, especially at low-field region. The reason why we used the reciprocal of non-local.

We have also measured the differential conductance characteristics, *dI/dV,* in two different measurement configurations with varying magnetic field. Figure 3(a) and 3(b) show the three-dimensional surface plot of differential conductance curves as a function of magnetic field for the four-probe and the non-local measurement configurations, respectively. Note that the four-probe differential conductance curves show an energy gap structure of size below 1 meV modulated by magnetic field. This energy gap modulation with the magnetic field is reflected on the oscillating patterns of the four-probe MR data. For $H = -0.6$ T where the four-probe MR exhibited its global maximum as shown in Figure 1, the differential conductance curve showed the largest gap structure. For $H = -3.2$ T where the four-probe MR showed its local minimum, on the other hand, the differential conductance curve exhibited a suppressed energy gap structure. Three representative differential conductance curves at $H = -0.6$ T, -2.0 T, and -3.2 T are plotted in Figure 4. We emphasize that the four-probe and the non-local differential conductance curves are also out of phase as well as the four-probe and the non-local MR. As shown in Figure 3, the four-probe and the non-local differential conductance curves exhibit dramatically contrasting behaviours: the conductance dip in four-probe differential conductance curves corresponds to the conductance peak in non-local ones.

It is well known that the conductance of a sample with arbitrary shape can be described by Landauer-Büttiker formula in the limit where electrons transverse the sample without suffering phase-destroying events [5]. With the lead voltage $\{V_j\}$ and the transmission coefficients $\{T_{ij}\}$ given, the current flow I_i through the lead i can be written as

$$I_i = G_0 \sum_j T_{ij} V_j \qquad (1)$$

where

$$T_{ii} = -T_i = -\sum_{i \neq j} T_{ij} = -\sum_{i \neq j} T_{ji} \qquad (2)$$

In the ideal contact limit, T_{ij} vanishes if $|i\text{-}j| \geq 2$. Assuming that all the contacts in our sample are nearly ideal, we obtain after some algebra the four-probe resistance $R_{14,23}$ given by

$$R_{14,23} = \frac{1}{G_0 T_{32}} . \qquad (3)$$

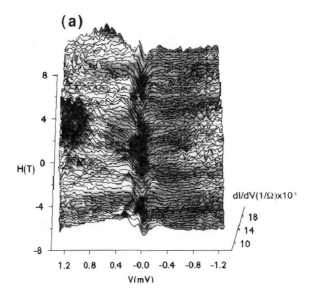

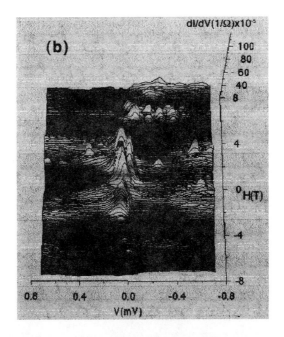

Figure 3. The perspective views of the differential conductance curves in the *V-H* plane for (a) four-probe (*I*:1,4 and *V*:2,3) and (b) non-local (*I*:3,4 and *V*:2,1) measurement configurations. The temperature was 200 mK.

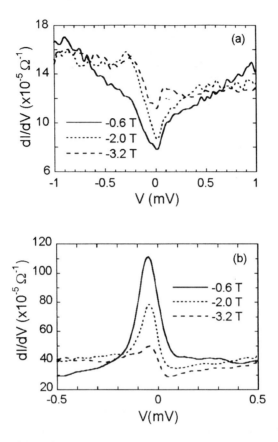

Figure 4. The differential conductance curves at magnetic fields $H = -0.6$ T, -2.0 T, and -3.2 T for (a) four-probe and (b) non-local measurement configurations. The temperature was 200 mK.

The non-local resistance $R_{34,21}$ vanishes in the ideal contact limit. In order to calculate the leading non-vanishing contribution to $R_{34,21}$, one should restore all T_{ij}. Assuming that T_{ij} for $|i\text{-}j| \geq 2$ are much smaller than other matrix elements, one obtains in the leading order

$$R_{34,21} \approx \frac{1}{G_0} \left| \frac{T_{14}}{T_{12}T_{34}} - \frac{T_{13}T_{24}}{T_{12}T_{34}} \frac{1}{T_{23}} \right|. \tag{4}$$

Now let us define τ_2, τ_3, τ_{23} by

$$T_{13} \equiv T_{12}T_{23}\tau_2, \quad T_{24} \equiv T_{23}T_{34}\tau_3, \quad T_{14} \equiv T_{12}T_{23}T_{34}\tau_{23} \qquad (5)$$

where τ_2, τ_3, τ_{23} are small by definition. Then one finds

$$R_{34,21} \approx \frac{1}{G_0} |\tau_{23} - \tau_3\tau_2| T_{32} \qquad (6)$$

assuming $T_{23} = T_{32}$ in the first approximation.

Note that T_{32} now appears in the numerator while it does in the denominator in case of $R_{14,23}$. In the leading order, $R_{34,21}$ is the reciprocal of $R_{14,23}$ multiplied by the correction factor $|\tau_{23} - \tau_2\tau_3|$. The experimental results suggested that the correction factor $|\tau_{23} - \tau_2\tau_3|$ could be insensitive to the magnetic field, especially at low field region. And if this is the case, $R_{34,21}(H)$ becomes almost symmetric with respect to the magnetic field reversal since $T_{32}(H) \approx T_{32}(-H)$ for the almost ideal contacts. This is consistent with the experimental result that the non-local MR is almost symmetric with respect to the magnetic field reversal.

In order to observe such dramatic non-local transport phenomena, all the contact must be close to the ideal ones but not perfectly. If they were perfect ideal contacts, no non-local transport phenomena would be observed. Actually because of the mismatch of the Fermi energy levels between the CNT and the Ti/Au electrodes, it is natural to consider the existence of the scattering barriers at the interface between the CNT and the electrodes. Such scattering barriers might produce the nonlocal MR and the MR asymmetry even in the absence of insulating barrier between the CNT and the electrodes. According to the Landauer-Büttiker formula, the non-local transport phenomena we have observed are inter-related to the MR asymmetry.

We have shown that four-probe differential conductance curve exhibited energy gap-like structure. As for the origin of this energy-gap like structure, one probable explanation is the energy quantization along the direction of the tube axis. The energy level spacing estimated using the formula [1] $\Delta E = hv_F/2L$ with $v_F = 8.13 \times 10^5$ m/sec and $L = 3.8$ μm is 0.44 meV which agrees well with the observed size of the energy-gap like structure [see Figure 4(a)]. Also should be addressed is the two distinct oscillation periods in the four-probe MR. Ajiki and Ando [8] predicted that the electronic band structure of CNT changes drastically with the magnetic flux threading through the tube with the period of $\Phi_0 = h/e$. If we assume the diameter of our CNT is 30 nm, the single flux quantum h/e through the cylinder corresponds to $H = 5.5$ T which is not far from the period of MR variation observed (Figure 1). The small oscillation period can be attributed to the Aharonov-Bohm(AB) effect combined with the CNT chirality. As pointed out by Bachtold *et al.*[3], if CNT has any chirality, conduction path can also have chirality and electrons prefer a helical trajectory. The electrons started at one end of the CNT will arrive at the other end after surrounding the CNT many times and picking up AB phase.

We have also measured the temperature dependence of the nonlocal MR with the magnetic fields applied in the perpendicular direction to the CNT axis. We could change the direction of applied magnetic fields by exchanging the sample stages in the room temperature. Figure 5 shows the conventional and nonlocal MR curves in three different temperatures of 80 mK, 4.1 K, and 18.7 K. It is clearly seen that nonlocal MR survives up to 18.7 K, although the amplitude of variation of MR is decreased as temperatures increased. As a result it is believed that electrons moved phase coherently over the whole length of CNT, 3.8 μm at least up to 18 K.

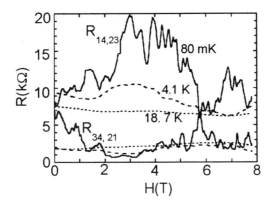

Figure 5. Conventional and nonlocal MR curves in various temperatures for magnetic fields perpendicular to the direction of the CNT axis. Solid , dashed, and dotted lines correspond to 80 mK, 4.1 K, and 18.7 K respectively.

In summary, we have observed non-local transport phenomena in a multi-wall CNT. The non-local MR measured was out of phase with respect to the conventional four-probe MR, which we explained in terms of the Landauer-Büttiker formula. Our observations indicate that the electron phase coherence length of multi-wall CNT exceeds the whole sample length.

ACKNOWLEDGMENTS

We would like to express sincere thanks to Prof. K.-J. Chang and H.-S. Sim at KAIST for fruitful discussions. This work was supported by MOST through Nano Structure Project and also by Korea Research Foundation Grant (KRF-99-015-DP0128) and also by KRISS (Project No. 00-0502-001).

REFERENCES

[1] C. Dekker, Physics Today, May, 22 (1999); M. Bockrath *et al.*, Nature **397**, 598 (1999); L. C. Venema, *et al.*, Science **283**, 52 (1999).
[2] S. Frank, P. Poncharal, Z. L. Wang, W. A. de Heer, Science **280**, 1744 (1998).
[3] A. Bachtold, *et al.*, Nature **397**, 673 (1999).
[4] C. Schönenberger, *et al.*, Appl. Phys. **A 69**, 283 (1999).
[5] M. Büttiker, Phys. Rev. Lett. **57**, 1761 (1986).
[6] J.-O. Lee *et al.*, Phys. Rev. B **61**, R16362 (2000).
[7] Y. Zhang *et al*, Science **285**, 1719 (1999).
[8] H. Ajiki, and T. Ando, J. of Phys. Soc. Jpn. **62**, 1255 (1993).

MACROSCOPIC QUANTUM PHENOMENA IN UNDERDAMPED JOSEPHSON JUNCTIONS

V. Corato, E. Esposito, C. Granata, B.Ruggiero, M. Russo,
and P. Silvestrini[*]

Macroscopic Quantum Coherence and Computing (MQC²) Group
Istituto di Cibernetica del CNR Via Toiano 6 I-80072 Arco Felice, Italy
MQC-INFN, Sezione Napoli, Via Cintia, I-80126, Naples, Italy
[*]e-mail:p.silvestrini@cib.na.cnr.it

Abstract   We present an experimental study of the supercurrent decay in extremely underdamped Josephson junctions. In the thermal regime, we present data on the rate of escape from the supercurrent state as function of the external energy both in quasi-stationary and non-stationary conditions for the system. We evaluate the relevant junction parameters and the effective dissipation and we get experimental evidence of energy level quantization at temperatures well above the classical-quantum crossover one. In the quantum limit we show results on macroscopic quantum tunneling in different extremely underdamped Josephson junctions. All measurements employ low noise electronics, as well as different techniques to reduce the effective dissipation by filtering the lines wiring the junction. The extremely low value of dissipation obtained in the observation of incoherent effects in underdamped Josephson junctions, is encouraging in view of experiments of coherent tunneling between energy levels in an rf-squid system.

Keywords: Josephson Effect, Macroscopic Quantum Tunneling and Coherence

1.   INTRODUCTION

Macroscopic quantum effects in Josephson systems have attrached great interest in the scientific community both for the physics involved and in view of applications [1,2]. The Josephson effect represents a very powerful tool to experimentally investigate a variety of interesting phenomena, including macroscopic quantum tunneling (MQT), and energy level quantization (ELQ) [3-7]. In this context the most fashinating topic is the observation of macroscopic quantum coherence (MQC) in Josephson systems[1,2,8]. This effect also has implications for quantum computing, because a coherent two-level system represents a single q-bit (Quantum BInary digiT), the elementary unit of quantum computer[9]. Recently important steps towards the realization of a q-bit based on a Josephson system have been reported in literature[10,11]. Such proposals are based on a nanometer-scale q-bit[10] or on a q-bit consisting of a micrometer-sized loop with three Josephson junctions[11].

In this paper, we show results on ELQ and MQT in extremely underdamped Josephson junctions with different classical-quantum crossover temperatures and in different regimes. In the thermal regime, we present data on the rate of escape from the supercurrent state as function of the external energy both in quasi-stationary and non-stationary conditions for the system. In the quasi-stationary case we evaluate the relevant junction parameters and the effective dissipation, in the non-stationary limit we get experimental evidence of energy level quantization at temperatures well above the classical-quantum crossover one. Furthermore, in the quantum limit we show results on macroscopic quantum tunneling in different extremely underdamped Josephson junctions.

All measurements have been performed at CNR-IC employing low noise electronics, and different techniques to reduce the effective dissipation by filtering the lines wiring the junction[12,6]. The extremely low value of dissipation obtained is encouranging in view of experiments of coherent tunneling between energy levels in a rf squid system [8], or to measure quantum decoherence between different fluxoid states of an rf-squid by using the procedure of "adiabatic inversion"[13].

2.   THEORETICAL BACKGROUND

The macroscopic quantum variable describing the junction dynamics is the phase difference φ between the macroscopic wave functions of the two superconductors forming the junction. Within the RSJ model the system dissipation is described by an effective resistance R[3]. In this way the junction behavior can be understood in analogy with the motion of a particle in a washboard potential : $U(\varphi)= - U_0 (\alpha \varphi + \cos \varphi)$, where $U_0=\hbar I_c/2e$, α is the bias current I normalised to the critical one, $\alpha=I/I_c$. The Josephson state corresponds to the particle trapped in one of the minima of the potential. Increasing the current, the barrier height is decreasing, and once is small enough we can have the escape from the potential well by thermal activation[14] or by quantum tunneling[1-4]. It is typically defined a crossover temperature between thermal and quantum regime[15] as $T_0=\hbar\omega_j/2\pi k$, where ω_j is the plasma frequency of the junction, $\omega_j=(2\pi I_c/\phi_0 C)^{1/2}(1-\alpha^2)^{1/4}$. Here ϕ_0 is the magnetic flux quantum, and C the junction capacitance.

The thermal activation theory for underdamped systems (namely hysteretical Josephson junction as used in the experiment) predicts, for the escape rate Γ out of the V=0 state of the junction, the following expression[14]:

$$\Gamma = A_t \frac{\omega_j}{2\pi} \exp\left[-\frac{E_b}{kT}\right] \tag{1}$$

where E_b is the barrier energy of the metastable state, which decreases with increasing the bias current : $E_b(\alpha)=U_0[2\alpha\sin^{-1}(\alpha)+2(1-\alpha^2)^{1/2}-\pi\alpha]$. The prefactor A_t depends on the damping η related, within the RSJ model, to a junction "effective" resistance R and capacitance C ($\eta=\frac{1}{RC}$)[3,14]. In underdamped junctions $((\omega RC)^{-1}<<1)$, A_t is responsible for two different damping regimes: the transition rate regime (TR) with A_t almost independent of η, and the extremely low damping regime (LD) with A_t proportional to η[12,14,16].

In the quantum regime we must consider N energy levels in each well of the potential describing the system, and transitions from the i-th (j-th) into the j-th (i-th) level will occur at a rate w_{ij} (w_{ji}), due to the interaction with the thermal bath. The escape process is well described by the kinetic equation for the probability ρ_j of finding the system in the j-th energy level E_j[15]:

$$\frac{\partial \rho_j}{\partial t} = \sum_i \left(w_{ji}\rho_i - w_{ij}\rho_{ji}\right) - \gamma_j\rho_j \qquad i,j=0, \ldots\ldots,N \tag{2}$$

where γ_j is the tunneling probability through the barrier which strongly depends on the energy level position E_j. The escape rate Γ is the sum of the contributions of tunneling from the various levels, $\Gamma= \Sigma_j\gamma_j\rho_j$, plus the thermal hopping. For $T<T_0$ the system is frozen in the ground state and the escape rate is just due to the tunneling. Experiments at $T< T_0$ on MQT and ELQ in moderate underdamped Josephson junctions have been performed[3]. For $T>T_0$, the thermal diffusion process from the bottom levels towards the top of the barrier must be taken into account. No quantum effects can be observed as long as the characteristic time of the escape process will be dominated by the thermal diffusion to reach the top of the barrier. This is the picture in quasi-stationary conditions[15], which are realised for a steady current or for sweeping frequency very low with respect to the characteristic time of the thermal diffusion(dN/dt<< w_{ij}).

The presence of discret energy levels can be revealed by a fast sweep of the external current, namely in non-stationary conditions[17]. A clear manifestation of the presence of quantized energy levels for $T > T_o$ have been recently reported[6,18].

3. EXPERIMENTS IN UNDERDAMPED JOSEPHSON JUNCTIONS IN THE THERMAL REGIME

In the experiments the decay from the metastable state is reveled by measuring the switching current from the V=0 state to the V≠0 state occurs randomly following a certain distribution P(I), related to escape rate $\Gamma(I)$ by the following relation:

$$P(I) = (dI/dt)^{-1} \Gamma(I) \exp\left[-\int_0^I (dI'/dt)^{-1} \Gamma(I) \, dI' \right] \tag{3}$$

where dI/dt is the bias current rate.

In our measurements we use high quality Nb-AlOx-Nb Josephson tunnel junctions which exhibited a very low leakage current ($V_m > 80$ mV), and a quite uniform critical current density J_C. Note that the intrinsic dissipation parameter depends on the critical current density, but not on the junction area A_j. In this section we present data on two Josephson juctions: $J_C \cong 50$ Acm^{-2} at T=4.2 K, $A_j \cong 7 \times 7$ μm^2, and C=2.5±0.5 pF for sample 1, $J_C \cong 700$ Acm^{-2} at T=4.2 K, $A_j \cong 5 \times 5$ μm^2, and C=1.5±0.3 pF for sample 2. Measurements were performed on a pumped liquid ^{4}He cryostat with three μ-metal shields and a copper layer to reduce electromagnetic noise. The sample was mounted on a chip carrier especially designed for the experiment, which contained an integrated low pass frequency filtering stage located very close to the junction. All the connections to room temperature went through manganine wires. It is of fundamental importance in our experiments to have a very low dissipation level taking into account that the effective resistance in this kind of experiment may be limited by any external shunting impedance[3]. In our case we had a 100 KΩ limiting resistor integrated in the chip carrier and located close to the junction, while a great care has been devoted to avoid stray capacitance, which may reduce the real part of the complex impedance. Measurements were performed following the standard time-fly technique[12]obtaining P(I) and $\Gamma(I)$. In Fig.1 we report the measured distribution width σ, $\sigma = (<I^2> - <I>^2)^{1/2}$, as function of temperature T for

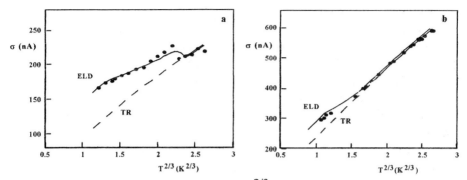

Figure1 Distribution width σ as function of $T^{2/3}$ for : a) sample 1 and b) sample 2. Experimental data (dots) are compared with the Buttiker-Harris-Landauer theoretical predictions (solid lines; ELD)[14]. In the theoretical curves the effective resistance is assumed to be the subgap resistance R_{qp}. The dashed lines are the predictions of the simple transition rate theory (TR). Note the different temperature range for the crossover between the two thermal regimes as expected for junctions with different dissipation level (at T=1.2 K: R=20 KΩ for sample 1 and R=2 KΩ for sample 2).

samples 1 and 2 together with the theoretical predictions for the transition rate(TR) and extremely low damping(ELD) regimes[14].

In the theoretical curves, the relevant junction parameters are independently determined from the I-V characteristics: the critical current has been measured as a function of the temperature, and the capacitance is measured from the Fiske step position. In our measurements, any residual amount of external noise is negligible(T_n=0), and the best fitting values of the resistance R are compatible with the subgap resistance independently measured on the I-V curve. *Therefore in our experimental setup the effective dissipation is dominated by intrinsic mechanism.* The experimental data show an evident crossover between the two regimes which well fits the theoretical predictions. We note that a higher level of dissipation correctly shifts the crossover region towards lower temperatures(Fig.1b). The agreement with the theory, obtained *without free fitting parameters*, is quite convincing for both the junctions used in the experiment, and this sets the relevant junction parameters useful for quantum measurements.

4. ENERGY LEVEL QUANTIZATION

Previous experiments show quantum effects in Josephson systems at temperatures below the crossover temperature T_0[3-5], in order to guarantee the condition that the thermal energy be very low with respect to the energy level spacing[3]. In a previous experiment[6], we find that, increasing the sweeping frequency of the external bias, the rate of escape from the supercurrent state is an oscillating function of the bias current at temperature above T_o. Those data were consistent with a quantum picture, assuming that the tunneling through the barrier can only occurr from quantized energy levels. However, the experimental resolution could not exclude other possibilities related to resonances at the plasma frequency. Very recently[18], the achieved time resolution and noise condition allowed us to observe the oscillatory behaviour of the rate of escape in Josephson junctions with significantly different expected spacing of the macroscopic energy levels, as well as for significantly different sweeping frequency of the external bias, before recovering the usual thermal activation at lower sweeping frequency. This guarantees at the first sight that the observed oscillation is definitely associated to the energy change of the external bias leading to the energy level quantization in Josephson junctions at high temperatures ($T>T_o$). The experimental setup to perform such kind of measurements is described elsewhere[6,18], and here we present some data on ELQ in Josephson junctions in non-stationary regime. In order to have different energy level spacing, Nb-AlO$_x$-Nb Josephson junctions with different critical current densities J_c, are chosen. The junction parameters independently measured at 1.3 K for sample 3 (sample 4) are: I_c=26 ± 0.5 µA (I_c=80 ± 1 µA) and C=3.3 ± 0.4 pF (C=1.2 ± 0.2 pF), with $J_c \cong 52$ A/cm^2 ($J_c \cong 700$ A/cm^2). The system dissipation is determined from the fitting of data in the pure thermal limit (see the previous section), namely from the low frequency measurements, and in this case results an effective resistance of R=20±10 kΩ (R=10±5 kΩ). This way of measuring the average dissipation of the whole system includes the contribution from the shunting (frequency dependent) impedance of the load line[3], as well as the intrinsic one (quasiparticle resistance)[12,16]. The thermal independent measurements allow us to determine all the relevant quantum parameters[15], namely energy levels, tunneling rates, thermal rates, and the plasma frequency, to compare data by theory. The crossover temperature has been also independently determined by measuring at low temperature (down to 50 mK) the well known transition of the quasi-stationary distributions in the quantum regime(see next section)[3]. All the measurements lead to the same result for the relevant junction parameters, confirming the correctness of our quantum picture of the junction. In Fig.2 we report both the experimental histograms P and escape rate Γ as functions of the external current I biasing the junction, obtained for sample 3 at T=1.3 K. The sweeping frequency is high enough to induce the observation of quantum effects, dI/dt>20 A/s. In Fig.3, P(I) and Γ(I) are reported at T=1.2 K for sample 4. The raw experimental data already give evidence of energy level quantization.

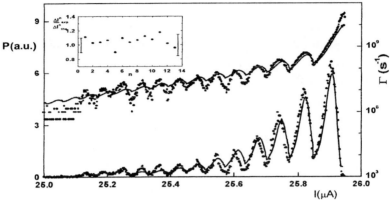

Figure 2 Switching current distribution P (left axis, lower curve) and escape rate of the supercurrent state Γ as a function of the current I (right axis, upper curves) for sample 3. Dots refer to data taken at T=1.3 K and dI/dt = 42 A/s. The solid lines are the theoretical predictions from non-stationary theory on *ELQ*[17] with the following junction parameters: I_c=26 μA, C=3.3 pF, R= 20 kΩ, dI/dt= 42 A/s and T=1.3 K. In the inset the experimental current spacing $\Delta I_{exp}=I_{min}^N-I_{min}^{N-1}$ normalized to the expected one[15], ΔI_{the}, is shown. The error bars include the uncertainty on the time resolution and the statistical one as well.

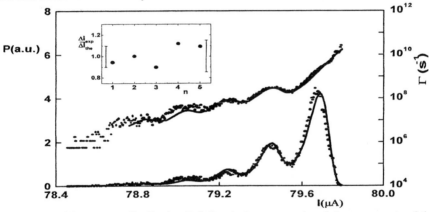

Figure 3 Switching current distribution P (left axis, lower curve) and the escape rate of the supercurrent state Γ as a function of the current I (right axis, upper curves) for sample 4. Dots refer to data taken at T=1.2 K and dI/dt = 102 A/s. The solid lines are theoretical predictions from non-stationary theory on *ELQ*[17] refer to: I_c=80 μA, C=1.2 pF, R= 10 kΩ, T=1.2 K, and dI/dt = 102 A/s. In the inset the experimental current spacing $\Delta I_{exp}=I_{min}^N-I_{min}^{N-1}$ normalized to the expected one[15], ΔI_{the}, is shown. The error bars include the uncertainty on the time resolution and the statistical one as well.

5. MACROSCOPIC QUANTUM TUNNELING

In this section we present data on two Josephson junctions measured in the quantum regime (T<T_0): $J_c \cong$ 140 Acm^{-2} at T=50 mK, $A_j \cong$6x6 μm^2, and C=2.0±0.4 pF-sample 5; $J_c \cong 490$ Acm^{-2} at T=50 mK, Aj$\cong$4x4 μm^2, and C=0.70±0.14 pF-sample 6. For the two junctions we expect a different crossover temperature: $T_0 \sim 125$ mK(sample 5) and $T_0 \sim 330$ mK (sample 6).

These measurements were performed on ^{3}He/^{4}He Oxford dilution cryostat having a base temperature of 20 mK. The chip carrier inserted into the mixing chamber, was especially designed for the experiment and contained a filtering stage located very close to the junction. We used thermocoaxes [19] as wires coming from the junction up to 4.2 K. The electrical lines between 4.2 K and room temperature were made of cryogenic coaxes[20] with the addition of a radio-frequency filtering stage at 4.2 K. The whole filtering system assures an attenuation of 200 dB at 50 GHz. In Fig.4 we report data and theoretical predictions in the thermal regime for the distribution width σ as function of temperature for the two samples.

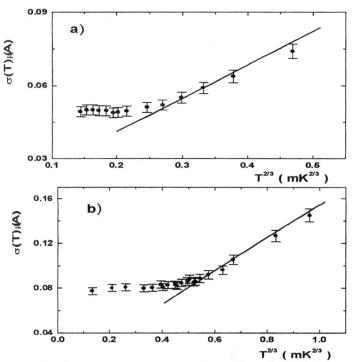

Figure 4 Distribution width σ as function of T $^{2/3}$ for Nb/AlO$_x$/Nb underdamped Josephson junctions. a) sample 5: $J_c \cong 140$ Acm^{-2} at T=50 mK, Aj$\cong$6x6 μm^2, C=2.0±0.4 pF, and dI/dt= 95 mA/sec ; b) sample 6: $J_c \cong 490$ Acm^{-2} at T=50 mK, Aj$\cong$4x4 μm^2, C=0.70±0.14 pF, and dI/dt= 57 mA/sec . Experimental data (dots) are compared with the theoretical predictions in the thermal regime[14](solid line). The theoretical curves are obtained for R=24 KΩ and C=1.6(sample 5), and for R=27 KΩ and C=0.6 (sample 6). The experimental extrapolation of the crossover temperatures for the two junctions gives: T_0=125 mK(sample 5), and T_0=330 mK(sample 6).

We note that for $T < T_0$ for both junctions we observe the expected saturation as due to MQT[3].

The experimental extrapolation of the crossover temperatures T_0 for the two different samples are in agreement with the expected ones within the experimental uncertainty. In the comparison between data and theory, we again use for the junction parameters the values independently determined from the current-voltage characteristics (I_c and C). We note that any residual amount of external noise is negligible in these measurements(T_n=0), and the fitting values of the resistance R are compatible with the subgap resistance independently measured on the I-V curve. *This gives an effective dissipation in the quantum regime due to intrinsic mechanisms.*

6. CONCLUSION

We have presented data and theoretical predictions on the switching current distribution P(I) and on the escape rate $\Gamma(I)$, as functions of the bias current I in different temperature ranges in respect to the classical-quantum crossover temperature T_0. For $T < T_0$, in quasi-stationary conditions (dI/dt up to 0.3 Asec^{-1}), we observe the expected saturation of the switching distribution width σ decreasing the temperature T, as due to MQT. For $T > T_0$, extending measurements at higher sweeping frequency (dI/dt up to 100 Asec^{-1}), we get evidence of ELQ at high temperature (up to 2 K) due to the transition of the system in non-stationary regime[4].

Our experiments extend the condition for which the thermal mixing of eingenstates does not destroy quantum effects. In fact, in the presence of low dissipation level for the system, this typically occurs when the thermal energy is much less than the quantum one[3]. Here we report data which realize this condition even if the rate of change of the external energy is fast with respect to the rate of thermally induced energy transitions between levels. This condition is realized at high temperatures by a fast sweeping of the external bias, provided that the lifetime of the quantum states is much longer than the characteristic time of the system (level width much smaller than the level separation).

The extremely low value of dissipation obtained is encouraging in view of experiments of coherent tunneling between energy levels in a rf squid [2,8]. In particular, work is in progress to measure the decoherence time in an rf-squid system by adiabatic inversion technique[13].

ACKNOWLEDGMENTS

The authors are grateful to D. Esteve for stimulating discussions. We thank to A.Andreone, A. Monaco, V.G.Palmieri, and G. Testa for help in the experimental work. Thanks are due to L. Serio and P. Alicante for technical assistance. This work is partially supported by INFN under MQC project.

REFERENCES

1. Leggett A.J., in: Kagan Yu. and Leggett A.J. (eds), *Quantum Tunneling in Condensed Media*, Elsevier Science, (1992) p. 2; Caldeira A.D. and Leggett A.J., Influence of Dissipation on Quantum Tunneling in Macroscopic Systems. *Phys. Rev. Lett.* **46** (1981) pp. 211-214.
2. Proceedings of Int. Workshop on "*Macroscopic Quantum Tunneling and Coherence*", Napoli, Italy 1998, Special Issue of *J. of Superconductivity* **12 (6)** (1999).
3. Martinis J.M., Devoret M.H. and Clarke J., Experimental Tests for the quantum behavior of a macroscopic degree of freedom: The phase difference across a Josephson junction. *Phys. Rev.* **B35** (1987) pp. 4682-4698; and references therein.
4. Schwartz D.B., Sen B., Archie C.N. and Lukens J.E., Quantitative Study of the Effect of the Environment on Macroscopic Quantum Tunneling. *Phys. Rev. Lett.* **57** (1985) pp. 1547-1550.
5. Rouse R., Han S. and Lukens J.E., Observation of Resonant Tunneling between Macroscopically Distinct Quantum Levels. *Phys. Rev. Lett.* **75** (1995) pp. 1614-1617.
6. Silvestrini P., Palmieri V.G., Ruggiero B., and Russo M., Observation of Energy Levels Quantization in Underdamped Josephson Junctions above the Classical-Quantum Regime Crossover Temperature. *Phys. Rev. Lett.* **79** (1997) pp. 3046- 3049.
7. Cosmelli C., Carelli P., Castellano M.G., Chiariello F., Diambrini Palazzi G., Leoni R., and Torrioli G., Measurements of the Intrinsic Dissipation of a Macroscopic System in the Quantum Regime. *Phys. Rev. Lett.* **82** (1999) pp. 5357- 5360.
8. Nakamura Y., Pashkin Yu. A. and Tsai J. S., Coherent control of macroscopic quantum states in a single-Cooper-pair box. *Nature* **398** (1999) pp. 786-788; Friedman J., Patel V., Chen W., Tolpygo S.K., and Lukens J.E.,Quantum Superposition of Distinct Macroscopic States. *Nature* **406**, (2000) pp. 43-46; Van der Wal C.H., ter Haar A.C.J., Wilhelm F.K., Schouten R.N., Harmans C.J.P.M., Orlando T.P., Lloyd S., and Mooij J.E., Quantum superposition of macroscopic fluxoid states *(this volume)*.
9. Averin D. V., Solid State qubits under control. *Nature* **398** (1999) pp. 748-749.
10. Makhlin Y., Schon G., and Schirman A., Josephson-junction qubits with controlled couplings. *Nature* **398** (1999) pp. 305-307.
11. Mooij J. E., Orlando T.P., Levitov L., Tian L., Van der Wal C., and Lloyd S., Josephson persistent current qubit. *Science* **258** (1999) pp.1036-1042.
12. Ruggiero B., Silvestrini P., Granata C., Palmieri V.G., Esposito A., and Russo M., Supercurrent decay in extremely underdamped Josephson junctions. *Phys. Rev.* **B57** (1998) pp. 134-137; Ruggiero B., Granata C., Esposito E., Russo M., and Silvestrini P., Extremely underdamped Josephson junctions for low noise applications. *Appl. Phys. Lett.* **75** (1999) pp.121-123.
13. Silvestrini P., and Stodolsky L., Study of Macroscopic Coherence and Decoherence in the squid by Adiabatic Inversion. *cond. mat./* **0004472** (2000*); Ruggiero B., Corato V., Esposito E., Granata C., Russo M., Silvestrini P., and Stodolsky L., Macroscopic Quantum Coherece in Josephson Systems. *Int.Jour. of Mod. Phys.(in press)*

14. Buttiker M, Harris E P, and Landauer R., Thermal activation in extremely underdamped Josephson-junction circuits. *Phys. Rev.* **B28** (1983) pp. 1268-1275;. Kramers H.A., Browian motion in a field of force and the diffusion model of chemial reactios. *Physica* **7** (1940) pp.284-304.

15. Larkin A.I., and Ovchinnikov Yu.N., Effect of level quantization on the lifetime of metastable states. *Sov. Phys. JETP* **64** (1986) pp. 185-189 [*Zh. Eksp. Teor. Fiz.* **91** (1986) pp. 318-325].

16. Silvestrini P., Liengme O., and Gray K.E., Current Distributios of Thermal Switching in Extremely Underdamped Josephson Junctions, *Phys. Rev.* **B37** (1988) pp.1525-1531, Silvestrini P., Pagano S., Cristiano R., Liengme O., and Gray K.E., Effect of Dissipation on Thermal Activation in an Underdamped Josephson Junction: First Evidence of Transition between Different Damping Regimes. *Phys. Rev. Lett.* **60** (1988) pp. 844-847.

17. Silvestrini P., Ruggiero B., and Ovchinnikov Yu. N., in *"Exploring the Quantum-Classical Frontier: Recent Advances in Macroscopic and Mesoscopic Quantum Phenomena"*, Friedman J.R., and Han S. eds., Plenum pub. NY, USA (in press).

18. Silvestrini P., Ruggiero B., Granata C., and Esposito E., Supercurrent Decay of Josephson Junctions in non-stationary conditions: experimental evidence of macroscopic quantum effects. *Phys. Lett.* **A267** (2000) pp. 45-51.

19. Zorin A.B., The thermocoax cable as the microwave frequency filter for single electron circuits. *Rev. Sci. Instrum.* **66** (1995) pp.4296-4300.

20. Glattli D.C., Jacques P., Kumar A., Pari P., and Saminadayar L., A noise detection scheme with 10 mk noise temperature resolution for semiconductor single electron tunneling devices *J.Appl.Phys.* **81** (1997) pp. 7350-7356.

SUPERCONDUCTING DEVICES TO TEST MACROSCOPIC QUANTUM COHERENCE ON THE FLUX STATES OF A RF SQUID

C. Cosmelli[1,2], P. Cappelletti[1], P.Carelli[3], M.G. Castellano[2,4], F. Chiarello[1,2], G. Diambrini Palazzi[1,2], R. Leoni[2,4], N. Milanese[1], G. Torrioli[2,4]

[1]*Dipartimento di Fisica, Universita` La Sapienza, 00185 Roma, Italy,* [2] *Istituto Nazionale di Fisica Nucleare, 00185 Roma, Italy,* [3] *Dipartimento di Energia Elettrica, Universita` dell'Aquila, Monteluco di Roio, 67040 L'Aquila, Italy,* [4] *Istituto di Elettronica dello Stato Solido, CNR, Via Cineto Romano, 00156 Roma, Italy.* [*]

Key words: Macroscopic Quantum Coherence, SQUID, Laser Switch.

Abstract: To perform tests of Macroscopic Quantum Coherence (MQC) on a SQUID system we have realised a set of chips containing all the different devices necessary for the measurements at a temperature of 10mK. In this paper we will report on the characteristics of all the devices in the chip as well as the measurement performed on each device. We will also describe briefly the measurements we plan to perform on the experimental apparatus.

INTRODUCTION

In 1980 T. Leggett [1] proposed a set of measurements on a rf SQUID to test one of the puzzling problems of physics: can a macroscopic object be described by means of Standard Quantum Mechanics (QM) theory, in contrast with the predictions given by the Macrorealism coming from Classical Mechanics (CM) interpretation? The proposed test was essentially

[*] Supported by Istituto Nazionale di Fisica Nucleare, Italy, under the MQC project.

a "macroscopic" dual of the experiments done by A. Aspect and co-workers to test Bell inequalities on polarised photons [2].

The Leggett proposal was to make the same kind of test on the two flux states associated to an rf SQUID polarised by half of the quantum flux.

The main difference with respect to the microscopic tests was that the usual Bell inequalities, were the measurements are done at the same time on two far away objects, in the SQUID are tested on the same object but at different times.

At the time of the first Leggett proposal the technology to construct high quality Josephson junctions was still not developed, so the first measurements made to test the Rabi oscillations of the rf SQUID (the first step on the overall set of measuring procedure) failed due to the very high level of dissipation of the device.

After the developing of trilayer technology to realise very high quality Nb/AlOx/Al Josephson junction, it was finally possible to try the Leggett experiment, and many groups started an experimental program to perform the measurements proposed by Leggett (ref. [3] and references therein).

Up to know the first indication of a superposition between two excited states in a SQUID system has been recently observed by the groups of Stony Brook and Delft [4][5]. The direct detection of the Rabi oscillations, however, has not jet observed, and no test on CM vs. QM predictions has yet been performed.

In this paper we will describe how we plan to realise an experiment were the Rabi coherent oscillations between flux states of an rf SQUID can be directly measured. We will report on all the characteristics of the different devices on our integrated chip were all the parts necessary to perform the measurements are integrated. We will describe briefly the technique proposed to realise the measurements [6], this technique being different from the first proposed by Leggett.

THE EXPERIMENTAL APPARATUS

The experimental SQUID system is composed by four different devices:

- The laser switch (for the state preparation)
- A double rf SQUID (the source of the MQC state)
- A shunted dc SQUID (the amplifier)

- An hysteretic dc SQUID (the switch for the state probing)

The laser switch has been realised on a separate chip, while the three other devices, arranged as shown in fig.1, are realised on the same chip starting from a trilayer Nb/AlOx/Nb with critical current of about 400A/cm^2.

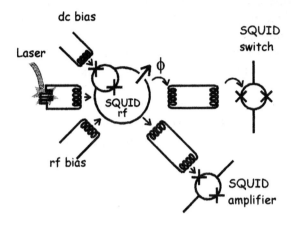

Figure 1. Scheme of the experimental apparatus

The first device at left is a laser switch, designed to give the starting time for the free oscillations of the flux. Then there is the double SQUID, an rf SQUID having the critical current tunable with an external flux. The lower dc SQUID is a standard SQUID amplifier to monitor the proper functioning of the rf SQUID, and it is supposed to be turned off when we want to do the true measurements. The last dc SQUID is a switch SQUID designed to perform stroboscopic Non Invasive Measurements (NIM) of the rf SQUID flux.

In the following paragraph we will describe the devices designed for this experiment and the test done on them.

THE LASER SWITCH

The first problem in measuring the Rabi oscillations in the time domain is the preparation of the state in one of the well and the knowledge of the time when the system begins the free evolution.

This task can be accomplished by means of a laser switch. This part of the experimental set up is composed by a superconducting Nb loop inductively coupled to the rf SQUID. A small part of the loop, designed as a meander line, can be illuminated by short pulses of laser light which drives the Niobium film into the normal state. The overall laser operation is the following: suppose to start with the rf SQUID potential well balanced by half quantum flux; then a superconducting permanent current is stored in the loop; this current makes an unbalance in the rf SQUID potential so that the flux state is stabilised in one of the two wells (the Left f.i.). A the time *t* a laser pulse is sent on the meander, driving it into the normal state, and the persistent current decays to zero in a time of the order of tenths of ns. The SQUID therefore will be balanced and begins the free oscillations starting from the Left well. The switch we have realised is done with 24 lines in the form of a meander having 3 μm width, 30nm thickness and total dimension of 150 μm.

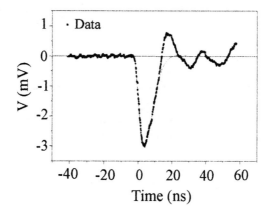

Figure 2. Voltage pulse read across the meander following the laser irradiation; the width of the pulse gives the time resolution of the system.

The decay time of a current stored in the loop (see fig.2) is approximately 10 ns, well within the time resolution required to measure oscillations in the range of few MHz. In the case of faster oscillations the time resolution of the switch can be easily reduced (by varying the loop parameters) to the value of fraction of ns, the intrinsic time response of the film being hundreds of ps.

THE DOUBLE SQUID

On of the major problems in realising the set up to measure tunnelling oscillations (coherent or incoherent) between the two SQUID wells is the right value of the potential well.

One solution can be the substitution of the Josephson junction in the rf SQUID with a very small hysteretic dc SQUID driven by an external magnetic field [7]. In this way the rf SQUID critical current can be tuned by the external flux and so the potential barrier, being the barrier height linearly dependent on the critical current.

This double SQUID however could not be necessarily equivalent to the single rf SQUID, due to the fact that now the SQUID dynamics is no more represented by a mono-dimensional potential, but by a two-dimensional potential. To test how the double SQUID behaves like a single SQUID we made escape measurements out of one of the flux states, calculating the SQUID escape temperature in function of the magnetic field applied to the small dc SQUID. The results shows that the two SQUIDs can behave similar but for a value of the factor $\beta = 2\pi L I_0 / \phi_0$ very small (smaller than 1), and for a value of the external magnetic flux far from $0.5\ \phi_0$.

THE SQUID SWITCH

The rf-SQUID magnetic flux state can be read-out by using an inductively coupled hysteretic dc-SQUID [8]. This device is like a threshold (or switch) detector, giving a two-state response to the input magnetic flux: no response if the input magnetic flux is below the fixed threshold, positive response otherwise.

This partial read-out is sufficient for a complete characterisation of the rf-SQUID flux state. In fact the rf-SQUID, when operating in the double well mode, presents two distinct and well separated flux states, peaked around the two minima of the potential.

The use of a threshold detector instead of the complete reading of the flux state with a magnetometer presents a series of advantages, for example a nominally zero back-action on the probed system (essential for MQC tests and for state preparations), a simpler circuitry, a smaller contribution to system noise, and a higher operation speed [8].

A hysteretic dc-SQUID is a superconducting ring interrupted by two non shunted Josephson junctions, that can be biased by a current and by a constant magnetic flux. Under appropriate conditions, it works like a single Josephson junction with critical current I_c modulated by the external magnetic flux ϕ_{dc}. For $I_c(\phi_{dc}) > I_{dc}$ the dc-SQUID remains in the superconducting state with zero voltage, and so the back-action on the observed system is neglectable. For $I_c(\phi_{dc}) < I_{dc}$ there is a transition to the normal state and the switch voltage reaches or excess the superconductor gap value (~ 2.7 mV in Nb).

One can use a dc-SQUID-switch inductively coupled to the rf-SQUID to read-out its flux state. This read-out is performed by applying appropriate flux and current biases to "turn-on" the switch, so that a measurement of the switch voltage will give $V_{dc} = 0$ (permanence in the superconducting state) if the measured flux is in the left state, and $V_{dc} \neq 0$ (transition to the normal state) otherwise.

The switching behaviour is a sharp step in the ideal case, but it is smeared out by thermal noise effects (and zero point fluctuations in the quantum regime).

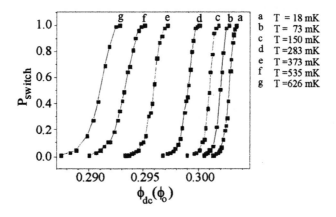

Figure 3. Switching probabilities at different temperatures for $\tau = 1$ms.

The important parameter to describe the switching is the escape rate Γ from the zero voltage state to the normal state. In the ideal case it must be zero below the threshold and infinity above it, but in real cases $\Gamma(\phi_{dc})$ grows exponentially as a function of the measured flux, from very small values (below the threshold) to very high values (above it).

The probability to have a switch to the normal state if the device is turned-on just for a short time τ is given by:

$$P(\phi_{dc}) = 1 - \exp\left[-\Gamma(\phi_{dc})\tau\right] \tag{1}$$

In fig. 3 there are reported switching probabilities as functions of the applied magnetic flux, measured at different temperatures on a dc-SQUID-switch with total inductance $L=5pH$ and maximum critical current $I_o=25\mu A$.

At 18 mK the step sharpness is sufficient to distinguish the two rf-SQUID flux states, but at high temperature (4.2 K) this is not possible.

However by using an indirect technique it is possible to test the correct working of the dc-SQUID-switch in probing the rf-SQUID state at high temperature too.

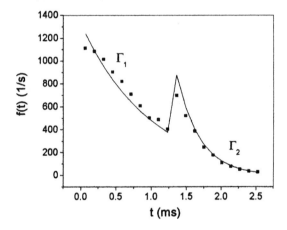

Figure 4. Distribution function of the switching probability as a function of time, with the rf-SQUID flux state changed after a time $\Delta t=1.5$ ms. Boxes are experimental points, while the line is the fitting curve.

The switch is turned-on and, after a time Δt, the rf-SQUID flux state is changed from a value ϕ_1 to a new value ϕ_2 by modifying the relative bias. This modification changes the switch rate in the dc-SQUID from Γ_1 to Γ_2. The switching probability distribution function at a given time t is then given by an exponential distribution with a double rate:

$$f(t) = \begin{cases} \Gamma_1 \exp(-\Gamma_1 t) & \text{for } t < \Delta t \\ \Gamma_2 \exp[-\Gamma_1 \Delta t - \Gamma_2 (t - \Delta t)] & \text{for } t > \Delta t \end{cases} \quad (2)$$

The time interval between the device turn-on and the switching can be experimentally measured by using flight-time techniques, and a set of N repetition of such measurements (in our case $N=1000$) allows the estimation of this distribution function. We report in fig. 3 the switching distribution function for a dc-SQUID (with $L=16$ pH and $I_o=20$ μA) reading the flux state in a coupled double rf-SQUID, with coupling constant $k = 0.011$, where the rf-SQUID flux state is changed after $\Delta t = 1.5$ ms.

MEASUREMENTS OF THE INTRINSIC DISSIPATION

The main limitation in performing a MQC experiment is the decoherence that the system experiments because of the coupling with the environment. In a rf-SQUID double well system the coherent oscillations damping rate due to the decoherence is given by:

$$\gamma_c = \frac{\Delta\phi^2}{2\hbar R} \frac{k_B T}{\hbar} \quad (3)$$

where $\Delta\phi$ is the distance between the two minima, T is the bath temperature, and R is an effective resistance related to the dissipation effects [9]. In planning a MQC experiment it is of extreme importance to know R for the considered system.

This is possible, for example, by studying Energy Level Quantization (ELQ) phenomena in Josephson junctions and rf-SQUIDs. The potential of such systems is characterised by a series of wells that can be modified by changing the applied bias. For example, it is possible to reduce the barrier separating one well from the next lower well. Each well presents a series of metastable energy levels, and at low temperature only few of them are populated. The level populations in a well change because of two causes: a) the tunnelling escape to the next well, and b) the thermal transitions to upper or lower levels in the same well (interlevel transitions). This second effect is related to temperature and dissipation, and depends on the effective resistance R. If one changes the potential by reducing the barrier height with a rate faster than the interlevel transition rates, it is possible to observe the effects of the depopulation of a single level at a time due to tunnelling, visible as peaks in the overall escape from the well. One can use a fit on this curve to estimate the effective resistance R.

In Figure 7 are reported escape rates for a Josephson junction (left) and for an rf-SQUID (right) as a function of the bias parameter x that changes the potential shape (the bias current for the junction, and the bias flux for the rf-SQUID), with a bath temperature $T \sim 30\ mK$ [10,11].

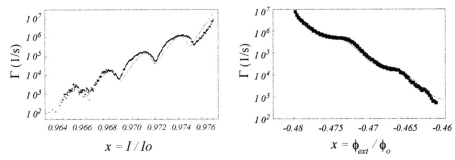

Figure 7. Escape rates in a Josephson junction (left) and in a rf-SQUID (right) as a function of the bias parameter x (bias current in the Josephson junction and bias flux in the rf-SQUID). One can notice the peaks due to ELQ effects.

The fits on these data give in both cases an effective resistance R of the order of some $M\Omega$, corresponding to a decoherence time for the considered system of the order of few μs, sufficient to perform a MQC experiment.

MQC PLANNED MEASUREMENTS

On the system we have described we plan to perform a series of measurements to test, as explained in the introduction, QM against CM predictions. The first set of measurements is the detection of time oscillations between the two flux states of the rf SQUID, and related measurements [3].

There are however two more possible measurements that can be performed to detect a non-classical behaviour of the system:

- By a series of subsequent switch of the dc-SQUID detector, that is switched off an on for a very short time, one can record a string of "Left" or "Right" measurements of the rf-SQUID flux state. By keeping only the couple of measurements where the first is a Non Invasive Measurements, and by varying the time delay between the two measurements, one con reconstruct the time oscillation of the probability to find the flux in the Left or Right well.
- One more possibility is to make the same measurement described above, but reducing the time delay well the supposed period of the Rabi oscillations. In this case, when the time delay becomes comparable to the decoherence time, one should observe the Quantum Zeno Effect, i.e. the starting flux state, should be "frozen" in the starting state without any decay to the mixed state.

REFERENCES

[1] A.J. Leggett, *Prog. Theor. Phys. Suppl.* 69, 80 (1980).

[2] A. Aspect et al., *Phys. Rev. Lett.,* 49, 1804 (1982).

[3] G. Diambrini, *Journal of Superconductivity,* 12, 767 (1999).

[4] J. R. Friedman, *these proceedings (2000).*

[5] C.H. Van der Wal, *these proceedings (2000).*

[6] L. Chiatti, C. Cosmelli, submitted to Int. J. of Modern Physics (2000).

[7] S. Han, J. Lapointe, J.E. Lukens, *Phys. Rev. Lett.* , 63, 1712 (1989).

[8] C. D. Tesche, *Phys. Rev. Lett.,* 64, 2358 (1990).

[9] U. Weiss, *Quantum Dissipative Systems, Ed. World Scientific, Singapore, (1999).*

[10] C. Cosmelli, P. Carelli, M. G. Castellano, F. Chiarello, G. Diambrini Palazzi, R. Leoni, G. Torrioli, *Phys. Rev. Lett.* 82, 5357 (1999),

[11] B. Ruggiero, M. G. Castellano, G. Torrioli, C. Cosmelli, F. Chiarello, V. G. Palmieri, C. Granata, P. Silvestrini, *Phys. Rev. B* 59, 177 (1999)].

BIEPITAXIAL YBa$_2$Cu$_3$O$_{7-x}$ GRAIN BOUNDARY JOSEPHSON JUNCTIONS: 0- AND π- RINGS FOR FUNDAMENTAL STUDIES AND POTENTIAL CIRCUIT IMPLEMENTATION

Francesco Tafuri
INFM Dip. Ingegneria dell'Informazione, Seconda Università di Napoli, Aversa Dip. Scienze Fisiche Università di Napoli "Federico II", 80125 Napoli (ITALY)

Franco Carillo, Filomena Lombardi, Fabio Miletto Granozio, Fabrizio Ricci, Umberto Scotti di Uccio and Antonio Barone
INFM-Dip. Scienze Fisiche dell'Università di Napoli "Federico II", 80125 Napoli (ITALY)

Gianluca Testa and Ettore Sarnelli
Istituto di Cibernetica del CNR, Via Toiano 6, Arco Felice (NA) (ITALY)

John R. Kirtley
IBM T.J. Watson Research Center, P.O. Box 218, Yorktown Heights, NY 10598,USA

Keywords: Josephson junctions, High critical temperature superconductors

Abstract   We present various concepts and experimental procedures to fabricate biepitaxial YBa$_2$Cu$_3$O$_{7-x}$ grain boundary Josephson junctions. The performance of our junctions indicates significant improvement in the biepitaxial technique. Different types of structures can be realized including "π-loops" in which one of the junctions is intrinsically π-phase shifted. Due to its versatility the biepitaxial technique offers the possibility to investigate basic aspects of the physics of grain boundary Josephson junctions and also the influence of π-loops in transport properties. Further, we discuss the possibility to produce circuits in which "0-" and "π-loops" are controllably located on the same chip.

1.   INTRODUCTION

The possibility of realizing electronic circuits in which the phase differences of selected Josephson junctions are biased by π in equilibrium is quite stimulating [1]. The concept of such π -phase shifts

83

was first introduced in the "extrinsic" case for junctions with ferromagnetic barriers [2] and in the "intrinsic" case for junctions exploiting superconductors with unconventional order parameter symmetries [3]. As a result of the possible $d_{x^2-y^2}$ order parameter symmetry of high critical temperature superconductors (HTS) [4], the presence of intrinsic π loops has also been considered for HTS systems [5]. This has been discussed recently in view of novel device concepts, and in particular for the implementation of a solid state qubit [1,6,7,8,9] and for Complementary Josephson junction electronics [10].

In this paper we discuss how $YBa_2Cu_3O_{7-x}$ (YBCO) structures made by the biepitaxial technique [11,12] can be successfully employed to produce arbitrary circuit geometries in which both "0" and π -loops are present, and possibly to obtain a doubly degenerate state [1,6]. Of course, great caution should be used because of the intrinsic nature of HTS and of stringent requirements on junctions parameters for practical applications of such devices.

Josephson junctions based on artificially controlled grain boundaries have been widely employed for fundamental studies on the nature of HTS [4,7,8]. Although the mechanism of high-T_C superconductivity and the influence of grain boundaries on the transport properties are not completely determined, reproducible and good quality devices are routinely fabricated. The bicrystal technique [13] typically offers junctions with better performances and allows in principle the realization of all different types of GBs ranging from [001] and [100] tilt to [100] twist boundaries. GB junctions based on the biepitaxial technique offer the advantage, with respect to the bicrystal technology, of placing the junctions on the substrate without imposing any restrictions on the geometry. We intend to show that significant improvements with respect to the original technique developed by Char et al. [12] are possible for performance of the biepitaxial junctions, and that the resulting devices have potential for applications. As a matter of fact, in traditional biepitaxial junctions, the seed layer used to modify the YBCO crystal orientation on part of the substrate produces an artificial 45° [001] tilt (c-axis tilt) GB. The nature of such a GB seems to be an intrinsic limit for some real applications. A convincing explanation has been given in terms of the d-wave nature of the order parameter and the intrinsic faceting of a grain boundary. More specifically the uncontrolled presence of π -loops leads to an extreme depression of the I_CR_N factor (I_C

is the critical current and R_N the normal state resistance) and several undesired effects for most applications [14].

We will show that the implementation of the biepitaxial technique [11] we developed to obtain 45° [100] tilt and twist (a-axis tilt and twist) GBs junctions makes such a technique interesting for both applications and fundamental studies. The phenomenology observed for the junctions based on these GBs and Scanning SQUID Microscopy investigations demonstrate the absence of π-loops, as we expect from their microstructure. As a consequence higher values of the $I_C R_N$ values, a Fraunhofer like dependence of I_C on the magnetic field and lower values of the low frequency flux noise, when compared with 45° c-axis tilt GBs, have been measured. These features are important tests to employ junctions for "traditional" applications in the field of superconducting electronics. Finally, we will also show that the same π-loops that make 45° [001] tilt (c-axis tilt) GBs undesirable for "traditional" applications, if suitably controlled, can be used for novel devices. We extended our process to other types of GB by using different seed layers to obtain junction configurations where π loops can be controllably produced. We shall not dwell on conceptual principles and actual feasibility of qubit devices. Instead we discuss the importance of the biepitaxial technique in having "0" and "π -" loops on the same chip. This makes the biepitaxial technique more versatile and promising for circuit design.

2.   CONCEPTS AND FABRICATION PROCEDURE

The biepitaxial technique allows the fabrication of various GBs by growing different seed layers and using substrates with different orientations. We have used MgO, CeO_2 and $SrTiO_3$ as seed layers. The MgO and CeO_2 layers are deposited on (110) $SrTiO_3$ substrates, while $SrTiO_3$ layers are deposited on (110) MgO substrates; in all these cases the seed layers grow along the [110] direction and have a thickness of 20-30 nm. Ion milling is used to define the required geometry of the seed layer and of the YBCO thin film respectively, by means of photoresist masks. YBCO films, typically 120nm in thickness, are deposited by inverted cylindrical magnetron sputtering at a temperature of 780 °C. YBCO grows along the [001] direction on MgO (substrates or seed layers) and on the CeO$_2$ (seed layers), while it grows along the [103]/[013] direction on SrTiO$_3$ (substrates or seed layers). In order to select the [103] or [$\bar{1}$03] growth and to ensure a better structural uniformity of the GB interface, we have also successfully employed vicinal substrates. Detailed structural investigations on these GBs,

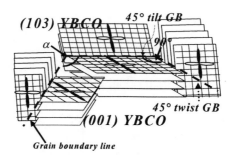

Figure1 A schematic representation of the artificial grain boundary structure. The boundary is obtained at the interface between the [001] oriented YBCO film grown on the [110] MgO seed layer and the [103] YBCO film grown on the bare [110] STO substrate. In this case the order parameter does not produce an additional π phase shift.

including Transmission Electron Microscopy (TEM) analyses, have been performed and the results have been presented elsewhere [11]. Depending on the patterning of the seed layer and the YBCO thin film, different types of GBs ranging from the two ideal limiting cases of 45° a-axis tilt and 45° a-axis twist have been obtained (see Figure 1). The intermediate situation occurs when the junction interface is tilted at an angle α different from 0 or π/2 with respect to the a- or b-axis of the [001] YBCO thin film. In all cases, the order parameter orientations do not produce an additional π- phase shift along our junction, in contrast with the 45° asymmetric [001] tilt junctions. As a consequence, no π loops should occur independently of the details of the interface orientation. The CeO$_2$ seed layer may produce a more complicated GB

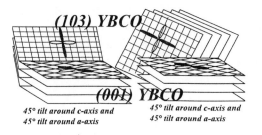

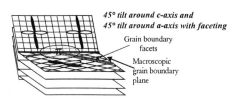

Figure 2 The CeO$_2$ seed layer produces an artificial GB that can be seen as a result of two rotations: a 45° [100] tilt or twist followed by a 45° tilt around the c-axis of the (001) film. For this junction configuration a d-wave order parameter symmetry would produce π-loops.

structure, in which a 45° c-axis tilt accompanies the 45° a-axis tilt or twist (see Figure 2a). In this case, as shown in Figure 2b, π loops should occur in analogy with the traditional biepitaxial junctions based on 45° c-axis tilt GBs. We found that the deposition conditions to select the uniform growth of YBCO 45° tilted around the c-axis of the (001) film are critical [15].

3.   EXPERIMENTAL RESULTS

3.1 TRANSPORT MEASUREMENTS

In this section we attempt to describe some of the main features of the transport properties of the biepitaxial junctions described above. Particular attention is given to the different phenomenology of the junctions based on 45° a-axis tilt or twist GBs (MgO seed layer) (BPMg) and of those based on GBs where a 45° c-axis tilt accompanies the 45° a-axis tilt or twist (CeO$_2$ seed layer) (BPCe) respectively. The differences seem to be strongly correlated to the absence or presence of π loops respectively. For each type of junction (BPMg and BPCe), we can have the tilt or the twist cases or intermediate situations where the junction interface is tilted at an angle α different from 0 or π/2 with respect to the a- or b-axis of the [001] YBCO thin film angle.

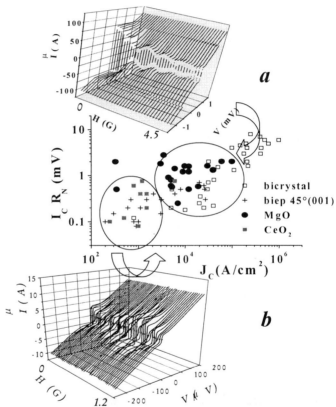

Figure3 I$_C$R$_N$ values vs J$_C$ for different types of junctions, including BPMg (full circles) and BPCe (full squares). In the insets a) and b) I-V curves are shown as a function of an externally applied magnetic field for the junctions BPMg and BPCe respectively (T = 4.2 K).

The first remarkable difference between the BPMg and BPCe junctions is the I$_C$R$_N$ value. For the former this value typically ranges from 1 mV to 2 mV at T = 4.2 K, while for the latter in average it is one order of magnitude less. This difference is clearly shown in Figure 3 , where the I_CR_N values are reported for several different BPMg and BPCe junctions as a function of their corresponding critical current density (J_C). This difference usually is mainly due to the higher values of the J_C measured in BPMg junctions. As a matter of fact, the normal state resistance values in the two types of junctions are close to each other. Data available from the literature and relative to other types of GB junctions are also reported in Fig.3 for a comparison. I_CR_N values of the BPCe junctions are close to those reported for conventional biepitaxials, while those of the BPCe junctions are of the same order of magnitude as in GB bicrystal and step edge junctions [11]. In the insets the typical dependencies of the critical current on the magnetic field for the two types of junctions are also shown, but we will comment on them below.

Let us first consider BPMg-type junctions. While the values of critical current density and normal state specific conductance in the tilt case are quite different from the twist case, the I_CR_N values are approximately the same for both. Moreover I_CR_N does not scale with the critical current density [11]. In the tilt cases $J_C \approx (0.5\text{-}10) \times 10^3$ A/cm^2 and normal state specific conductance $\sigma_N \approx (1\text{-}10)$ ($\mu\Omega$-cm^2)$^{-1}$ are measured at T = 4.2 K respectively. Twist GBs junctions are typically characterized by higher values of J_C in the range $(0.1\text{-}4.0) \times 10^5$ A/cm^2 and of σ_N in the range $(20\text{-}120)$ ($\mu\Omega$-cm^2)$^{-1}$ (at T = 4.2 K). A demonstration of the possibility of tailoring the critical current density and of the different transport regimes occurring in the tilt and twist cases has been given by measuring the properties of junctions with different orientations of the GB barrier on the same chip. By patterning the seed layer as shown in Fig.4a, we could measure the properties of a tilt junction and of junctions whose interface is tilted in plane by an angle α = 30°, 45° and 60° with respect to the a- or b-axis of the [001] YBCO thin film respectively. In all cases the order parameter orientations do not produce an additional π phase shift along our junction, in contrast with the 45° [001] tilt junctions, and no π loops should occur. We measured the expected increase of the critical current density with increasing angle, which corresponds on average to an increase of the twist current component. The values measured at T = 4.2 K are reported in Fig. 4a and range from the minimum value J_C = 3 x10^2 A/cm^2 in the tilt case to the maximum J_C = 10^4 A/cm^2 corresponding to an angle of 60°, for which the twist component is higher.

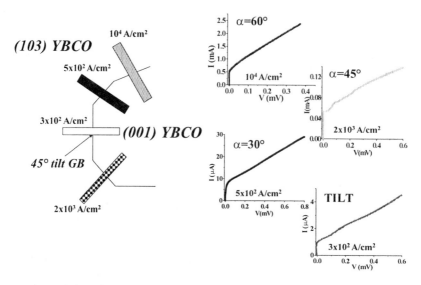

Figure 4 a) Scheme of the seed layer patterning, which allows the measurement on the same chip of the properties of a tilt junction and of junctions whose interface is tilted in plane of an angle α = 30°, 45° and 60° with respect the a- or b-axis of the (001) YBCO thin film respectively. b) The I-V characteristics (measured at T = 4.2 K) of the microbridges reported in a).

The values of normal state resistances are higher in the tilt case and decrease with increasing α, producing $I_C R_N$ values about the same for all the junctions independently of the angle α. In Figure 4b the I-V characteristics measured at T = 4.2 K, corresponding to the junctions of Figure 4a, are shown for approximately the same voltage range. Deviations from RSJ behavior appear for higher values of the critical current density (α = 60°). This possibility of modifying the GB macroscopic interface plane by controlling the orientation of the seed layer's edge is somehow equivalent to the degree of freedom offered by bicrystal technology to create symmetric or asymmetric GBs, with the advantage of placing all the junctions on the same substrate. The 45° a-axis tilt and twist GBs and the intermediate situations can represent ideal structures to investigate the junction physics in a wide range of configurations. The anisotropy of the (103) films and the possibility to select the orientation of the junction interface by suitably patterning the seed layer, and eventually the use of other seed layers which produce different YBCO in plane orientations, allow the fabrication of different types of junctions and the investigation of different aspects of HTS junction phenomenology.

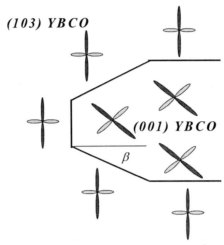

Figure5 Scheme of the seed layer patterning for the structure in Figure 2, which allows the measurement on the same chip of the properties of a tilt and of a twist junction and of junctions whose interface is tilted in plane of an angle β.

Let us consider BPCeO-type junctions. In this case we expect significant effects on the transport properties related to the presence of junctions with π shift phase and enhanced by the faceting. These were confirmed by the experimental results. Current vs voltage measurements showed for instance the expected decrease of the critical current density due to the presence of negative Josephson current (π phase shift). In addition, preliminary measurements showed that the critical current densities in the tilt and twist cases are quite similar $J_C \approx (1\text{-}10) \times 10^2$ A/cm^2 at T = 4.2 K. They also gave clear evidence that the tilt and twist cases have lower values of the critical current density when compared with intermediate situations, we characterize by an angle β (see Fig.5). For angles $\beta = 15°$ and $75°$ we measured values of the critical current density from 5 to 10 times higher than the tilt and twist cases in various samples. These results are quite consistent with the picture of a predominantly d-wave order parameter symmetry in presence of a random faceting. The tilt and the twist cases represent the most unfavorable situation for the Josephson current because of the higher negative contributions. From this point of view they resemble more the (001) 45° asymmetric GBs case. Intermediate situations with $\beta \neq 0, 45°, 90°$ are more similar to the (001) 45° symmetric GBs configuration. In this case, as experimentally confirmed, we expect a lower contribution due to negative currents. A maximum of the critical current should occur for angles β = 22.5°, 67.5°. The measured values of normal state specific conductance for β = 15° and 75° were close to those corresponding to the tilt and twist (σ_N in the range (1-5) ($\mu\Omega$-cm^2)$^{-1}$ at T = 4.2 K). The highly resistive barriers are mainly due to the microstructure of the grain boundary.

Different behaviors of the critical current as a function of an externally applied magnetic field are observed for BPMg and BPCe junctions respectively. In both cases we measured symmetric patterns around zero magnetic field and clear modulations of the critical current I_C. For BPMg we observe I_C modulations following the usual Fraunhofer-like dependence (see inset a of Fig.3), and in all samples the absolute maximum of I_C occurs at H=0. The presence of the current maximum at zero magnetic field is consistent with the fact that in our junction configuration the order parameter orientations do not produce an additional π phase shift. For BPCe I_C modulations do not follow the Fraunhofer-like dependence (see inset b of Fig.3), and in all samples the absolute maximum of I_C occurs at symmetric values of H $\neq$ 0. This was more evident for narrower microbridges and lower values of the critical current. This behavior is similar to what observed in 45° [001] tilt GB junctions [14,16,11] and consistent with the presence of π phase shift along the grain boundary. In both cases the value of the magnetic field between two adjacent minima of the critical current is consistent with the classical expression $H = \dfrac{\Phi_o}{2W\left(t + \lambda_R + \lambda_L\right)}$, W being the junction width, t the barrier thickness and $\lambda_{R(L)}$ the penetration depth in the R(L) electrode respectively, rather than with the expression taking into account flux focussing effects H $\propto$ 1/W^2 [17]. We notice that, due to the anisotropic structure of our junctions, we use the London penetration depth λ_c along the c-axis which is one order of magnitude larger that the λ_{a-b} value ($\lambda_c \approx 4$ μm), as resulting from Scanning SQUID measurements (see next sub-section).

Fiske steps as a function of H have been observed for junctions characterized by Frauhnofer-like pattern, giving some evidence of a dielectric-like behavior of some of the layers at the junction interface. We already reported about this work elsewhere [18]. The Fiske steps do not depend on the use of a particular substrate, since they have been observed in junctions based both on SrTiO$_3$ and MgO substrates. Typical values of the ratio between the barrier thickness t and the relative dielectric constant ε_r range from 0.2 nm to 0.7 nm. These values are derived by considering the c-axis penetration depth inferred from SSM measurements (see section III.b). Resonances, which manifest themselves as steps in I-V characteristics, have been also observed in junctions characterized by an anomalous magnetic pattern (BPCe). Further investigations on the nature of such resonances are in progress.

We conclude this section by reporting on the characterization of BPMg dc-SQUIDs. These SQUIDs exhibit very good properties, and noise levels which are among the lowest ever reported for biepitaxial junctions [19]. The noise spectral densities of the same dc-SQUID have been measured at T = 4.2 K and T = 77 K using standard flux-locked-loop modulated electronics. The energy resolution $\varepsilon = S_\Phi/2L$ (with S_Φ being the magnetic-flux-noise spectral density) at T = 4.2 K and T = 77 K is reported in Fig. 6. At T = 4.2 K and 10 kHz, a value of $S_\Phi = 3~\mu\Phi_0/(Hz)^{1/2}$ has been measured, corresponding to an energy resolution $\varepsilon = 1.6 \times 10^{-30}$ J/Hz. This value is the lowest reported in the literature for YBCO biepitaxial SQUIDs. Moreover, the low frequency 1/f flux noise spectral density at 1 Hz is more than one order of magnitude lower than the one reported for traditional biepitaxials, as is also evident from the comparison with data at T = 4.2 K of Ref. [20]. The lower values of low frequency noise are consistent with the absence of π - loops on the scale of the faceting for these types of GBs.

3.2. SCANNING SQUID MICROSCOPY ON BIEPITAXIAL JUNCTIONS WITH MgO SEED LAYER

Figure 7 is a scanning SQUID microscope [21] image of a 200x200 μm^2 area along a grain boundary separating a (001) region from a (103) region (as labeled in the figure) of a thin YBCO biepitaxial film grown as described above. The position of the grain boundary is indicated by the dashed line. The image was taken at 4.2 K in liquid helium with an octagonal SQUID pickup loop 4 microns in diameter after cooling the sample in a few tenths of

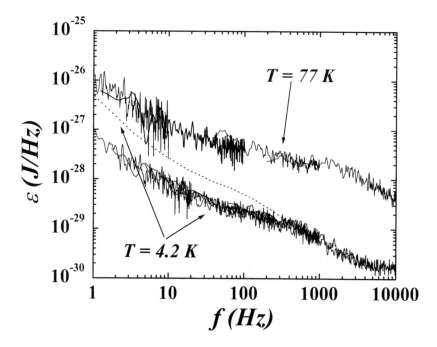

Figure 6 Magnetic flux noise spectral densities of a [100] tilt biepitaxial SQUID at T = 77 K and T = 4.2 K. The SQUID, with an inductance L=13 pH, was modulated with a standard flux-locked-loop electronics. The right axis shows the energy resolution. Data at T = 4.2 K are compared with results on SQUIDs based on [001] tilt biepitaxial junctions from Ref. [20].

a μT externally applied magnetic field normal to the plane of the sample. The grey-scaling in the image corresponds to a total variation of $0.13\Phi_o$ of flux through the SQUID pickup loop. Visible in this image are elongated interlayer Josephson vortices in the (103) area to the right magnetization in the (001) area to the left, of the grain boundary. Fits to the interlayer vortices give a value for the c-axis penetration depth of about 4 μm. Apparently localized regions of magnetic flux with random magnitudes and orientations are spontaneously generated in the (001) film, regardless of the value of external field applied [22]. Temperature dependent scanning SQUID microscope imaging shows that this spontaneous magnetization, which appears to be associated with defects in the film, arises when the film

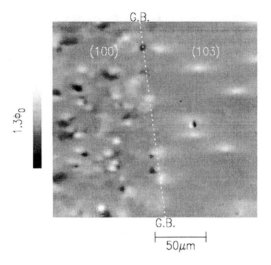

Figure7 Scanning SQUID microscope image of a 200x200 μm^2 area along a grain boundary separating a (100) region from a (103) region of a thin YBCO biepitaxial film grown. The position of the grain boundary is indicated by the dashed line.

becomes superconducting [22]. Although it is difficult to assign precise values of total flux to the apparently localized vortices, since they are not well separated from each other, fits imply that they have less than Φ_o of total flux in them, an indication of broken time-reversal symmetry [23]. Although there is apparently some flux generated in the grain boundary region, the fact that these SQUIDs have relatively low noise seems to indicate that this flux is well pinned at the temperatures at which the noise measurements were made. These results are consistent with the absence of π loops along the grain boundary.

4. BIEPITAXIAL JUNCTIONS FOR EXPERIMENTS ON THE SYMMETRY OF THE ORDER PARAMETER AND FOR A DEVELOPMENT OF A DOUBLY DEGENERATE SYSTEM

The particular junction configurations investigated in this work allow some consideration of the possible impact of these types of junctions on the study of the Josephson effect and the order parameter symmetry in YBCO and on the development of concepts for devices [1,6,7,9,10]. We first recall that the biepitaxial technique can provide circuits composed completely of junctions without any π -loops. By varying the interface orientation with respect to the [103] electrode orientation, the junction properties can be adjusted. On the other hand the traditional biepitaxial technique [12], producing 45° [001] tilt GBs or the types of junctions described in this work by using CeO_2 (see Figure 8 top view and inset of Figure 8 three-dim. view), can controllably generate π -loops on macroscopic scales. In the

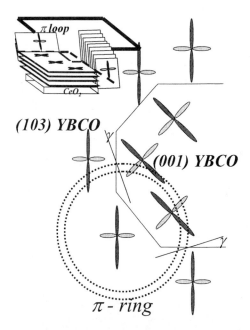

Figure 8 Top view of a π-SQUID based on GBs resulting from two rotations: a 45° [100] tilt or twist followed by a 45° [001] tilt. Inset) 3-dimensional view

schemes of Figure 8 we use a corner geometry with a 90° angle. This angle γ can be obviously tuned to enhance the effects related to the phase shift (see

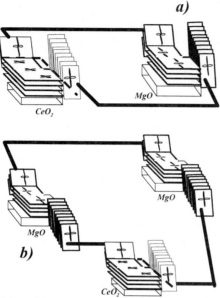

Figure 9 Scheme of the qubit structure proposed in Ref. [1] designed using the biepitaxial grain boundaries proposed in the paper.

top view Figure 8) and this change is particularly easy to realize by using the biepitaxial technique. In this section we focus our attention mainly on the feasibility of the biepitaxial junctions to obtain the doubly degenerate state required for a qubit. Of course, great caution should be used because of the intrinsic nature of HTS, probably characterized by high dissipation, and of stringent requirements on junctions parameters for practical applications of such devices. In Ref. [1a] the design is based on quenching the lowest order coupling by arranging a junction with its normal aligned with the node of the d-wave order parameter, thus producing a doubly periodic current-phase relation. It has been shown that the use of π phase shifts in a superconducting phase qubit also provides a naturally bistable device and does not require external bias currents and magnetic fields [1b]. The direct consequence is the quietness of the device over other designs. A π junction provides the required doubly degenerate fundamental state, which also manifests itself in a doubly periodic function of the critical current density as a function of the phase [8]. This principle has been used in a small inductance five junction loop frustrated by a π -phase shift [1b]. This design provides a perfectly degenerate two-level system and offers some advantages in terms of fabrication ease and performance. HTS may represent a natural solution for the realization of the required π-phase shift due to the pairing symmetry of the order parameter and, therefore, due to the possibility of producing π phase shifts. Experimental evidence of YBCO π-SQUIDs has been given by employing the bicrystal technique on special tetracrystal substrates [7]. Our technique allows the realization of circuits where π-loops can be controllably located in part of the substrate and separated from the rest of the circuit based on "0"-loops, i.e. junctions where no additional π phase shifts arise. This can be easily made by depositing the MgO and CeO_2 seed layers on different parts of the substrate. Obviously, part of the substrate will be partly not covered by any seed layer.

As a test to show how the biepitaxial junctions could be also useful possibly for device implementation of circuits based on π-loops without the topological restriction imposed by the bicrystal technique, we refer to the structures proposed in Ref. [1] for quantum computation as exemplary circuits. The first is composed of a s-wave (S)- d-wave (D)- s-wave (S') double junction connected with a capacitor and an ordinary "0" Josephson junction based on s-wave superconductors (the S-D'-S junction generates the doubly degenerate state). The second consists of a five junction loop with a π junction. Our technique would combine the possibility of placing the ordinary "0" junctions corresponding to the MgO seed layer and to exploit the possible doubly degenerate state of asymmetric 45° GB junctions corresponding to the CeO_2 seed layer to replace the S-D-S' system or the π junction respectively. In Figures 9a and 9b we show how devices for instance such as those proposed in Ref. [1] could be obtained by employing the biepitaxial technique respectively. The biepitaxial technique can offer possible alternatives for the realization of the structures above as extensively discussed in Ref.[9].

Finally we also notice that our experiments based on SSM [22] raise the possibility of intentionally introducing time-reversal symmetry breaking effects by, for example, photolithographically patterning small defects in high-T_C samples. In perspective this could be applicable to (e.g.) the fabrication of elements for quantum computation, opening new perspectives in the design of such devices without necessarily using Josephson junctions.

5.   CONCLUSIONS

The performance of the biepitaxial junctions based on 45° a-axis tilt or twist GBs (MgO seed layer) demonstrates that significant improvements in the biepitaxial technique are possible, and the resulting devices have potential for applications. This is due to the absence of π-loops in these grain boundary junctions. The use of a CeO$_2$ rather than a MgO seed layer can produce π-loops in the same junction configurations. The advantage of placing junctions in arbitrary locations on the substrate without imposing any restrictions on the geometry, and the ease of obtaining different device configurations by suitably patterning the seed layer, make the biepitaxial technique competitive for the testing of new concept devices, such as those based on π -loops. Some simple examples of situations in which π-loops can be suitably produced in specific locations of a more complicated circuit have also been discussed.

ACKNOWLEDGMENTS

This work has been partially supported by the projects PRA-INFM "HTS Devices" and by a MURST COFIN98 program (Italy). The authors would like to thank Dr. E. Ilichev and Dr. A. Golubov for interesting discussions on the topic.

REFERENCES

[1]   L.B. Ioffe, V.B. Geshkenbein, M.V. Feigel'man, A.L. Fauchere and G. Blatter, Nature 398, 679 (1999); G. Blatter, V. B. Geshkenbein and L. B. Ioffe, Cond. Mat. 9912163 (1999).

[2]   L.N. Bulaevskii, V.V. Kuzii and A.A. Sobyanin, JETP Lett. 25, 290 (1977).

[3]   V.B. Geshkenbein, A.I. Larkin and A. Barone, Phys. Rev. B 36, 235 (1986).

[4]   C.C. Tsuei and J.R. Kirtley, to be published in Review of Modern Physics (2000); C.C. Tsuei, J.R. Kirtley, C.C. Chi, L.S. Yu-Jahnes, A. Gupta, T. Shaw, J.Z. Sun and M.B. Ketchen, Phys. Rev. Lett. 73, 593 (1994); J.R. Kirtley, C.C. Tsuei, J.Z. Sun, C.C. Chi, L.S. Yu-Jahnes, A. Gupta, M. Rupp and M.B. Ketchen, Nature 373, 225 (1995); D.A. Wollman, D.J. Van Harlingen, D.M. Ginsberg and A.J. Leggett, Phys. Rev. Lett. 73, 1872 (1994).

[5]   M. Sigrist and T.M. Rice, J. Phys. Soc. Jap. 61, 4283 (1992).

[6]   A.M. Zagoskin, Cond. Mat. 9903170 (1999).

[7]   R. Schulz, B. Chesca, B. Goetz, C.W. Schneider, A. Shmehl, H. Bielefeldt, H. Hilgenkamp and J. Mannhart, Appl. Phys. Lett.76, 912 (2000).

[8]   E. Il'ichev, V. Zakosarenko, R.P.J. Ijsselsteijn, H.E. Honig, V. Schultze, H.G. Meyer, M. Grajcar and R. Hlubina, Phys. Rev. B 60, 3096 (1999).

[9]   F. Tafuri, F. Carillo, F. Lombardi, F. Miletto Granozio, F. Ricci, U. Scotti di Uccio, A. Barone, G. Testa, E. Sarnelli and J.R. Kirtley, Phys. Rev. B in press (2000)

[10] E. Terzioglu and M.R. Beasley, IEEE Trans. Appl. Supercond. 8, 48 (1998).

[11] F. Tafuri, F. Miletto Granozio, F. Carillo, A. Di Chiara, K. Verbist and G. Van Tendeloo, Phys. Rev. B 59, 11523 (1999).

[12] K. Char, M.S. Colclough, S.M. Garrison, N. Newman and G. Zaharchuk, Appl. Phys. Lett. 59, 773 (1991).

[13] D. Dimos, P. Chaudari, J. Mannhart and F.K. LeGoues, Phys. Rev. Lett. 61, 219 (1988).

[14] J. Mannhart, H. Hilgenkamp, B. Mayer, Ch. Gerber, J.R. Kirtley, K.A. Moler and M. Sigrist, Phys. Rev. Lett. 77, 2782 (1996).

[15] U. Scotti di Uccio, F. Lombardi, F. Ricci, E. Manzillo, F. Miletto Granozio, F. Carillo and F. Tafuri unpublished (2000).

[16] Copetti C.A., F. Ruders, B. Oelze, Ch. Buchal, B. Kabius and J.W. Seo, Physica C 253, 63-70 (1995).

[17] P.A.Rosenthal, M.R. Beasley, K. Char, M.S. Colclough and Z. Zaharchuk, Appl. Phys. Lett. 59, 3482 (1991)·

[18] F. Tafuri, B. Nadgorny, S. Shokhor, M. Gurvitch, F. Lombardi, F. Carillo, A. Di Chiara and E. Sarnelli, Phys. Rev. B 57, R14076 (1998).

[19] G. Testa, E. Sarnelli, F. Carillo and F. Tafuri, Appl. Phys. Lett. 75, 3542 (1999).

[20] A.H. Miklich, J. Clarke, M.S. Colclough, and K. Char, Appl. Phys. Lett. 60, 1989 (1992).

[21] J.R. Kirtley, M.B. Ketchen, K.G. Stanwiasz, J.Z. Sun, W.J. Gallagher, S.H. Blanton and S.J. Wind, Appl. Phys. Lett.66, 1138 (1995).

[22] F. Tafuri and J.R. Kirtley, Phys. Rev. B in press (2000).

[23] D.B. Bailey, M. Sigrist and R.B. Laughlin, Phys. Rev. B 55, 15239 (1997); M. Sigrist, Progr. Theor. Physics 99, 899 (1998).

THE SUPERCONDUCTING SINGLE ELECTRON TRANSISTOR: *IN SITU* VARIATION OF THE DISSIPATION

J.B. Kycia,[1,2]J. Chen,[1,2]R. Therrien,[1,2]Ç. Kurdak,[1,2,*]K.L. Campman,[3]
A.C. Gossard,[3] and John Clarke[1,2]

[1]*Department of Physics, University of California, Berkeley, CA 94720;*

[2]*Materials Sciences Division, Lawrence Berkeley National Laboratory, Berkeley, CA 94720;*

[3]*Department of Physics, University of California, Santa Barbara, CA 93106*

Abstract   We have fabricated a superconducting single electron transistor (sSET) on a GaAs/AlGaAs substrate containing a two-dimensional electron gas (2DEG) about 100 nm below the surface. The island separating the two junctions, each of resistance 18 kΩ, and the leads connected to the sSET are capacitively coupled to the 2DEG. Depleting the 2DEG by means of a voltage applied to a back gate enables us to decrease the dissipation experienced by the sSET *in situ*. The measured minimum zero-bias conductance, G_o^{min}, of the sSET follows the approximate power law dependences $G_o^{min} \propto T^\alpha g^\beta$, where T is the temperature and $g \equiv R_K/4R_g$; here $R_K = h/e^2$ and R_g is the resistance per square of the 2DEG. We find $\alpha \approx -1$ and $\beta \approx 0.2 - 0.4$, depending on the temperature. These results are in qualitative agreement with the theory of Wilhelm *et al.* However, the fact that β, in particular, has a temperature dependence suggests that one cannot separate the temperature and dissipation dependences with the functional form $T^\alpha g^\beta$.

1.   INTRODUCTION

There is considerable interest in the study of macroscopic quantum states of magnetic flux and electric charge in superconducting circuits, particularly with regard to their coherent superposition. This research is driven partly by a longstanding fundamental interest in the coherence of macroscopic states [1] and partly by the potential use of these elements as qubits in quantum computers [2]. In the flux case, the element consists of a superconducting loop containing one or more Josephson junctions, and the two quantum states involve supercurrents in opposite direction [1, 3]. Very recently, the coherent superposition of such states has been observed, manifested by the level splitting induced by coherent macroscopic quantum tunneling between quantum states [3-5]. In the charge case, the two-state system consists of a Cooper pair "box" onto which Cooper pairs can tunnel from a superconducting reservoir coupled to it by a Josephson junction [6-8]. Quantum oscillations between two charge states differing by a charge $2e$ have been demonstrated [6]. However, in order to obtain decoherence times of sufficient duration for either flux-state or charge-state devices to be of interest in quantum computing, the dissipation which they experience will have to be drastically reduced. In the case of Josephson junctions, the effects of dissipation [1] have been extensively studied, in most detail by experiments in which the environment could be varied by an *in situ* technique [9, 10]. On the other hand, in the case of the charge-state devices, although the effect of the environment on tunneling rates in small junctions has long been formulated [11], to our knowledge there are no previous experiments in which one could vary the environment *in situ* while keeping all the other parameters of the system fixed.

In this paper, we describe an experiment in which a superconducting single-electron transistor (sSET) is fabricated on a GaAs/AlGaAs substrate containing a two-dimensional electron gas (2DEG) just below the surface. By applying a voltage to a back gate on the reverse side of the substrate, we increase the resistance of the 2DEG, thereby reducing the dissipation in the sSET *in situ* [12]. This experiment enables us to make accurate measurements of the effects of dissipation on the zero-bias conductance of the sSET.

The sSET consists of two small area tunnel junctions, each with capacitance C and resistance R_N, separated by a small superconducting island [13]. The charge on the island can be varied by means of a voltage applied to a gate coupled to it via a capacitance C_g. In the absence of dissipation, the behavior of the sSET is determined by two characteristic energies. The first is the charging energy required to place an electron on the island, $E_c = e^2/2C_\Sigma$, where $C_\Sigma = 2C + C_g$. The second is the Josephson coupling energy of each tunnel junction, $E_J = \Delta R_K/8R_N$, where Δ is the superconducting energy gap and $R_K = h/e^2 \approx 25.8$ kΩ is the resistance quantum. The excess charge Q on the island and the quantum mechanical phase ϕ of the superconducting order parameter of the island are canonically conjugate variables satisfying the uncertainty relationship $\Delta Q \Delta \phi \geq 2e$; ΔQ and $\Delta \phi$ are fluctuations in Q and ϕ. At zero temperature, $\Delta \phi$ is small if $E_J \gg E_c$, and the device exhibits high conductance at zero bias. Conversely, if $E_c \gg E_J$, the charge fluctuations ΔQ are small, representing localized charge, and the device exhibits a Coulomb blockade with a low zero-bias conductance. Increasing the temperature increases the magnitude of both ΔQ and $\Delta \phi$. Under all circumstances, providing E_c, $E_J \gg k_B T$, the zero-voltage conductance is periodic in the gate voltage V_g, which controls the effective charging energy, with a period of e or $2e$ [14].

In the current experiment, we are particularly concerned with the effect of adding a dissipative element in parallel with the sSET. The dissipation tends to reduce $\Delta\phi$ and thus to increase the zero-voltage conductance. Our experiment enables us to study the effect of varying this dissipation.

2.   EXPERIMENTAL CONFIGURATION

The sSET was fabricated on a $GaAs/Al_{0.3}Ga_{0.7}As$ heterostructure in which a 2DEG is located below the surface. The heterostructure was grown on a GaAs substrate using molecular beam epitaxy and consists of the following layers: 500 nm of GaAs, 104 nm of $Al_{0.3}Ga_{0.7}As$, and 6 nm of GaAs. The $Al_{0.3}Ga_{0.7}As$ is selectively doped with Si donors situated 40 nm from the lower $GaAs/Al_{0.3}Ga_{0.7}As$ interface, at which the 2DEG forms. The substrate is placed on a metallic back gate. We bias the back gate negatively with respect to the 2DEG with a large voltage to reduce the sheet density n_S of electrons in the 2DEG, and thereby change its resistance per square, R_g. To reduce the required voltage, we thinned the substrate to about 250 μm. Four leads were attached to the 2DEG with indium-tin contacts annealed through the upper $Al_{0.3}Ga_{0.7}As$ and GaAs layers, and the resistance per square was measured using the van der Pauw method. The resistance varied from 160 to 600 Ω per square as the back gate voltage was varied from 0 to -300 V.

The configuration of the sSET, fabricated with double-angle evaporation through a shadow mask made with electron-beam lithography [15], is shown in Fig. 1(a). The first Al layer is 20 nm thick, and the second, deposited after oxidation of the first layer, is 35 nm thick. The central island is 0.5 x 2 μm^2, and its capacitance to the 2DEG is $C_g = 1.8$ fF. We determined C_g by driving the Al films into the normal state by means of a 0.3 T magnetic field and measuring the $1e$-periodicity of the zero-voltage conductance with respect to a gate voltage applied to the 2DEG. From scanning-electron micrographs, we estimate the two Al-$Al_x O_y$-Al junctions formed by the overlap of the narrow Al lines to have areas of 90 x 90 nm^2. Using a specific capacitance of 45 fF$/\mu$m^2 [16], we estimate $C \approx 0.37$ fF and hence $E_c/k_B \approx 360$ mK. The normal state resistances R_N of the junctions at 4.2 K, assumed to be equal, are 18 kΩ, leading to $E_J/k_B \approx 380$ mK. The narrow leads from the island are each connected to a lead, of width 10 μm and length 760 μm, that connects the sSET to separate current and voltage contacts to which Cu wires are bonded with pellets of In. Each of the Al leads forms a lossy transmission line with the 2DEG.

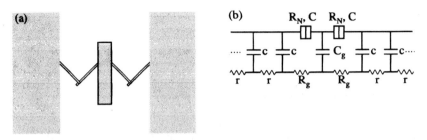

Figure 1   (a) Configuration of an sSET. The central 0.5 μm x 2 μm island has an Al-Al$_x$O$_y$-Al tunnel junction on each side formed by the overlap of the two perpendicular, narrow lines. Each of the junctions is coupled to a 10 μm-wide Al strip that forms a transmission line with the 2DEG below it. Only a small portion of this line is shown. (b) Circuit model for the sSET, with junction resistance R_N and capacitance C, capacitively coupled to a 2DEG of resistance R_g per square. The capacitance between the central island and the 2DEG is C_g. The capacitors c and resistors r represent transmission lines; the inductance of these lines is negligible at frequencies of interest.

Figure 1(b) shows the circuit representation of the sSET, its Al-film leads, and the 2DEG. Each junction is shunted by the resistance of the 2DEG between the island and the wide lead, with a value of approximately R_g, in series with C_g and the impedance of the transmission line, $Z_t(\omega)$. This transmission line has an inductance per unit length $l_t \approx 25$ nH/m, a capacitance per unit length $c_t \approx 10$ nF/m (assuming a dielectric constant of 12 for the substrate), and a resistance per unit length $r_g = R_g/w$ that can be varied from about 16 to 60 MΩ/m. The line impedance is dominated by resistive losses for all frequencies below $r_g/2\pi l_t \approx (1-4) \times 10^{14}$ Hz. This frequency is much higher than the plasma frequency of the junction, $(\Delta/2\hbar R_N C)^{1/2} \approx 60$ GHz, and the characteristic frequency $1/2\pi R_N C \approx 24$ GHz [17]. Thus, we can safely neglect propagating modes in the transmission line. The attenuation length for incident signals $(2/\omega c_t r_g)^{1/2}$ is shorter than the length of the line for frequencies below a few megahertz, so that we can regard the line as infinite for frequencies of interest. Thus, the real part of the line impedance $\text{Re}[Z_t(\omega)] = (r_g/2\omega c_t)^{1/2}$.

The sSET was cooled to temperatures as low as 18 mK in a dilution refrigerator surrounded by a screened room. Because noise coupled to the sSET from sources at higher temperatures can raise the effective noise temperature it experiences to values substantially above the bath temperature, we heavily filtered all four leads coupled to it, following the prescription of Vion *et al.* [18]. For each lead, there were four Cu-powder filters [9] at the base temperature, a Cu-powder filter and RC-filters at 4.2 K, and LC-filters at room temperature. The four Cu-filters attached to the mixing chamber comfortably exceeded the criterion [18] that the attenuation should exceed 200 dB at frequencies above 400 MHz. The four leads to the 2DEG were similarly heavily filtered.

3. RESULTS

A representative current-voltage characteristic of the sSET is shown in Fig. 2 for T $= 40$ mK and $R_g = 160$ Ω/square. There is a supercurrent-like branch for currents below about 1.5 nA. As the current is increased, the voltage switches to a current step at a voltage $2\Delta/e$, and subsequently to a second step at $4\Delta/e$. We determined the zero-bias conductance in a four-terminal arrangement by applying a small oscillating current at a few hertz and measuring the voltage with a lock-in detector. The amplitude of the excitation current was typically 30% of the switching current. To within the accuracy of the measurement, 2%, we obtained the same conductance values when the amplitude of the current was varied from 0.24 nA to 0.48 nA.

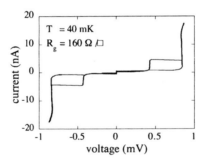

Figure 2 Current vs. voltage for sSET at 40 mK with 2DEG resistance of 160 Ω/square.

When we swept the voltage of the 2DEG relative to the sSET, the zero-bias conductance G_o oscillated with a period e/C_g, as shown in Fig. 3 for four temperatures and $R_g = 160$ Ω/square. As we expect, G_o increases monotonically as we lower the temperature. Generally, the amplitude of the conductance oscillations also increases as we lower the temperature; however, for reasons we do not understand, there is no discernible oscillation at 100 mK. We always observe a periodicity of $1e$ for all temperatures and levels of dissipation. This may be due, at least in part, to the fact that the heavy filtering at low frequencies forces us to sweep the 2DEG voltage relatively slowly, increasing the probability that a single electron will be transferred to the island at a voltage $e/2C_g$ [14]. Thus, the sSET never reaches the degeneracy point at which charge states differing by one Cooper pair are degenerate, and at which G_o is a maximum. For this reason, to avoid the issue of $1e$ or $2e$ periodicity, for our subsequent analysis we measured the minimum zero-bias conductance, G_o^{min}. This corresponds to the 2DEG voltage at which there is a maximum in the effective charging energy, the energy change for a Cooper pair tunneling onto the island.

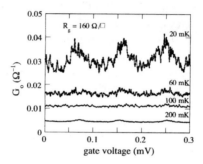

Figure 3 Zero-bias conductance G_o vs. gate voltage at four temperatures for $R_g = 160\ \Omega$ per square.

Figure 4(a) shows G_o^{min} vs. T for two values of the dissipation parameter $g \equiv R_K/4R_g$; the higher value of g represents increased dissipation. For a given value of g, G_o^{min} increases rapidly as the temperature is lowered, while at a given temperature, G_o^{min} is higher at the higher value of g.

Figure 4(b) shows G_o^{min} vs. g for five temperatures. At each temperature, G_o^{min} increases with g, while at fixed g, G_o^{min} increases rapidly with decreasing T, as in Fig. 4(a).

We now compare our results with theoretical predictions. Wilhelm *et al.* [19] have calculated the transport properties of an sSET in the configuration of our experiment. Using a perturbative expansion of the Josephson term in a charge basis, they calculate the values of G_o^{min} arising from the first and second order terms in the expansion, corresponding to sequential tunneling and co-tunneling, respectively. In the linear regime, both terms have the same temperature and dissipation dependence, and follow the power law

$$G_o^{min} \propto T^\alpha g^\beta. \tag{1}$$

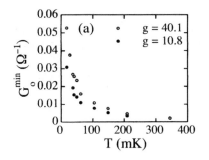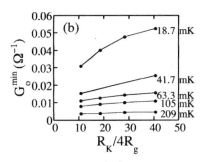

Figure 4 (a) G_o^{min} vs. T for two values of $g \equiv R_K/4R_g$; (b) G_o^{min} vs. $R_K/4R_g$ for five values of T.

Wilhelm *et al.* calculate the values of α and β for two different model systems. The first is a lumped-circuit model in which they assume that the transmission line on each side of the sSET can be replaced by a single, large capacitor. Thus, the dissipation is assumed to arise solely from the resistance of the 2DEG between the edges of the island and the edges of the two transmission lines. Although this model is unlikely to be a good approximation to the experimental configuration, it has the virtue of simplicity, in that the real part of the impedance shunting each junction is $R_g/[1 + (\omega R_g C)^2]$, corresponding to that of an ohmic shunt. For this model, Wilhelm *et al.* find $\alpha = -2$ and $\beta = 1$. They subsequently calculate the exponents for the much more complicated case involving the lossy transmission lines, and find $\alpha = -5/3$ and $\beta = 1/3$.

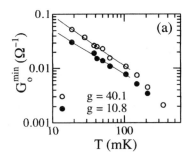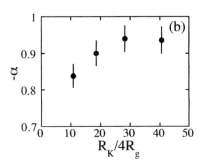

Figure 5 (a) G_o^{min} vs. T for two values of $g \equiv R_K/4R_g$. Lines are least square fits to the data below 100 mK; (b) α vs. g, where $G_o^{min} \propto T^\alpha$.

In Fig. 5(a), we plot G_o^{min} vs. T on a log-log scale for the two values of g shown in Fig. 4(a). At temperatures below about 100 mK, that is, well below E_c/k_B and E_J/k_B, G_o^{min} appears to follow a power law dependence on T for fixed g. However, as seen in Fig. 5(b), the measured values of $|\alpha|$ are substantially smaller than the prediction for either the lumped-circuit or transmission line model. Moreoever, α may depend weakly on g, varying from -0.84 ± 0.04 for $g = 10.8$ to -0.93 ± 0.04 for $g = 40.1$.

In Fig. 6(a), we plot G_o^{min} vs. g on a log-log scale for the five values of T shown in Fig. 4(b). We see that G_o^{min} has a relatively weak power law dependence on g. The values of β, plotted in Fig. 6(b), vary from about 0.2 at the highest temperature to about 0.4 at the lowest temperature. These values of β are markedly below the value of unity predicted by the lumped-circuit model, but embrace the value of $1/3$ predicted by the transmission line model. The temperature dependence of β is not predicted by the theory. Our data suggest that the dependences of G_o^{min} on temperature and dissipation are not separable.

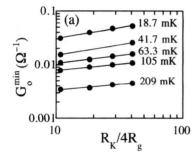 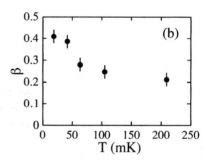

Figure 6 (a) G_o^{min} vs. $g \equiv R_K/4R_g$ at five temperatures. Lines are least square fits to the data; (b) β vs. T, where $G_o^{min} \propto g^\beta$.

4. CONCLUDING REMARKS

The temperature and dissipation dependence of the minimum zero-bias conductance, G_o^{min}, of an sSET capacitively coupled to the variable resistance of a 2DEG scales approximately as $T^\alpha g^\beta$. The fact that β is temperature dependent, and α is possibly dissipation dependent, implies that we cannot separate the dependences of G_o^{min} on T and g with this functional form. The exponent α is approximately -0.9, while β increases from about 0.2 to about 0.4 as the temperature is reduced from about 200 mK to about 20 mK. These values are qualitatively in accord with the predictions of Wilhelm *et al.* [19]. The measured value of $|\alpha|$ is roughly one-half that predicted by either the lumped-circuit model or the transmission line model. The value of β is much less than that predicted by the lumped-circuit model; this is not surprising, since this model is not a good approximation to the experimental configuration.

The more appropriate transmission line model, on the other hand, yields a value of β that is reasonably close to the value measured at the lowest temperatures, although it does not capture the observed temperature dependence. Clearly, more work is required to reconcile the experimental observations and theoretical predictions.

Two further points are in order. First, the fact that G_o^{min} scales as g^β, where $\beta \approx 0.2 - 0.4$, implies that reducing the environmental dissipation coupled to an sSET – or presumably to any other single-charge device – does not reduce the effect of dissipation on the characteristics of the device as rapidly as one might like, for example, in order to increase the decoherence time. Second, the 2DEG provides a convenient means of varying the dissipation experienced by a wide variety of superconducting or normal circuits. In the current experiment, the dissipative element was capacitively coupled to the device. Alternatively, one could couple directly to the 2DEG by means of vias in the upper layer of an InAs heterostructure.

Acknowledgment

We thank F. Wilhelm, G. Zimanyi and G. Schön for a copy of their unpublished manuscript, and them and A. Rimberg for many illuminating discussions. This work was supported by the Director, Office of Science, Office of Basic Energy Sciences, Materials Sciences Division of the U.S. Department of Energy under contract number DE-AC03-76SF00098, by the University of California Campus-Laboratory Collaboration Program under grant number UCCLC-24796, by the AFOSR under grant number F49620-94-1-0158, and by QUEST, an NSF Science and Technology Center.

References

[*] Current address: Department of Physics, University of Michigan, Ann Arbor, MI 48109

[1] A.J. Leggett, S. Chakravarty, A.T. Dorsey, Matthew P.A. Fisher, Anupam Garg, and W. Zwerger, Rev. Mod. Phys. **59**, 1 (1987).

[2] G.P. Berman, G.D. Doolen, V.I. Tsifrinovich, Superlattices Microstruct. **27**, 89 (2000).

[3] Jonathan R. Friedman, Vijay Patel, W. Chen, S.K. Tolpygo, and J.E. Lukens, Nature **406**, 43 (2000).

[4] J.E. Mooij, T.P. Orlando, L. Levitov, Lin Tian, Caspar H. van der Wal, and Seth Lloyd, Science **285**, 1036 (1999).

[5] C.H. van der Wal, A.C.J. ter Haar, F.K. Wilhelm, R.N. Schouten, C.J.P.M. Harmans, T.P. Orlando, S. Lloyd, and J.E. Mooij, International Conference on Macroscopic Quantum Coherence and Computing, Naples, Italy, June 14-17, 2000 (unpublished).

[6] Y. Nakamura, Yu. A. Pashkin, and J.S. Tsai, Nature **398**, 786 (1999).

[7] Alexander Shnirman, Gerd Schön, and Ziv Hermon, Phys. Rev. Lett. **79**, 2371 (1997).

[8] V. Bouchiat, D. Vion, P. Joyez, D. Esteve, and M.H. Devoret, Physica Scripta T **76**, 165 (1998).

[9] J.M. Martinis, M.H. Devoret, and John Clarke, Phys. Rev. B **35**, 4682 (1986).

[10] E. Turlot, D. Esteve, C. Urbina, J.M. Martinis, M.H. Devoret, S. Linkwitz, and H. Grabert, Phys. Rev. Lett. **62**, 1788 (1989).

[11] G.L. Ingold and Yu V. Nazarov, in *Single Charge Tunneling*, ed. H. Grabert and M.H. Devoret (Plenum Press, New York, 1992), p. 48; D. Vion, M. Gotz, P. Joyez, D. Esteve, and M.H. Devoret, Phys. Rev. Lett. **77**, 3435 (1996).

[12] A.J. Rimberg, T.R. Ho, Ç. Kurdak, John Clarke, K.L. Campman, and A.C. Gossard, Phys. Rev. Lett. **78**, 2632 (1997).

[13] D.V. Averin and K.K. Likharev, in *Mesoscopic Phenomena in Solids*, ed. B.L. Altshuler, P.A. Lee, and R.A. Webb (Elsevier, Amsterdam, 1991), p. 173.

[14] M.T. Tuominen, J.M. Hergenrother, T.S. Tighe, and M. Tinkham, Phys. Rev. Lett. **69**, 1997 (1992); T.M. Eiles, J.M. Martinis, and M.H. Devoret, Phys. Rev. Lett. **70**, 1862 (1993); A. Amar, D. Song, C.J. Lobb, and F.C. Wellstood, Phys. Rev. Lett. **72**, 3234 (1994).

[15] G.J. Dolan, Appl. Phys. Lett. **31**, 337 (1977).

[16] L.J. Geerligs, M. Peters, L.E.M. de Groot, A. Verbruggen, and J.E. Mooij, Phys. Rev. Lett. **63**, 326 (1989).

[17] M.H. Devoret and H. Grabert, in *Single Charge Tunneling*, ed. H. Grabert and M.H. Devoret (Plenum Press, New York, 1992),p. 1.

[18] D. Vion, P.F. Orfila, P. Joyez, D. Esteve, and M.H. Devoret, J. Appl. Phys. **77**, 2519 (1995).

[19] F. Wilhelm, G. Zimanyi, and G. Schön, unpublished.

[20] C. Nguyen, J. Werking, H. Kroemer, E. L. Hu, Appl. Phys. Lett. **57**, 87 (1990).

SUPERCONDUCTING ELECTROMETER FOR MEASURING THE SINGLE COOPER PAIR BOX

A. Cottet,[1] A. Steinbach,[1] P. Joyez,[1] D. Vion,[1] H. Pothier,[1] D. Esteve,[1] and M. E. Huber[2]

[1]*Service de Physique de l'Etat Condensé, CEA-Saclay, F-91191 Gif-sur-Yvette, France*
[2]*Department of Physics, University of Colorado at Denver, Denver, CO 80217, USA*

Abstract   We discuss for the single Cooper pair box the contributions to relaxation and to decoherence of the electromagnetic environment, of the offset charg enoise, and of a measuring Single Electron Transistor. We show that a single Cooper pair transistor can also be used for that purpose. Experimentally, we have operated such a device by measuring the variations of its critical supercurrent with the gate voltage using a SQUID series array amplifier. We describe the characteristics of this new electrometer and compare different schemes for measuring the critical current.

1. INTRODUCTION

The giant leap that quantum mechanics could bring to computing science [1] motivates an intense research of systems suitable for implementing quantum bits (qubits) and quantum algorithms. The requirements are formidable: the quantum states of the elementary qubits should be manipulable at will without significant loss of coherence over times much longer than the duration of elementary transformations, the couplings between qubits should be fully controllable, and the state of a qubit should be readable reliably. Furthermore, the implementation of the error correcting codes necessary to fight the unavoidable residual decoherence would require to perform measurements on some qubits during the computation process in order to perform adequate correction manipulations [2]. At the present time, quantum entanglement up to four qubits [3] and operation of elementary quantum gates [4] have already been demonstrated in quantum-optics based systems. Although less developed, microfabricated solid state systems are more appealing because they could be integrated on a large scale far more easily. The most advanced results reported so far with solid state systems have been obtained on qubits based on flux-states of small superconducting loops [5], and on charge states of small superconducting islands [6, 7]. In particular, Rabi precession between the two states of a charge qubit has been demonstrated in the single Cooper pair box over a few tens of oscillations [8], and longer coherence times are expected. The understanding of all decoherence sources which limit the duration of coherent oscillations in this system is thus an important issue. In this work, we discuss the influence on the single Cooper pair box of the residual dissipation in the box circuit, of the offset charge noise, and of the measuring system. In order to reduce the back-action of the measuring apparatus, we consider a new type of electrometer based on the superconducting version of the single electron transistor [7]. Finally, we report the first electrometry measurements performed with such an electrometer, and we discuss the different possible measuring set-ups.

2. DESCRIPTION OF THE SINGLE COOPER PAIR BOX

The single Cooper pair box [6], described in Fig. 1, consists of a single supercon-ducting island connected to a voltage source U through a small capacitor C_g on one side and through a small Josephson junction [9], with capacitance C_J and Joseph-son energy E_J, on the other side. When the superconducting gap Δ in the junction electrodes is larger than the charging energy $E_c = e^2/2C_\Sigma$ (with $C_\Sigma = C_J + C_g$), two charge states $|n\rangle$ and $|n + 1\rangle$ differing by one Cooper pair in the island form, close to their electrostatic energy degeneracy point, a two-level-system well decoupled from other degrees of freedom. The effective spin 1/2 hamiltonian of this two-level system writes:

$$H_0 = -\tfrac{1}{2}\vec{B} \cdot \vec{\sigma}$$

(1)

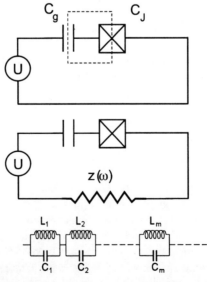

Figure 1. Top: Schematic circuit of the single Cooper pair box. The dashed line encloses the box island. Middle: Realistic circuit with a residual series impedance. Bottom: Representation of the electromagnetic modes coupled to the box.

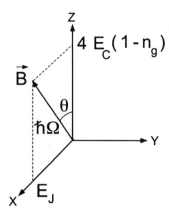

Figure 2. Fictitious spin 1/2 representation of the Cooper pair box. The vector $\vec{B}$ is the effective field acting on the spin.

where σ is a vector of Pauli operators, and B a fictitious field with components $\{E_J, 0, 4E_c(1 - n_g)\}$, with $n_g = C_g U/e$ (see Fig. 2). The ground and excited states of the above hamiltonian, $|\sigma_B = 1\rangle$ and $|\sigma_B = -1\rangle$, are the two states of the qubit. Their energy difference is $\hbar\Omega = E_J/\sin\theta$, where θ is the angle between B and the z axis. This description is however oversimplified, and the qubit is coupled to other degrees of freedom, both at thermal equilibrium and out of thermal equilibrium. We will consider in this work the effect of residual electromagnetic dissipation in the Cooper pair box circuit, the effect of moving charges in the neighborhood of the box island, and the effect of an electrometer measuring the state of the qubit from the electrostatic potential of the box island.

3. RELAXATION AND DECOHERENCE INDUCED BY THE ELECTROMAGNETIC ENVIRONMENT

The electromagnetic degrees of freedom of the box circuit can be modeled by inserting a small series impedance $z(\omega)$, as shown in Fig. 1. This impedance incorporates the effect of the voltage source and wiring impedances. Its effect is to couple the box to a set of bosonic electromagnetic modes with frequencies $\{\omega_j\}$ through the hamiltonian:

$$h = -\sqrt{\pi}\sum_j {}'\omega_j \sqrt{\frac{Z_j}{R_K}} i(b_j - b_j^\dagger)\sigma_z, \qquad (2)$$

where Z_j, b_j and $b_j^\dagger$ are the impedance, the annihilation and creation operators of mode j, respectively, and $R_K = h/e^2$ is the resistance quantum. These modes, which form the electromagnetic environment of the charge qubit, are assumed at thermal equilibrium. Any summation $\sum_j f(\omega_j)Z_j$ over the set of modes is performed in the following way

$$\sum_j f(\omega_j) Z_j = \frac{2}{\pi} \int_0^\infty f(\omega) \operatorname{Re} Z(\omega) \frac{d\omega}{\beta}, \tag{3}$$

with, in the weak coupling regime $\kappa = V_g/(C_J + C_g) \ll 1$ relevant for the experiments, $\operatorname{Re} Z(\omega) = \kappa^2 \operatorname{Re} z(\omega)$. The coupling hamiltonian [2] induces transitions between the box and the modes of the environment. The standard second order perturbation theory yields for the downward and upward transition rates $\Gamma_\downarrow$ and $\Gamma_\uparrow$ between the excited state and the ground state of the box:

$$\Gamma_{\downarrow(\uparrow)} = \frac{(2\pi)^2 \sin^2 \theta \, S(+(-)\Omega)}{\hbar R_K}, \tag{4}$$

where $S(\omega$ is the spectrum of the voltage fluctuations across the effective impedance $Z(\beta)$

$$S(\omega) = \frac{\hbar|\omega|}{2\pi} \left(\coth\left(\frac{\hbar\omega}{2k_B T}\right) + 1 \right) \operatorname{Re} Z(\omega). \tag{5}$$

The total relaxation rate $\Gamma_1^{ENv} = \Gamma_\downarrow + \Gamma_\uparrow$ is the decay rate of the diagonal part of qubit matrix density. In the low temperature regime $T \ll \hbar\Omega/k_B$, one has $\Gamma_\downarrow \simeq 0$, and the total relaxation rate is simply:

$$\Gamma_1^{ENV} \simeq 4\pi\Omega \sin^2 \theta \, \kappa^2 \operatorname{Re} z(\Omega)/R_K \tag{6}$$

Numerically, one gets $\Gamma_1^{ENV} = 50$ kHz for the boxparameters $\Omega/2\pi = 10$ GHz ($\hbar\Omega/k_B \simeq 0.5$ K), $\kappa = 2.5\%$, $\operatorname{Re} z(\Omega) = r = 5\Omega$, and $\theta = \pi/4$.

We now discuss the decay of the coherence amplitude $\langle X'|X'(t)\rangle$ of a state prepared at $t = 0$ as $|X'\rangle = 1/\sqrt{2}(|\sigma_B = 1\rangle + |\sigma_B = -1\rangle)$. Although on-resonance modes exchanging energy with the box contribute to the decay of coherence, out-of-resonance oscillators also contribute because they get entangled with the box. The coherence amplitude $\langle X'|X'(t)\rangle$ picks an extra decay factor [11]:

$$A(t) = \exp\left(4\cos^2 \theta \cdot \operatorname{Re}|J(t)]\right) \tag{7}$$

where $J(t)$ is the phase correlation function which appears in the theoryof Coulomb blockade [12]:

$$J(t) = 2 \int_0^\infty \frac{d\omega}{\omega} \frac{\operatorname{Re} Z(\omega)}{R_K} \frac{\exp(-i\omega t) - 1}{1 - \exp(-\hbar\omega/k_B T)} \tag{8}$$

In the simple case when $z(\omega) = r$, the function $J(t)$ can be calculated exactly. The long time behavior of $\operatorname{Re} J(t)$ is:

1. At zero temperature: $\operatorname{Re} J(t) \sim -2\frac{\kappa^2 r}{R_K}(\gamma + \ln(t/\kappa r C_j))$

2. At finite temperature: $\operatorname{Re} J(t) \sim -2\frac{\pi\kappa^2 r}{R_K}(K_B T t/\hbar)$

At zero temperature, $A(t)$ follows a power-law with a small exponent, and decoherence is weak. At temperatures $T > 10$ mK, the classical regime is almost reached, and $A(t)$ decays exponentially at a rate Γ_2^{ENV}:

$$\Gamma_2^{ENV} B = 8\pi \left(\frac{k_B T}{\hbar}\right) \frac{\kappa^2 r}{R_K} \cos^2 \theta. \tag{9}$$

This result can be easily retrieved by performing the following semi-classical average:

$$A(t) = \left\langle \exp i \int_0^t f(\Omega(t') - \overline{\Omega}) dt' \right\rangle \tag{10}$$

where $\Omega(t)$ is now the time dependent transition frequency modulated by the thermal fluctuations of the voltage across the impedance $z(\beta)$. In the classical regime, this entanglement is dominated by a random phase factor between the two states of the coherent superposition. It is worth noticing that the entanglement between the qubit and its environment is not an irreversible process by itself, and that one could in principle recover to some extent the loss of coherence due to low frequency modes using echo or even more sophisticated pulse techniques, analogous to those developped in nuclear magnetic resonance (NMR). If the impedance $z(\omega)$ is frequency independent, the ratio between the relaxation rate and the decoherence rate due to thermally excited oscillators reduces to:

$$\frac{\Gamma_1^{ENV}}{\Gamma_2^{ENV}} = \frac{\hbar \Omega}{2k_B T} \tan^2 \theta. \tag{11}$$

Since the experiments are performed in the low temperature regime $k_B T \ll \hbar \Omega$, one has $\Gamma_1^{ENV}/\Gamma_2^{ENV} \gg 1$ in practice. In this case, the electromagnetic environment of the qubit relaxes the whole qubit density matrix at the rate Γ_1^{ENV}.

4. RELAXATION AND DECOHERENCE INDUCED BY THE OFFSET CHARGE NOISE

It is well known that the island of a Single Electron Transistor (SET) [13] is subject to an offset charge noise with a $1/f$ spectrum [14]. This noise is attributed to a set of charges randomly fluctuating between two positions in the junction barriers and/or in the insulators close to the SET island. Occasionnaly, slow two-level fluctuators (TLF) have been directly observed over long times. The $1/f$ character of this TLF noise has been probed up to about 10 MHz. Its typical intensity is $S_q(f) = e^2 B/f$, with $B \approx 10^{-7}$. Like the electromagnetic environment, the charge noise induces relaxation and decoherence on the box quantum states. Although the spectral density ofthe charge noise has only been measured at frequencies much smaller than Ω, recent experiments on the single electron pump [15] have provided experimental evidence that the charge noise extends up to 100 GHz, i.e. well above Ω. Amazingly, the spectral density estimatedfrom the measured transition rate of otherwise forbidden transitions falls rather close to the extrapolated value of the $1/f$ spectrum, with $B' \approx 10^{-8}$. This noise should result in upward as well as downward transitions at a rate Φ_1^{TLF}:

$$\Gamma_1^{TLF} = \frac{2\pi B'}{\Omega} \left(\frac{E_c}{\hbar}\right)^2 \sin^2 \theta. \tag{12}$$

The estimated value is about $(1 \times \sin^2 \theta)$, significantly larger than the estimated contribution of the box circuit electromagnetic environment.

The decay rate of the coherence amplitude picks an extra contribution from the low-frequency offset charge fluctuations. Because of divergences inherent to the $1/f$ spectrum, the experimental protocol has to beprecisely defined. When a measurement, performed during a short time τ averaged over a time $t_{av} \gg$ the transition frequency has drifted away from its initial value and one finds for the decay function:

$$A'(\tau) \simeq \exp -[8B\cos^2 \theta (E_c\tau/\hbar)^2 \ln (t_{av}/\tau)]. \tag{13}$$

The contributions of low frequency TLF to the decay of a coherence signal can be suppressed using an echo pulse sequence similar to those used in NMR to compensate for inhomogeneous line-broadening. In this case, the decay function does not depend on the averaging time and writes:

$$A'(\tau) \simeq \exp -\left[8B\,\text{vos}^2\,\theta\,(E_c\tau/\hbar)^2 \frac{\log(2)}{2}\right]. \tag{14}$$

The decay is still fast, and coherence is lost after a time $\sim(20/\cos \theta)$ ns for $B = 10^{-7}$, $E_c = 0.5\,K_B$ K. The $1/f$ noise is thus a serious limitation, and its reduction to a level significantly smaller than commonly achieved is an important issue.

5. RELAXATION AND DECOHERENCE INDUCED BY A MEASURING SET ELECTROMETER

The state of the box can be measured either by measuring the electrostatic potential of the island [6], or by connecting it to an extra small probing tunnel junction which directly exchanges electrons with it [7]. Whereas this latter method results in a destructive measurement of the qubit state, the first one discussed here could in principle allow quantum non demolition (QND) measurements. For that purpose, the measuring electrometer should be able to distinguish both states of the qubit without inducing transitions between them. Repeated measurements of the qubit state should then give the same answer. An important characteristic of any measuring electrometer is thus the amount of information it can provide before irreversible transitions induced by the measuring system or by other relaxation mechanisms occur. Figure 3 shows a measuring set-up with a SET electrometer capacitively coupled through C_c to the box island, a set-up which has been used to measure the island potential in the box ground state [6]. The measuring process has been theoretically investigated in great detail for this set-up [16]. The back-action ofthe SET results from the fluctuations of the electrostatic potential of its own island while the current is flowing. When one electron enters or exits the island, the voltage varies by $\delta V = e/V_{SET}$, where C_{SET} is the capacitance of the SET island. Typically, δV is a few hundreds of microvolts. At a practical

SET electrometer

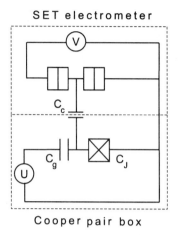

Cooper pair box

Figure 3. Schematic circuit of a single Cooper pair box electrostatically coupled to a measuring SET. The SET measures the potential of the box island. The voltage fluctuations of its island produce a back-action noise on the box, which induces relaxation and decoherence.

working point, the correlation time of these voltage fluctuations is $\tau_c \approx 1/R_{ET}C_{SET} \approx e/I$, where R_{SET} is the tunnel resistance of the SET junctions, and I is the average current. The spectrum is thus lorentzian with a cut-off frequency τ_c^{-1}. Its detailed form depends on the biasing point, and is precisely known for SETs with junction resistances larger than R_K [17]. Low frequency fluctuations result in decoherence, and a SET at the threshold can reach the quantum limit in the sense that the qubit state is decohered just on the time scale needed to measure it [16]. However, fluctuations at the qubit transition frequency Ω induce transitions at a significant rate when $\Omega\tau_c < 1$. In this regime, the induced relaxation rate follows from Eq. (4) using the spectral density of the SET island voltage and the relevant coupling factor $k_{SET} = C_c/C_J \ll 1$ between the box and the SET:

$$\Gamma_1^{SET} \simeq \frac{(2\pi)^2 \sin^2\theta\, \kappa_{SET}^2\, (\delta V/2)^2\, \tau_c}{\hbar R_K}. \tag{15}$$

More precise estimates can be obtained to take into account the precise working point of the SET and the effect of the rf drive in the case of an rf-SET [18]. The question thus arises if a SET is able to measure the box state before the induced relaxation has destroyed the qubit. A figure of merit can be defined as $f^{SET} = 1/(\Gamma_2^{SET} \cdot \tau_m)$, τ_m being the minimum time necessary to perform a measurement of the box state. Assuming that the measurement accuracy is solely limited by the SET intrinsic noise and not by the electronic amplifiers measuring the SET current, one finds $f^{SET} \approx \cot^2\theta$. A usual SET can thus perform a single measurement of the qubit [16], but not by a large margin. Note that a SET operated in the strong tunneling regime could possibly achieve a better performance. From the experimental point of view, the sensitivity of the best rf-SET is still presently limited by the noise of the microwave amplifier at the carrier frequency but the intrinsic limit is not beyond reach. In the following, we examine another type of electrometer based on the superconducting version of the SET, the SSET.

6. THE SUPERCONDUCTING SET ELECTROMETER

The SSET is almost equivalent to a single small Josephson junction whose critical current $I_c(n_g)$ is periodically modulated by the gate charge [8]. The only difference with a single junction is that the current-phase is not strictly sinusoidal, and that higher energy states can be excited. In the case when the gate charge is modulated at frequencies smaller than the band gap, the adiabatic approximation holds, and the SSET behaves as a tunable junction. The modulation pattern is determined by the ratio E_c/E_J', where $E_J' = I_0\hbar/2e$ is the Josephson energy of each junction with critical current I_0. For practical values $E_c/E_J' \sim 1$, the maximum critical current of a SSET, obtained for a reduced gate charge $n_g = 1$ [mod 2], is of the order of $I_0/2$, and the slope dI_c/dn_g is of the order of I_0 at practical working points. The predicted critical current for a SSET is compared in Fig. 4 with the average maximum supercurrent measured in a current-biased set-up [8]. In this case, the current-voltage $I - V$ characteristic is hysteretic and, upon ramping the bias current, the junction switches out of the zero-voltage-state at a switching current I_S. The values of I_S are distributed with an histogram whose average value and width depend on the electromagnetic impedance as seen from the SSET [19]. The average value is smaller than the critical current but almost scales with the predicted variations for $I_c(n_g)$. Depending on the biasing circuitry impedance, the critical current of a SSET can be measured in two different ways, which yields to two very different types of electrometers.

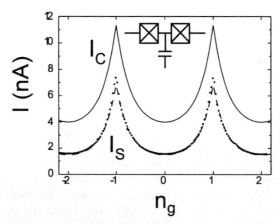

Figure 4. Straight line: predicted variations for the critical current of a SSET with $E_c = 0.66\,k_B\,$K and $E_J' = 0.50\,k_B\,$K. Dots: Measured average switching current I_S for this SSET in a moderate damping circuit at $T = 20$ mK. Dashed line: theoretical predictions.

6.1. Electrometry Based on the Switching of an ac-Shunted SSET

The variations of the average switching current with the gate charge can be used for electrometry. The obtention of narrow switching histograms suitable for electrometry requires to damp the dynamics of the phase across the junction. Switching histograms with an average close to the critical current and a relative width smaller than 10^{-3}} have been obtained in the case of a single junction using an RC ac-shunting circuit [19]. However, the bias current ramp-rate required for measuring a box is larger than used in the previous experiment [19], and we have found that histograms get wider when the ramp rate increases. Experimentally, we have measured switching histograms of an RC-shunted SSET using ramping speeds up to $0.2 \times I_S/\mu s$. Preliminary results show switching histograms narrow enough to discriminate the two box states within a measuring time of 1 μs.

6.2. Electrometry Based on the dc-Shunted SSET

When a Josephson junction is shunted by a small enough resistor, its $I - V$ characteristic is no longer hysteretic. However, its measurement is far more difficult than in the unshunted case because the voltage across the junction is too small to be measured using room temperature amplifiers. Recently, our group has used the SQUID series array amplifiers developed at NIST-Boulder [20] to measure the full $I - V$ characteristic of a small Josephson junction. The array we have used consists of 100 dc-SQUIDs in series, and delivers a signal large enough to be amplified by room temperature amplifiers without degradation [21]. Closed loop operation of the arrays is possible within a few MHz bandwidth. The experimental results on single junctions [22], in excellent agreement with the calculated $I - V$ characteristics [26], show that the classical regime for the phase dynamics was indeed reached. This result made possible the design of an electrometer based on the measurement of a shunted SSET [23]. For that purpose, we have implemented the set-up schematically described in Fig. 5, in order to measure the current I_R through the shunting resistance R_S of a SSET. More precisely, a fraction of this current flows in the input coil of a SQUID array amplifier. In order to avoid any high frequency resonance, an extra ac-shunt has been placed on-chip across the SSET, but careful mounting of discrete components should however be sufficient to obtain an adequate environment.

Sensitivity and Bandwidth. The sensitivity of the electrometer is set by the intrinsic SSET sensitivity $dI_c/dn_g \approx I_0$, by the ratio $dI_R/dI_c \lesssim 1$ at the bias point of the device, and by the noise of the SQUID array referred to its input $S_I \approx 3$ pA/$\sqrt{Hz}$. The intrinsic noise floor which results from the thermal fluctuations of I_R is much smaller and will be considered later on. The sensitivity, expressed in $e/\sqrt{Hz}$ is thus $s \approx S_I/I_0$. Although a larger critical current I_0 should result in better sensitivity, too large values of I_0 result in a strong renormalisation of the island charging energy due to virtual quasiparticle tunneling and correlatively to a strong reduction of the modulation depth of the critical current. In practice, we have

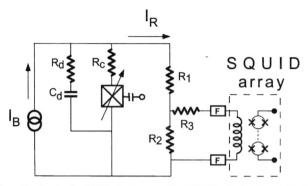

Figure 5. Schematic circuit of a SSET electrometer. The SSET is displayed as a tunable junction. A fraction ~1/3 of the dc-current in the shunt resistance passes in the input coil of a 100 SQUID series array. R_d and C_d form an on-chip damping circuit, R_c is a contact resistance, and R_1, R_2 and R_3 are surface-mounted components. The low-pass filters F prevent the Josephson oscillations in the SQUIDs from disturbing the SSET.

found that the optimal value of I_0 is in the range 20–40 nA , which would result in a sensitivity $s \approx 10^{-4}e/\sqrt{\text{Hz}}$ for a maximal coupling between the SSET and the array. This figure is significantly worse than the sensitivity $s \approx 7\ 10^{-6}e/\sqrt{\text{Hz}}$ already achieved with the rf-SET, but two-stage SQUID amplifiers might allow to improve the present sensitivity by about one order of magnitude. In this design, the SSET is connected to the input coil of a single dc-SQUID, which is itself in series with the input coil of a SQUID array. The bandwidth is limited by the input circuit and by the electronics backing the array. As seen from the array input coil, the SSET behaves as a source with a resistive impedance which is of the order of the series resistance in the input coil circuitry, \$10 Ω in our case. For an input coil inductance of \$200 nH, the resulting bandwidth is about 10 MHz, but can be made larger if needed.

Experimental Results. We have fabricated SSETs using 3 angle deposition through a shadow mask [24]. The two first aluminum layers form the junction electrodes, and the third gold layer forms normal wires which help eliminating spurious quasiparticles in the superconducting electrodes, and connect the SSET to an on-chip *RC* damping circuit fabricated by optical lithography. The SSET junction have an area of 80×130 nm^2 and a tunnel resistance of 7.5 kΩ. The samples were mounted in a shielded box fitted with coaxial connections to the SQUID series array box, and microfabricated RC filters [25] were installed on the bias and gatelines of the electrometer. As shown in Fig. 6 for a series of bias current values, the current through the shunt resistor of the SSET is modulated by the gate voltage, with a measured period corresponding to 2e. At larger bias currents, the modulation depth decreases progressively. We have also determined the *IV* characteristic of the SSET for different gate voltages. The extremal *IVs*, obtained for $n_g = 0$ and $n_g = 1$, are shown in Fig. 7. The overall agreement between the experimental results and the theoretical predictions [26] for a tunable Joseph-

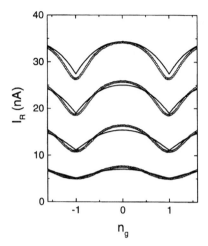

Figure 6. Open symbols: Variations of the current through the shunt resistor I_R with the gate charge n_g at $T = 40$ mK and for bias currents $I = 11, 21, 31, 40$ nA, from bottom to top. Full lines: Theoretical predictions at $T = 200$ mK based on the effective junction model.

son junction is satisfactory. The parameters are $Ec = 0.55 \ k_B$ K, and $I_0 = 40$ nA which are in good agreement with the values estimated from the area and tunnel resistance of the junctions, and from the superconducting gap $\Delta = 180 \ mueV$. This corresponds to a ratio $Ec/E_J' = 0.55$. The damping resistance as seen from the SSET is $R_{eff} = 38 \ \Omega$, larger than the estimated value $25 \ \Omega$ taking into account all components in the circuit. The effective temperature of 200 mK needed to fit the data is larger than the fridge temperature 40 mK because filtering between the SQUID array and the SSET has been greatly reduced compared to previous experiments on a single junction. This excess noise temperature is however tolerable for electrometry applications, and can be reduced if needed. We have also applied to the SSET gate a $0.2e$ step and recorded the response of the electrometer. In order to be sure that the measured signal originates from the SSET

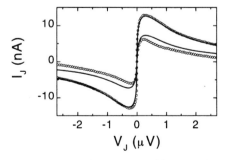

Figure 7. Open symbols: Extremal $I - V$ characteristics of the SSET obtained for $n_g = 0$ and $n_g = 1$, at $T = 40$ mK. Full lines: Theoretical predictions at $T = 200$ mK.

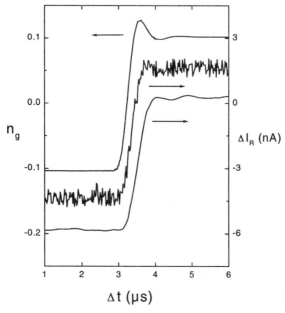

Figure 8. From top to bottom: Applied charge step at the gate of the SSET electrometer; electrometer response averaged over 4000 steps and measured within 10 MHz and 1 MHz bandwidths, respectively. Curves have been offsetted for clarity.

and not from a direct coupling to the SQUID electronics, we have substracted the traces with the SSET on and off. The applied step is compared in Fig. 8 with the electrometer response for two different output bandwidths, averaged over 4000 traces. The results demonstrate that the overall system has a bandwidth of about 1 MHz, probably limited by the extra filtering installed on the SQUID array lines. The noise level corresponds to the estimated one of $3.10^{-4}e/\sqrt{\text{Hz}}$, taking into account that only a fraction of the modulated current goes through the array input coil, and numerical factors. Although a faster response would be convenient, it would be of little use for measuring the state of a Cooper pair box if the sensitivity is not improved. Indeed, with the achieved noise level, one would need a typical time of a few μs to perform a measurement of the box state, assuming a coupling factor $\kappa_{\text{SSET}} = 2.5\%$. Improving the sensitivity is thus mandatory, and two stage SQUID amplifiers will be tested in the future for that purpose.

Back-Action Noise. The fluctuations of the island voltage V are very different from those in the SET because the SSET island is not sequentially charged and discharged. Since the island voltage varies periodically with the superconducting phase across the SSET, the island voltage fluctuations follow the phase fluctuations. Assuming that the box transition frequency is lower than the SSET band gap, the dynamics of the SSET can be calculated using an adiabatic approximation, and we find for the relaxation rate Γ_1^{SSET} due to the measuring SSET:

$$\Gamma_1^{SSET} \sim \frac{R_s I_c^2}{\hbar \Omega} \kappa_{SSET}^2 \sin^2 \theta. \tag{16}$$

For the electrometer we have operated, this rate would be of the order of 1 MHz, assuming again $\kappa_{SSET} = 2.5\%$. The charge sensitivityof the SSET as an electrometer is limited by the thermal fluctuations ofthe voltage at the working poin [23, 27]. The intrinsic figure of merit of the shunted SSET for the measurement of the box is:

$$f^{SSET} = \frac{1}{\Gamma_1^{SSET} \tau_m} \sim \frac{\hbar \Omega}{k_B T} \cot^2 \theta, \tag{17}$$

where τ_m is again the minimum time needed to discriminate the two states of the box. Although the factor $\hbar \Omega / k_B T$ can be large, the practical figure of merit of the present electrometer would be smaller than one because the measuring time would be limited by the SQUID array and not by the intrinsic noise of the shunted SSET. The decoherence results from the low frequency fluctuations of the island voltage. Although the spectrum of the phase fluctuations is known [27], the complicated relation between the phase across the SSET and the voltage in the island does not allow to deduce the island voltage fluctuation spectrum except in some limits. In the non-running state, the fluctuations of the phase are small, and a perturbative calculation of around the average value leads to:

$$\Gamma_2^{SSET} = \frac{4 k_B T}{h} \frac{\kappa_{SSET}^2 R_K}{r} \left(\frac{E_C}{E_J} \right)^2 g^2 \cos^2 \theta, \tag{18}$$

where

$$g = \frac{I_c}{\sqrt{I_c^2 - I_{SSET}^2}} \frac{\partial (C_{SSET} V / e)}{(\arcsin (I_{SSET} / I_c))} \tag{19}$$

is an increasing function of the dc current $I_{SSET} < I_c$ with $g(0) = 0$. For our electrometer biased at $I_{SSET} = I_c / 2$, the decoherence rate of a box coupled to it with $\kappa_{SSET} = 2.5\%$ would be of the order of a few MHz. This implies that, for a SSET used in the switching mode, the bias current has to be reduced close to zero during box manipulations. When the ratio EC/E_J' is larger than one, the variations of the island voltage are almost proportional to $\cos(\delta)$, and the spectrum can be calculated numerically along the lines of refs. [27] in all regimes. We find that the spectrum of the island voltage fluctuations has a peak at the Josephson frequency, and that the induced decoherence rate is the largest when the current through the electrometer is close to its maximum value. Quantitatively, the SSET wehave operated would result in a decoherence rate $\Gamma_2^{SSET} \approx 0.3$ MHz} at the working point used for electrometry.

Alternate High Bandwidth Set-up Using a SQUID Array. In the shunted SSET set-up, the bandwidth is ultimately limited by the time constant of the *RL* circuit at the input of the array. This limitation is not mandatory, and an alternate high bandwidth set-up is possible. For that purpose, a SSET, directly connected in parallel with the input coil of a SQUID, is current biased well below its critical current. Due to the residual resistance in the coil wiring, the dc bias current flows through the SSET only, and the phase difference accross the SSET adjusts to accomodate it. When the critical current of the SSET is ac-modulated, its effective inductance is varied accordingly, and the distribution of the bias current between the SSET and the SQUID input coil is modulated. In this set-up, the SSET behaves as a charge to current transducer, and the sensitivity, bandwidth and back-action are entirely determined by the current measuring stage.

7. CONCLUSIONS

We have evaluated in the single Cooper pair box the relaxation and decoherence rates due to electromagnetic dissipation in the box circuit itself and to external charges moving in the neighborhood of the box. The offset charge noise is likely the dominant source of relaxation and of decoherence, which rises an important problem. We have also evaluated the relaxation and decoherence rates due to a measuring electrometer coupled to the box island, and discussed in particular set-ups based on unshunted or shunted SSETs. Experimentally, we have operated an electrometer based on the continuous measurement of a shunted SSET with a SQUID series array. The sensitivity achieved by this system was about 3.10^{-4} $e/\sqrt{\text{Hz}}$, limited by the SQUID array noise, within a few MHz bandwidth. Such an electrometer would be useful in Cooper pair box experiments only if its sensitivity is significantly improved. Other measuring strategies can however be used for that purpose. In particular, the switching of an unshunted SSET is a simple one because it does not require a cold amplifier and is well suited for pulsed operation.

ACKNOWLEDGMENTS.   We thank Michel Devoret for stimulating discussions on quantum limited measurements.

REFERENCES

1. P. W. Shor, in *Proceedings of the Symposium on the Foundations of Computer Science* (IEEE Computer Society Press, New York,1994); L. K. Grover, Phys. Rev. Lett. **79**, 325 (1997).
2. P. Shor, Phys. Rev. A 52, R2493 (1995); A. M. Steane, Nature, **399**, 124 (1999).
3. W. Lange and H. J. Kimble, Phys. Rev. A **61**, 63817 (2000); C. A. Sackett, D. Kielpinski, B. E. King, C. Langer, V. Meyer, C. J. Myatt, M. Rowe, Q. A. Turchette, W. M. Itano, D. J. Wineland, and C. Monroe, Nature **404**, 256 (2000).
4. C. Monroe, D. M. Meekhof, B. E. King, W. M. Itano, and D. J. Wineland, Phys. Rev. Lett. **75**, 4714 (1995).
5. C. H. van der Wal, A. C. ter Haar, F. K. Wilhelm, R. N. Schouten, C. J. P. M. Harmans, T. P. Orlando, Seth Lloyd, and J. E. Mooij, to be published.
6. V. Bouchiat, D. Vion, P. Joyez, D. Esteve, and M. H. Devoret, Physica-Scripta **76**, 165 (1998) and J. Superconductivity, **12**, 789, (1999).
7. Y. Nakamura, Y. A. Pashkin, and J. S. Tsai, Physica-B **280**, 405 (2000) and this book; Y. Nakamura, J. S. Tsai, J. Superconductivity, **12**, 799 (1999) and J. Low. Temp. Phys. **118**, 765 (2000).

8. P. Joyez, P. Lafarge, A. Filipe, D. Esteve and M. H. Devoret, Phys. Rev. Lett. **72**, 2548 (1994).
9. A. Barone and G. Paternò, *Physics and Applications of the Josephson Effect* (Wiley, New York, 1992).
10. A. J. Leggett, in *Chance and Matter* (North-Holland, Amsterdam, 1987)
11. A. Cottet *et al.*, unpublished.
12. G. Ingold and Yu. V. Nazarov, in *Single Charge Tunneling*, edited by H. Grabert and M. H. Devoret (Plenum Press, New York, 1992).
13. T. A. Fulton and G. J. Dolan, Phys. Rev. Lett. **59**, 109 (1987).
14. A. B. Zorin, F. J. Ahlers, J. Niemeyer, T. Weimann, H. Wolf, V. A. Krupenin, and S. V. Lotkhov, Phys. Rev. B **53**, 13682 (1996); V. A. Krupenin, D. E. Presnov, M. N. Savvateev, H. Scherer, A. B. Zorin, and J. Niemeyer, Conference on Precision Electromagnetic Measurements Digest, IEEE, **140** (1998).
15. M. Covington, Mark W. Keller, R. L. Kautz, and John M. Martinis, Phys. Rev. Lett. **84**, 5192 (2000).
16. A. Shnirman and G. Schon, Phys. Rev. B **57**, 15400 (1998); Y. Makhlin, G. Schon, and A. Shnirman, Physica-B **280**, 410 (2000).
17. B. Starmark, T. Henning, T. Claeson, P. Delsing, A. N.Korotkov, J. Appl. Phys. **86**, 2132 (1999). D. Averin, this book.
18. R. J. Schoelkopf, P. Wahlgren, A. A. Kozhevnikov, P. Delsing and D. E. Prober, Science **280**, 1238 (1998).
19. D. Vion, M. Gotz, P. Joyez, D. Esteve, and M. H. Devoret, Phys. Rev. Lett. **77**, 3435 (1996); P. Joyez, D. Vion, M. Gotz, M. H. Devoret, and D. Esteve, J. Superconductivity **12**, 757 (1999).
20. R. P. Welty and J. M. Martinis, IEEE Trans. Mag. **27**, 2924 (1991).
21. M. E. Huber *et al.*, Applied Superconductivity **5**, 425 (1998).
22. A. Steinbach, P. Joyez, D. Esteve, M. H. Devoret and M. E. Huber, to be published.
23. A. B. Zorin, Phys. Rev. Lett. **76**, 4408 (1996); A. B. Zorin *et al.*, J. Superconductivity **12**, 747 (1999).
24. G. J. Dolan and J. H. Dunsmuir, Physica B **152**, 7 1(988).
25. D. Vion, P. F. Orfila, P. Joyez, D. Esteve, and M. H. Devoret, J. Appl. Phys. **77**, 2519 (1995).
26. Yu. M. Ivanchenko and L. A. Zil'berman, Soviet Phys. JETP **28**, 1272, (1969); V. Ambegaokar and B. I. Halperin, Phys. Rev. Lett. **22**, 1364 (1969).
27. A. V. Vystavkin, V. N. Gubankov, L. S. Kuzmin, K. K. Likharev, V. V. Migulin, and V. K. Semenov, Rev. Phys. Appl. **9**, 79 (1974); W. T. Coffey, Yu. P. Kalmykov, and J. T. Waldron, *The Langevin Equation* (World Scientific, 1996).

ADIABATIC TRANSPORT OF COOPER PAIRS IN ARRAYS OF SMALL JOSEPHSON JUNCTIONS

J.P. PEKOLA, J.J. TOPPARI, N. KIM, M.T. SAVOLAINEN, L. TASKINEN, AND K. HANSEN
Department of Physics, University of Jyväskylä, P.O. Box 35 (Y5), FIN-40351 Jyväskylä, Finland

Keywords: Josephson junctions, quantum bits, Cooper pair pump

Abstract   We present a quantitative theory of Cooper pair pumping in gated one-dimensional arrays of Josephson junctions. The pumping accuracy is limited by quantum tunneling of Cooper pairs out of the propagating potential well and by direct supercurrent flow through the array. Both corrections decrease exponentially with the number N of junctions in the array, but give a serious limitation of accuracy for any practical array. We also consider the Cooper pair trap and how it can be used to test the results experimentally.

PACS numbers: 73.23.Hk, 74.50.+r

1.   INTRODUCTION

When a potential well propagates adiabatically along an electron system which is effectively one-dimensional it carries with it additional electron density and induces a dc electric current through the system. Such a pumping effect is observed in mesoscopic systems ranging from small metallic tunnel junctions in the Coulomb blockade regime [1-4], to semiconductor quantum dots [5] and one-dimensional ballistic channels [6]. The propagation of the potential well is arranged either through the propagation of an acoustoelectric wave [2, 6] or by phase-shifted gate voltages [1, 3, 5]. Of particular interest is the pumping regime when the potential well carries a quantized number m of electrons so that the induced current I is related to the frequency f, with which the well crosses the system, by the fundamental relation $I = mef$. A well with a definite number of electrons can be created either by the Coulomb interaction, as, for instance, in the Coulomb blockade pumps [1-4], or it can be caused by the discrete nature of single-particle states inside the well [7, 8]. In the case of Coulomb blockade pumps, the precision of the pumped charge is reaching a level sufficient for metrological applications [3, 4]. Different sources of inaccuracy in the pumps have been discussed in the literature [3,9-12].

Until recently the pumping effect has been studied almost exclusively in normal systems where transport is due to individual electrons. A timely motivation for studying Cooper pair transfer comes from quantum computation, where pumping can play an important role in the dynamics of quantum logic gates [13]. This work presents a quantitative theory of Cooper pair pumping in one-dimensional arrays of superconducting tunnel junctions. In particular, we find fundamental corrections to the quantized pumping regime and show that they are unexpectedly large in arrays with a small number of junctions. These large quantum corrections explain the fact that the first experiment with pumping of Cooper pairs failed to demonstrate accurate pumping [14]. We also suggest that one can use a Cooper pair trap instead of a pump to experimentally measure the transport probabilities.

2.   COOPER PAIR PUMP

First we present the general expression for the charge transferred through an array of N superconducting tunnel junctions in the Coulomb blockade regime by adiabatic pumping of Cooper pairs [15]. In the standard model such arrays are characterized by two energies, the charging energy H_C as a system of capacitors, and the energy associated with tunneling [16]. The array is assumed uniform, so $C_1 = C_2 = \cdots = C_N \equiv C$

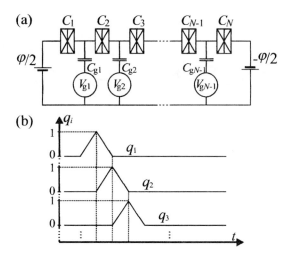

Figure 2.1. **(a)** A schematic drawing of a gated Josephson array of N junctions. In pumping Cooper pairs gate voltages V_{gi} are operated cyclically. C_i are the capacitances of the junctions, and C_{gi} are gate capacitances. In a uniform pump $C_i \equiv C$ for all $i = 1, 2, \ldots, N$. **(b)** A train of gate voltages to carry a charge in a pump. Here $q_i = -C_{gi} V_{gi}/2e$.

(fig. 2.1). We also assume that the characteristic energy $E_C \equiv (2e)^2/2C$ of a Cooper pair in the array and the temperature $(k_B T)$ are both much lower than the superconducting energy gap of the electrodes. The relation for E_C replaces the usual condition in the normal Coulomb blockade where the junction resistances should be larger than the quantum resistance. With these conditions fulfilled quasiparticle tunneling is exponentially suppressed, while the tunneling energy of Cooper pairs in junction i reduces to a constant $E_{Ji}/2$. (E_{Ji} is the Josephson coupling energy.) In this work, the bias voltage is set to zero, and thus a constant Josephson phase difference φ is fixed across the array. We can then treat the two external electrodes of the array as one, so that effectively the array forms a loop and φ plays the role of external flux threading it. Then, the Hamiltonian of the N-pump is [15]:

$$H = H_C(n - q) - \sum_{k=1}^{N} \frac{E_{Jk}}{2} \left(|n\rangle\langle n + \delta_k | e^{i\varphi/N} + h.c. \right) . \qquad (2.1)$$

The array $n \equiv \{n_1, n_2, \ldots, n_{N-1}\}$ represents the number n_i of Cooper pairs on the islands and $q \equiv \{q_1, q_2, \ldots, q_{N-1}\}$ the charge, normalized by $2e$, induced on each island by the gate voltage V_{gi} (fig. 2.1). The term δ_k describes the change of n due to tunneling of one Cooper pair in the kth junction. The charging energy of the homogeneous array can be written as

$$H_{\mathrm{C}} = \frac{E_{\mathrm{C}}}{N} \left[\sum_{k=1}^{N-1} k(N-k)u_k^2 + 2 \sum_{l=2}^{N-1} \sum_{k=1}^{l-1} k(N-l)u_k u_l \right], \qquad (2.2)$$

where $u_k \equiv n_k - q_k$. We will also need the current operator of the kth junction:

$$I_k = \frac{ieE_{\mathrm{J}k}}{2\hbar} \left(|n\rangle\langle n + \delta_k| e^{i\varphi/N} - h.c. \right). \qquad (2.3)$$

2.1   ADIABATIC APPROXIMATION

There are two mechanisms of Cooper pair transport in the array. One is the direct supercurrent through the whole array and the other is the pumping, i.e. the charge transfer in response to the adiabatic slow variation of the induced charges q_i. To derive the general expression for the total charge Q transferred during one pumping period we introduce the basis of instantaneous eigenstates $\{|m_{(t)}\rangle\}$ with eigenenergies $\{E_m^{(t)}\}$ of the full Hamiltonian (2.1) for a given $q(t)$. Assuming slowly varying gate voltages we may solve the time-dependent Schrödinger equation with the initial condition $|\psi(t_0)\rangle = |m_{(t_0)}\rangle$ to obtain [18]

$$|\psi_{(t_0+\delta t)}\rangle = e^{-iE_m^{(t_0)}\delta t/\hbar}|m_{(t_0)}\rangle + |\delta m_{(\delta t)}\rangle. \qquad (2.4)$$

Here the term $|\delta m_{(\delta t)}\rangle$ is a correction to the state $|m_{(t_0)}\rangle$ due to the change in gate charges q. The amount of charge that passes through the junction k during a short time interval δt is then

$$
\begin{aligned}
\delta Q_k &= \int_{t_0}^{t_0+\delta t} \langle\psi_{(t)}|I_k|\psi_{(t)}\rangle dt \\
&= \delta t \langle I_k\rangle_{|m_{(t_0)}\rangle} - 2\hbar \sum_{l(\neq m)} \mathrm{Im}\left[\frac{\langle m|I_k|l\rangle\langle l|\delta m\rangle}{E_l - E_m} \right] \qquad (2.5)
\end{aligned}
$$

where we have neglected the term quadratic in $|\delta m\rangle$ and oscillatory terms by assuming that the inequality $\delta t \gg \hbar/(E_l - E_m)$ holds for all l.

For a closed path γ the transferred charge must be equal for all N junctions so the total amount of charge, Q, transferred through the array over a pumping period τ is then given by $Q = Q_k = \int_0^\tau \langle I_k\rangle_{|m_{(t_0)}\rangle}dt + Q_{\mathrm{P}}$. The first term gives the charge transferred via direct supercurrent. The second term, the charge transfer induced by gates, can be written as

$$\frac{Q_{\mathrm{P}}}{-2e} = \frac{\hbar}{e} \oint_\gamma \sum_{l(\neq m)} \mathrm{Im}\left[\frac{\langle m|I_k|l\rangle\langle l|dm\rangle}{E_l - E_m} \right], \qquad (2.6)$$

where $|dm\rangle$ is the differential change of $|m\rangle$ due to a differential change of the gate voltages dq (see [15]).

2.2   HOMOGENEOUS PUMP

In the regime of accurate pumping the main contribution to Q comes from the induced charge transfer Q_P while the supercurrent gives only small corrections limiting the pumping accuracy [15]. The necessary condition for this regime to exist is $E_J \ll E_C$, which we assume from now on. We consider two different pumping paths around the degeneracy point, which occurs when $q_k = 1/N$ for all k. At this degeneracy point the energies of several charge states coincide as illustrated in fig. 2.2(a) for $N = 3$. If the pumping process is slow, the Cooper pair is transported adiabatically between the islands by the usual two-state level-crossing transitions that shift it along the array following the gate voltages. One Cooper pair is then transported through the array per cycle corresponding to a q-space trajectory circling once around the degeneracy point. One more condition necessary for accurate pumping is that the probability of the Landau-Zener transitions to the excited states is negligible and the array remains in the minimum-energy state throughout the cycle. This condition limits the rate of pumping, $1/\tau$, by the relation, $\hbar/\tau \ll E_J^2/E_C$. However, even then, i.e. in the regime of the present work, the pumping is not accurate due to the nonvanishing E_J/E_C.

For the trajectory illustrated in fig. 2.1(b) we obtain by perturbation theory in E_J and by eq. (2.6) [15]:

$$\frac{Q_P}{-2e} = 1 - \frac{N^{N-1}(N-1)}{(N-2)!} \left(\frac{E_J}{2E_C} \right)^{N-2} \cos\varphi. \qquad (2.7)$$

Thus the probability of Cooper pair tunneling limiting the pumping accuracy decreases with increasing N.

For $N = 3$, the triangular pumping trajectory in the (q_1, q_2) plane shown in fig. 2.2 corresponds to triangular gate voltages. Another pumping scheme in the $N = 3$ pump [1, 14] is provided by harmonic gate voltages, corresponding to a circular trajectory around the degeneracy point $q_1 = q_2 = 1/3$. In this case it is possible to calculate the pumped charge directly from eq. (2.6). For $\varphi = 0$ we obtain [15]:

$$\frac{Q_P}{-2e} = 1 - \frac{3}{2} \left(\frac{1}{3\sqrt{2}\delta} + \frac{1}{2 - 3\sqrt{2}\delta} + \frac{1}{\frac{3}{\sqrt{5}}\delta} + \frac{1}{1 - \frac{3}{\sqrt{5}}\delta} \right) \frac{E_J}{E_C}, \qquad (2.8)$$

where $\delta \equiv [(q_1 - 1/3)^2 + (q_2 - 1/3)^2]^{1/2}$ is the radius of the trajectory. The results of eqs. (2.8) and (2.7) for $N = 3$ almost coincide for the optimum radius of $\delta \simeq 0.3$. It should be noted that the quantum inaccuracy

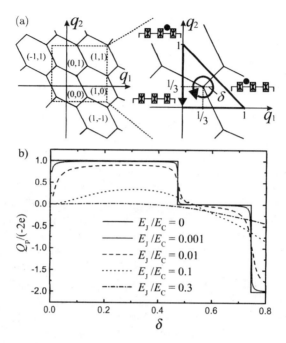

Figure 2.2. (a) The states with minimum charging energy of the uniform $N = 3$ pump on the (q_1, q_2) plane. The vector (n_1, n_2) denotes the stable configuration inside each hexagon. A circular path with radius δ, and the triangular path of fig. 2.1(b) are shown. Only states contributing to the inaccuracy in charge transport in the leading order are shown. (b) Numerically calculated quantum inaccuracies of a uniform 3-pump for different values of E_J/E_C. The analytical result of eq. (2.8) is exact in the limit of small E_J/E_C.

in pumping is very significant: it is more than 20 % at $E_J/E_C = 0.03$ (practically, E_J is limited from below by the temperature, while the maximum E_C is limited by physical dimensions of the fabrication.) The accurate coherent pumping is thus practically impossible in the $N = 3$ pumps. Figure 2.2 shows Q_P calculated numerically from eq. (2.6) for $\varphi = 0$ (no direct supercurrent present) as a function of δ. For small radii the charge is quadratic in δ, $Q_P = \pi\delta^2 (8E_C/27E_J)^2$, as can be derived from eq. (2.6). At large δ the pumped charge in fig. 2.2 starts to decrease since the trajectory approaches another degeneracy point at $q_1 = q_2 = 2/3$.

2.3   FURTHER RESULTS

To obtain quantitatively more precise results than by the first order perturbation theory one can use the renormalisation method (see, e.g., [17]). With this method it is possible to calculate the higher order corrections to the pumping inaccuracy in case of homogeneous arrays, inhomogeneity of the array, and nonideal pumping sequences [18].

In fig. 2.3 the pumped charge Q_P for $N = 3$ is shown as a function of the phase difference φ. Values calculated with renormalisation and numerical results are in good agreement and they clearly indicate that the deviations from the leading order result, $[Q_P/(-2e) = 1 - 9(E_J/E_C)\cos\varphi]$ are important also for finite ϕ.

Also the inhomogeneity of the array can be treated with renormalisation theory. To do this we define the inhomogeneity index of the array

$$X_{\text{inh}} = \sqrt{\frac{1}{N} \sum_{k=1}^{N} \left(\frac{C - C_k}{C}\right)^2}, \qquad (2.9)$$

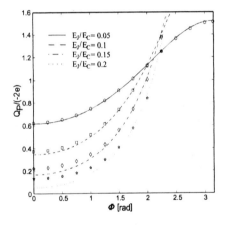

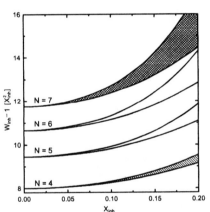

Figure 2.3. The pumped charge $Q_P/(-2e)$ as a function of φ for some values of E_J/E_C and $N = 3$. Curves denote renormalised values and symbols numerical values which were obtained for a 41-state basis. The points calculated by eq.(2.7) for $\varphi = 0$ and $E_J/E_C = 0.05$ and 0.1 are 0.55 and 0.10, respectively. The pumped charge is symmetric in φ and its period is 2π.

Figure 2.4. The limits for the ratio W_{inh} as a function of X_{inh} for array lengths $N = 4$ to $N = 7$. For small values of X_{inh}, $W_{\text{inh}} \approx 1 + a_N^{(\text{inh})} \cdot X_{\text{inh}}^2$, where the N-dependent constant $a_N^{(\text{inh})}$ can be evaluated [18].

where the "average" capacitance C is given by $C = N/\sum_{k=1}^{N} C_k^{-1}$. To make a comparison easier we consider W_{inh}, defined as the ratio between the inhomogeneous and homogeneous inaccuracies. The effects due to inhomogeneity can be parametrised by obtaining limits for W_{inh} as a function of X_{inh}. In fig. 2.4 we graphically present these limits for $(W_{\mathrm{inh}} - 1)/X_{\mathrm{inh}}^2$ as a function of X_{inh} in the cases $N = 4$ to $N = 7$.

3.   COOPER PAIR TRAP

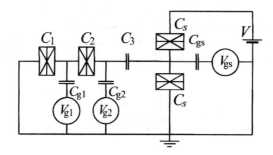

Figure 3.1. Schematic picture of a Cooper pair trap with two Josephson junctions. The voltage V should be switchable for measurements as described in the text.

To experimentally measure charge transport it is necessary to consider a Cooper pair trap coupled to an SET-electrometer instead of the straightforward pump array experiment discussed above. The last island of the trap, consisting of two or more Josephson junction in series, is capacitively coupled to the island of the SET (see fig. 3.1). Cooper pairs are transported to the trap by pumping with gate voltages V_{gi}. At the end of the pumping cycle one can activate the SET to measure if the Cooper pair has passed the array.

The trap differs from the three junction pump in that there is no Josephson coupling over the last element where a capacitance has been substituted for a Josephson junction. The change will cause hysteresis, e.g. when moving on the trajectory $q = (0,0) \leftrightarrow q = (0,1)$ (see fig. 2.2) in a two junction trap. The Hamiltonian of the homogeneous two junction trap can be written in the basis of charge states $\{|00\rangle, |10\rangle, |01\rangle\}$ as

$$H = \begin{pmatrix} E_{\mathrm{C1}} & -\frac{E_J}{2} & 0 \\ -\frac{E_J}{2} & E_{\mathrm{C2}} & -\frac{E_J}{2} \\ 0 & -\frac{E_J}{2} & E_{\mathrm{C3}} \end{pmatrix},$$ (3.1)

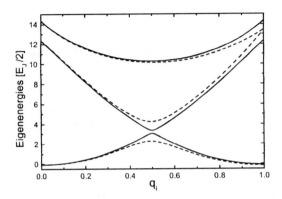

Figure 3.2. Eigenenergies of the two junction Cooper pair trap in the basis of charge states $\{|00\rangle, |10\rangle, |01\rangle\}$ plotted as a function of the normalised gate voltage q_1 (dashed) and q_2 (solid). The other gate voltage is kept at constant value $q_i = 0$ so the x-axis corresponds to the vertical and horizontal trajectories in the triangular pumping path shown in fig. 2.2. The value of E_J/E_C is 0.1.

where the E_{Ci} represent the charging energy of the three basis states. Zeros at the far off-diagonals correspond to the absence of direct coupling between the left and right side of the pumping line.

Also the energy spectrum of the trap differs from the one of the three junction pump. This can clearly be seen in fig 3.2 where the eigenenergies of the trap are plotted as a function of two different trajectories in the (q_1, q_2)-plane. From the figure it is clear that the system is not as symmetric as the pump and there is still a small gap between the un-coupled states $|00\rangle$ and $|01\rangle$. This gap is due to higher order tunnelling events.

Acknowledgments. This work has been supported by the Academy of Finland under the Finnish Centre of Excellence Programme 2000-2005 (Project No. 44875, Nuclear and Condensed Matter Programme at JYFL) and EU (contract IST-1999-10673). We thank D.V. Averin and M. Aunola for fruitful collaboration on the subject.

References

[1] H. Pothier, P. Lafarge, C. Urbina, D. Esteve and M.H. Devoret, Europhys. Lett. **17**, 249 (1992).

[2] J.P. Pekola, A.B. Zorin and M.A. Paalanen, Phys. Rev. B **50**, 11255 (1994).

[3] M.W. Keller, J.M. Martinis, N.N. Zimmerman and A.H. Steinbach, Appl. Phys. Lett. **69**, 1804 (1996); M.W. Keller, J.M. Martinis, and R.L. Kautz, Phys. Rev. Lett. **80**, 4530 (1998).

[4] S. V. Lotkhov, S. A. Bogoslovsky, A. B. Zorin and J. Niemeyer *The single electron R-pump: first experiment*, cond-mat/0001206

[5] L.P. Kouwenhoven, A.T. Johnson, N.C. Van der Vaart, C.J.P.M. Harmans and C.T. Foxon, Phys. Rev. Lett. **67**, 1626 (1991).

[6] V.I. Talyanskii, J.M. Shilton, M. Pepper, C.G. Smith, C.J.B. Ford, E.H. Linfield, D.A. Ritchie and G.A.C. Jones, Phys. Rev. B **56**, 15180 (1997).

[7] D.J. Thouless, Phys. Rev. B **27**, 6083 (1983).

[8] Q. Niu, Phys. Rev. Lett. **64**, 1812 (1990).

[9] H.D. Jensen and J.M. Martinis, Phys. Rev. B **46**, 13407 (1992).

[10] D.V. Averin, A.A. Odintsov and S.V. Vyshenskii, J. Appl. Phys. **73**, 1297 (1993).

[11] J.M. Martinis, M. Nahum and H.D. Jensen, Phys. Rev. Lett. **72**, 904 (1994).

[12] L. R. C. Fonseca, A. N. Korotkov and K. K. Likharev, Appl. Phys. Lett. **69**, 1858 (1996).

[13] D.V. Averin, Solid State Commun. **105**, 659 (1998).

[14] L.J. Geerligs, S.M. Verbrugh, P. Hadley, J.E. Mooij, H. Pothier, P. Lafarge, C. Urbina, D. Esteve and M.H. Devoret, Z. Phys. B **85**, 349 (1991).

[15] J.P. Pekola, J.J. Toppari, M. Aunola, M.T. Savolainen and D.V. Averin, Phys. Rev. B **60**, R9931 (1999).

[16] D.V. Averin and K. K. Likharev, in: *Mesoscopic Phenomena in Solids*, ed. by B. L. Altshuler, P. A. Lee, and R. A. Webb (Elsevier, Amsterdam, 1991).

[17] P.J. Ellis and E. Osnes, Rev. Mod. Phys. **49**, 777 (1978).

[18] M. Aunola, J.J. Toppari and J.P. Pekola, *Arrays of Josephson junctions in an environment with vanishing impedance*, Phys. Rev. B, *in print*.

ENTANGLED STATES IN A JOSEPHSON CHARGE QUBIT COUPLED TO A SUPERCONDUCTING RESONATOR

O. Buisson[1] and F.W.J. Hekking[2]

[1] *Centre de Recherches sur les Très Basses Températures, laboratoire associé à l'Université Joseph Fourier, C.N.R.S., BP 166, 38042 Grenoble-cedex 9, France.*

[2] *Laboratoire de Physique et Modélisation des Milieux Condensés & Université Joseph Fourier, C.N.R.S., BP 166, 38042 Grenoble-cedex 9, France.*

buisson@labs.polycnrs-gre.fr and hekking@belledonne.polycnrs-gre.fr

Abstract   We study the dynamics of a quantum superconducting circuit which is the analogue of an atom in a high-Q cavity. The circuit consists of a Josephson charge qubit coupled to a superconducting resonator. The Josephson charge qubit can be treated as a two level quantum system whose energy separation is split by the Josephson energy E_j. The superconducting resonator in our proposal is the analogue of a photon box and is described by a quantum harmonic oscillator with characteristic frequency ω_r. The coupling between the charge qubit and the resonator is realized by a coupling capacitance C_c. We have calculated the eigenstates and the dynamics of the quantum circuit. Interesting phenomena occur when the Josephson energy equals the oscillator frequency, $E_j = \hbar\omega_r$. Then the quantum circuit is described by entangled states. We have deduced the time evolution of these states in the limit of weak coupling between the charge qubit and the resonator. We found Rabi oscillations of the excited charge qubit eigenstate. This effect is explained by the spontaneous emission and re-absorption of a single photon in the superconducting resonator.

1.   INTRODUCTION

Recently, a substantial interest in the theory of quantum information and computing [1] has revived the physical research on quantum systems. The elementary unit of quantum information is a two-state system, usually referred to as a quantum bit (qubit). Basic operations are realized by preparation and manipulation of, as well as a measurement on, entangled states in systems which consist of several coupled qubits. However, the fabrication of physical systems which would enable the actual implementation of quantum algorithms is far from being realized in the near future and a substantial amount of fundamental research is still needed.

During the past five years, great progress has been made in the manipulation of entangled states in systems consisting of up to four qubits based on ion traps [2] and atoms in a high-Q cavity [3]. These two experiments demonstrate clearly and unambiguously the possibility to coherently control the entangled states of a limited number of qubits, as well as to perform a quantum measurement on them. In spite of this success, it seems quite difficult to realize circuits consisting of the large number of ion traps or atoms in a cavity necessary for quantum computation.

It has been suggested that small solid state devices fabricated using nanolithography technologies are promising for quantum circuit integration. However, the coherent manipulation of entangled states as well as the realization of quantum measurements remain fundamental issues to be investigated. One of the main challenges is to gain control over all possible sources of decoherence. At present the best candidates for the implementation of quantum gates based on solid state devices are circuits using small Josephson junctions. In the superconducting state, such circuits contain less intrinsic sources of decoherence. Indeed it has been experimentally demonstrated that a single Cooper pair box is a macroscopic two level system which can be coherently controlled [4, 5, 6]. At about the same time, theoretical works have proposed the use of the Cooper pair box as a qubit (the so-called Josephson charge qubit) in the context of quantum computers. In particular, systems consisting of several charge qubits with controlled couplings have been discussed, the quantum measurement problem has been addressed and the decoherence time has been estimated [7, 8]. More recently, qubits based on superconducting loops containing small Josephson junctions have been proposed (Josephson phase qubits) [9, 10, 11, 12] and are currently studied. But up to now, the existence of entangled states, which are at the heart of quantum information processing, has been demonstrated neither for charge qubits nor for phase qubits.

In this article we propose to study one of the simplest Josephson circuits in which entangled states can be realized. It consists of a charge qubit coupled to a superconducting resonator and can be described theoretically by a two level system coupled to a quantum harmonic oscillator. After a description of this quantum circuit in the next section, the Hamiltonian describing it will be derived in Sec. 3. In Sec. 4, the time evolution of the eigenstates is obtained and we demonstrate the existance of entanglement. In the last Section, we discuss the dynamics of the quantum circuit for typical experimental values of the system parameters.

2. QUANTUM CIRCUIT

The Josephson circuit we study hereafter is depicted in Fig. 1. It consists of three different elementary circuits: a Cooper pair box, an LC-resonator and a "coupling" capacitor.

For small enough junction capacitance C_j, gate capacitance C_g, and coupling capacitance C_c, the charging energy of the box is large compared to thermal fluctuations and the excess charge of the box is quantized. On the other hand, we assume the charging energy to be smaller than the superconducting gap Δ, such that no quasiparticles are present in the box. Thus the excess charge is entirely due to the presence of Cooper pairs and charge quantization occurs in units of $2e$. The gate voltage V_g is used as an external control parameter. When the gate charge $N_g = -C_g V_g/e$ is equal to unity, the Cooper pair box can be viewed as a macroscopic two-level quantum system whose energy separation is split by the Josephson energy E_j. The two eigenstates $|-\rangle$ and $|+\rangle$ correspond to a coherent superposition of the two different charge states of the box [4, 5, 6]. When $N_g \approx 1$, the Cooper pair box will be referred to as a Josephson charge qubit.

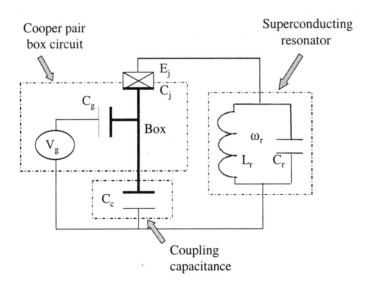

Cooper pair
box circuit

Superconducting
resonator

Figure 1 The quantum circuit.

The LC-resonator system can be described by a quantum harmonic oscillator whose characteristic frequency is given by ω_r. This system is the analogue of a high-Q cavity.

The capacitance C_c plays a crucial role in our proposed circuit since it couples the charge qubit and the resonator to each other. These two circuits are no longer independent and the system must be considered in its totality. Thus the proposed quantum circuit of Fig. 1 realizes the simple situation in which a two level system is coupled to a harmonic oscillator. In spite of its simplicity, such a system describes a great variety of interesting situations which were extensively studied in the past [13].

3.   HAMILTONIAN

The circuit depicted in Fig. 1 can be characterized mechanically by two generalized coordinates, ϕ_j and ϕ_r. These coordinates are associated with the voltage drop δV_j over the junction and δV_r over the resonator, respectively, according to the Josephson relation $\phi_i = 2e\delta V_i t/\hbar$ ($i = j, r$). We seek the Lagrangian $\mathcal{L}(\phi_j, \phi_r, \dot\phi_j, \dot\phi_r) = T - V$ describing the dynamics of these variables. The potential energy V is a function of the coordinates only,

$$V(\phi_j, \phi_r) = -E_j \cos\phi_j + \frac{E_r}{2}\phi_r^2, \tag{1.1}$$

where $E_r = (1/L_r)(\hbar/2e)^2$ is the energy associated with the inductance L_r of the resonator. The kinetic energy T is quadratic in the velocities $\dot\phi_j$ and $\dot\phi_r$. It is just the free electrostatic energy stored in the capacitators present in the circuit. This free energy can be written as

$$T = \frac{1}{2}\left[C_{\Sigma j}(\hbar\dot\phi_j/2e)^2 + C_{\Sigma r}(\hbar\dot\phi_r/2e)^2 + 2C_{\Sigma c}(\hbar/2e)^2\dot\phi_j\dot\phi_r\right]. \tag{1.2}$$

Here we introduced the capacitances $C_{\Sigma c} = C_c + C_g$, $C_{\Sigma j} = C_j + C_{\Sigma c}$, $C_{\Sigma r} = C_r + C_{\Sigma c}$. Note that the Lagrangian contains an interaction between the resonator and the junction: $\mathcal{L}_{int} = C_{\Sigma c}(\hbar/2e)^2\dot\phi_j\dot\phi_r$. The effective coupling between these two parts of the circuit is determined by the sum of the gate capacitance and the coupling capacitance. We also note that the equations of motion $d(\partial\mathcal{L}/\partial\dot\phi_i)/dt + \partial\mathcal{L}/\partial\phi_i = 0$ express current conservation in the circuit.

Through the Josephson junction, Cooper pairs can tunnel from or onto the island. The number of excess Cooper pairs on the island, n, depends on the gate voltage. Charge neutrality leads us to relation

$$2ne = C_{\Sigma j}(\hbar\dot\phi_j/2e) + C_{\Sigma c}(\hbar\dot\phi_r/2e) + N_g e. \tag{1.3}$$

It is always possible to add a term to the Lagrangian which is a total time derivative. Let us add the term $\hbar\dot{\phi}_j N_g/2$. As a result, the momenta conjugate to ϕ_j and ϕ_r are

$$
\begin{aligned}
p_j &= \partial\mathcal{L}/\partial\dot{\phi}_j = (\hbar/2e)[C_{\Sigma j}(\hbar\dot{\phi}_j/2e) + C_{\Sigma c}(\hbar\dot{\phi}_r/2e) + N_g e] = \hbar n, \\
p_r &= \partial\mathcal{L}/\partial\dot{\phi}_r = C_{\Sigma r}(\hbar/2e)^2\dot{\phi}_r + C_{\Sigma c}(\hbar/2e)^2\dot{\phi}_j.
\end{aligned}
$$

Note in particular that the momentum p_j is proportional to n.

The Hamiltonian is obtained with the help of the Legendre transform $H = p_j\dot{\phi}_j + p_r\dot{\phi}_r - \mathcal{L}$. We find $H = H_j + H_r + H_c$, where

$$
\begin{aligned}
H_j &= E_{C,j}(2n - N_g)^2 - E_j\cos\phi_j, & (1.4) \\
H_r &= E_{C,r}(2p_r/\hbar)^2 + E_r\phi_r^2/2, & (1.5) \\
H_c &= -E_{C,c}(2n - N_g)(2p_r/\hbar). & (1.6)
\end{aligned}
$$

The charging energies $E_{C,i}$ ($i = j, r, c$) appearing here are given by $E_{C,i} = e^2/2C_{i,\text{eff}}$, with $C_{j,\text{eff}} = C_j + (1/C_{\Sigma c} + 1/C_r)^{-1}$, $C_{r,\text{eff}} = C_r + (1/C_{\Sigma c} + 1/C_j)^{-1}$, and $C_{c,\text{eff}} = [C_{\Sigma c} + (1/C_j + 1/C_r)^{-1}][(C_j + C_r)/2C_{\Sigma c}]$. The Hamiltonian equations of motion, $\dot{p}_i = -\partial H/\partial\phi_i$, $\dot{\phi}_i = \partial H/\partial p_i$ lead us again to current conservation.

In order to obtain the quantum mechanical Hamiltonian $\hat{H}$, we replace p_i, ϕ_i by the corresponding operators, with $[p_k, \phi_m] = (\hbar/i)\delta_{k,m}$. In particular, in ϕ−representation we have $p_k = (\hbar/i)\partial/\partial\phi_k$. Note also that $[n, \phi_j] = -i$ and $n = -i\partial/\partial\phi_j$. Below we discuss the various contributions to $\hat{H}$ in some detail.

Josephson junction. The commutation relation $[n, \phi_j] = -i$ implies $[n, e^{i\phi_j}] = e^{i\phi_j}$. Using the basis states $|n\rangle$, where n corresponds to the number of excess Cooper pairs on the island, we thus have $e^{i\phi_j}|n\rangle = |n + 1\rangle$. Similarly, $e^{-i\phi_j}|n\rangle = |n - 1\rangle$. Therefore we can write $\hat{H}_j$ as

$$
\begin{aligned}
\hat{H}_j &= E_{C,j}\sum_n(2n - N_g)^2|n\rangle\langle n| \\
&\quad - \frac{E_j}{2}\sum_n(|n + 1\rangle\langle n| + |n - 1\rangle\langle n|). \quad (1.7)
\end{aligned}
$$

If the gate-voltage is such that $N_g \simeq 1$, the states with $n = 0$ and $n = 1$ are almost degenerate. At low temperatures, the Hamiltonian $\hat{H}_j$ involves only these two states, and thus can be presented as a matrix

$$
\hat{H}_j \simeq \begin{pmatrix} E_{C,j}N_g^2 & -E_j/2 \\ -E_j/2 & E_{C,j}(2 - N_g)^2 \end{pmatrix}. \quad (1.8)
$$

This matrix can be diagonalized. The eigenvalues are

$$
E_{\mp} = E_{C,j}[1 + (\delta N_g)^2] \mp \frac{1}{2}\sqrt{(\delta E_g)^2 + E_j^2}, \quad (1.9)
$$

where $\delta N_g = N_g - 1$ and $\delta E_g = -4E_{C,j}\delta N_g$. The corresponding eigenstates are

$$|-\rangle = \alpha|0\rangle + \beta|1\rangle \qquad (1.10)$$
$$|+\rangle = \beta|0\rangle - \alpha|1\rangle \qquad (1.11)$$

where $\alpha^2 = 1 - \beta^2 = [1 + \delta E_g/\sqrt{(\delta E_g)^2 + E_j^2}]/2$.

Resonator. Since the LC-circuit constitutes just a harmonic oscillator with a characteristic frequency $\omega_r = \sqrt{1/L_r C_{r,\text{eff}}}$, the Hamiltonian $\hat{H}_r$ can be written in the standard way

$$\hat{H}_r = \hbar\omega_r(a^\dagger a + 1/2), \qquad (1.12)$$

where

$$\phi_r = 2\sqrt{\frac{E_{C,r}}{\hbar\omega_r}}(a^\dagger + a), \qquad (1.13)$$

$$p_r = \frac{i\hbar}{4}\sqrt{\frac{\hbar\omega_r}{E_{C,r}}}\left(a^\dagger - a\right). \qquad (1.14)$$

Coupling term. The coupling term can also be written using the operators $a, a^\dagger$:

$$\hat{H}_c = -i\frac{E_{C,c}}{2}\sqrt{\frac{\hbar\omega_r}{E_{C,r}}}(2n - N_g)\left(a^\dagger - a\right). \qquad (1.15)$$

Note that the characteristic coupling energy is $E_c = \sqrt{\hbar\omega_r/E_{C,r}}E_{C,c}/2$.

A general analysis of the Hamiltonian $\hat{H}$ is beyond the scope of the present paper and will be presented elsewhere [14]. In the next section we will discuss an explicit matrix form of $\hat{H}$, which can be obtained under certain simplifying conditions which are nevertheless experimentally relevant.

4.   EIGENSTATES AND ENTANGLEMENT

Throughout this section we will work in the zero-temperature limit. We are interested in the situation $N_g \simeq 1$, such that we have to consider the charge qubit states $|-\rangle$ and $|+\rangle$ only. Furthermore, as far as the resonator is concerned, we will consider $\hbar\omega_r = E_j$ and work only with the ground state $|0\rangle$ and the first excited state $|1\rangle$. In the limit of weak coupling, $E_c \ll \hbar\omega_r$, the coupled system can be characterized by the four basis states $|-,0\rangle, |-,1\rangle, |+,0\rangle, |+,1\rangle$. The Hamiltonian matrix for this low-energy subspace reads

$$
\hat{H} = \begin{pmatrix}
E_0 & iE_\beta & 0 & -2i\alpha\beta E_c \\
-iE_\beta & E_1 & 2i\alpha\beta E_c & 0 \\
0 & -2i\alpha\beta E_c & E_2 & iE_\alpha \\
2i\alpha\beta E_c & 0 & -iE_\alpha & E_3
\end{pmatrix}, \tag{1.16}
$$

which is a hermitian matrix describing the two-level system coupled to the lowest states of the resonator. Here, $E_0 = E_- + \hbar\omega_r/2$, $E_1 = E_- + 3\hbar\omega_r/2$, $E_2 = E_+ + \hbar\omega_r/2$, $E_3 = E_+ + 3\hbar\omega_r/2$, $E_\alpha = E_c(2\alpha^2 - N_g)$, and $E_\beta = E_c(2\beta^2 - N_g)$.

Suppose that the system has been prepared in the state $|\psi(t=0)\rangle = |+, 0\rangle$ at time $t = 0$. This situation can be achieved by a suitable manipulation of the gate voltage V_g at times prior to $t = 0$ [6]. At times $t > 0$ we keep V_g fixed such that $N_g = 1$. The time evolution of $|\psi(t)\rangle$ describing the system at $t > 0$ is governed by the Hamiltonian (1.16). Thus we have $\alpha^2 = \beta^2 = 1/2$, and hence $E_\alpha = E_\beta = 0$. Moreover, as $\hbar\omega_r = E_j$, we have $E_1 = E_2 = E_{C,j} + E_j \equiv \bar{E}$: without coupling, the state $|+, 0\rangle$ would be degenerate with the state $|-, 1\rangle$. Thus the Hamiltonian takes the simple form

$$
\hat{H} = \begin{pmatrix}
E_0 & 0 & 0 & -iE_c \\
0 & \bar{E} & iE_c & 0 \\
0 & -iE_c & \bar{E} & 0 \\
iE_c & 0 & 0 & E_3
\end{pmatrix}. \tag{1.17}
$$

Note in particular that the state $|+, 0\rangle$ couples to the state $|-, 1\rangle$; as a result, the degeneracy between them is lifted and the states become entangled. The precise form of the entanglement is governed by the central 2×2 block of the matrix (1.17). The eigenstates of this block are

$$
\begin{aligned}
|\chi_1\rangle &= [|-, 1\rangle + i|+, 0\rangle]/\sqrt{2}, \\
|\chi_2\rangle &= [|-, 1\rangle - i|+, 0\rangle]/\sqrt{2},
\end{aligned}
$$

corresponding to the eigen energies $\bar{E} - E_c$ and $\bar{E} + E_c$, respectively. These two excited eigenstates thus correspond to a maximum entanglement of charge qubit and resonator states, induced by the capacitive coupling between them.

The time evolution of $|\psi(t)\rangle$ is given by

$$
|\psi(t)\rangle = \frac{1}{\sqrt{2}i} \left[e^{-i(\bar{E}-E_c)t/\hbar}|\chi_1\rangle - e^{-i(\bar{E}+E_c)t/\hbar}|\chi_2\rangle \right]. \tag{1.18}
$$

We see that the state $|\psi(t)\rangle$ oscillates coherently between $|\chi_1\rangle$ and $|\chi_2\rangle$, *i.e.* between $|-, 1\rangle$ and $|+, 0\rangle$. In fact, these so-called quantum Rabi oscillations can be interpreted as the spontaneous emission and re-absorption of one excitation quantum by the resonator. An interesting quantity is the probability $P_1(t)$ to find the harmonic oscillator in the state $|1\rangle$ after a certain time t. This probability shows Rabi oscillations as a function of t with frequency $2E_c/\hbar$,

$$P_1(t) = |\langle 1, -|\psi(t)\rangle|^2 = \frac{1}{2}[1 - \cos(2E_c t/\hbar)]. \qquad (1.19)$$

Since these Rabi oscillations are characteristic for the entanglement realized in the system, their measurement would provide direct evidence of the presence of the entangled states $|\chi_1\rangle$ and $|\chi_2\rangle$. We will discuss the feasibility of such a measurement in the next Section.

5.   DISCUSSION

For the numerical estimates presented below we will consider typical parameters related to an aluminium superconducting circuit [4, 6]. The Josephson charge qubit has the following characteristics: $E_j = 26.1\mu eV$, $E_{C,j} = 70\mu eV$, $\Delta = 240\mu eV$. As for the resonator, we take $L_r = 90pH$ and $E_{C,r} = 12neV$, as a result $\hbar\omega_r = 26.1\mu eV$. Finally, the coupling capacitance is chosen to be of the same order of C_j, $C_c = 0.5fF$, yielding $E_c = 256neV$. Note that the coupling energy is indeed much smaller than the Josephson energy, which in turn is equal to the excitation energy of the resonator.

Using the above paramaters, we have plotted $P_1(t)$, Eq. (1.19), as a function of time in Fig. 2. We clearly see the Rabi oscillations with perdiodicity $T_{Rabi} = 8$ ns.

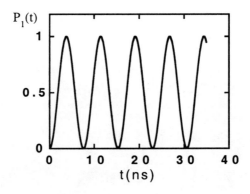

Figure 2 Probability $P_1(t)$ to find the system, prepared in the state $|+, 0\rangle$ at $t = 0$, in the state $|-, 1\rangle$ after a time t.

In order to be able to observe these oscillations, we need to satisfy various conditions. First of all, in order to avoid the presence of quasi-particles, the temperature must be much lower than the gap Δ. Secondly, it is necessary to have a decoherence time which is longer than T_{Rabi}. In our system, the decoherence time will be the shorter of the lifetime τ_r of the excited state of the resonator and the decoherence time τ_{qubit} of the charge qubit. A Q-factor of about 500 is quite realistic for a superconducting LC resonator. This yields a lifetime $\tau_r >$ 10 ns. As for the qubit, the experiment by Nakamura [6] has indicated that $\tau_{\text{qubit}} > 2\text{ns}$. Finally, a measurement of the number of excitations should be performed on the resonator. This can be done, *e.g.*, along the lines of Ref. [15], where the discrete, oscillator-like energy levels of an underdamped Josephson junction were measured.

Acknowledgments.   We thank F. Balestro, M. Doria, G. Ithier, and J.E. Mooij for invaluable discussions.

References

[1] A. Steane, Rep. Prog. Phys. **61**, 117 (1998).

[2] C.A. Sackett *et al.*, Nature **404**, 256 (2000).

[3] S. Haroche, Physics Today **36**, 51 (1998).

[4] V. Bouchiat *et al.*, Physica Scripta T**76**, 165 (1998).

[5] Y. Nakamura, C. D. Chen, and J. S. Tsai, Phys. Rev. Lett. **79**, 2328 (1997).

[6] Y. Nakamura, Yu. A. Pashkin, and J. S. Tsai, Nature **398**, 786 (1999).

[7] A. Shnirman, G. Schön, and Z. Hermon, Phys. Rev. Lett. **79**, 2371 (1997).

[8] Y. Makhlin, G. Schön, and A. Shnirman, Nature **398**, 305 (1999).

[9] J. E. Mooij *et al.*, Science 285,1036 (1999).

[10] M.V. Feigel'man *et al.* Journal of Low Temperature Physics **118**, 805 (2000).

[11] C.H. van der Wal and J. E. Mooij (unpublished).

[12] J.E. Lukens and J.R. Friedman (unpublished).

[13] M. Brune et al., Phys. Rev. Lett. **76**, 1800 (1996).

[14] G. Ithier (unpublished).

[15] J.M. Martinis, M.H. Devoret and J. Clarke, Pohys. Rev. Lett. **55**, 1543 (1985).

COOPER PAIR TUNNELING IN CIRCUITS WITH SUBSTANTIAL DISSIPATION: THE THREE-JUNCTION R-PUMP FOR SINGLE COOPER PAIRS

A. B. Zorin
S. A. Bogoslovsky*
S. V. Lotkhov
J. Niemeyer

Physikalisch-Technische Bundesanstalt

38116 Braunschweig

Germany

Abstract   We propose a circuit (we call it R-pump) comprising a linear array of three small-capacitance superconducting tunnel junctions with miniature resistors ($R > R_Q \equiv h/4e^2 \cong 6.5$ kΩ) attached to the ends of this array. Owing to the Coulomb blockade effect and the effect of dissipative environment on the supercurrent, this circuit enables the gate-controlled transfer of individual Cooper pairs. The first experiment on operating the R-pump is described.

Keywords: Josephson tunneling, Coulomb blockade, effect of electromagnetic environment

INTRODUCTION

The Coulomb blockade effect in circuits with small-capacitance tunnel junctions provides the means of manipulating single charge quanta (see, for example, the review paper by Averin and Likharev [1]). If periodic signals of frequency f are applied to the gate electrodes of the circuit, a train of single charges q can move across an array of junctions so that charge pumping, giving rise to the current $I = qf$, is achieved. It was experimentally proven that the normal-state

---

*Permanent address: Laboratory of Cryoelectronics, Moscow State University, 117899 Moscow, Russia

metallic circuits enable single electrons ($q = e$) to be effectively pumped at frequencies f of about several MHz [2]. Moreover, the accuracy of single-electron pumping can nowadays meet the requirements of fundamental metrology, *viz.* $\delta I/I \simeq 10^{-8}$ [3].

Unlike pumping of electrons, the pumping of Cooper pairs ($q = 2e$) in superconducting circuits has not been that successful so far. The only experiment had been carried out in 1991 by Geerligs *et al.* [4] with a three-junction Al sample. Although this experiment did evidence a pumping of the pairs, the pumping was strongly disturbed by several factors: the Landau-Zener transitions, Cooper pair co-tunneling, quasiparticle tunneling, etc. As a result, the shape of the current plateau at $I = 2ef$ in the I-V curve was far from being perfect.

Recently, in their theoretical paper Pekola *et al.* [5] concluded that pumping of Cooper pairs with reasonable accuracy in a three-junction array was impossible. For practical values of parameters (the ratio of the Josephson coupling energy to the charging energy $\lambda \equiv E_J/E_c = 0.01 - 0.1$) they evaluated the inaccuracy of pumping $\delta I/I$ to be as much as 9%-63%. This is because of intensive co-tunneling of pairs in the short arrays, i.e. the process of tunneling simultaneously across several (≥ 2) junctions. To suppress the co-tunneling and, in doing so, to improve the characteristics of the pump they proposed to considerably increase the number of junctions $N(\gg 3)$ and, hence, of the gates ($N-1$). However, such modification would make operation of the circuit more complex.

In this paper we propose an alternative way to improve Cooper pair pumping in a three-junction array. We modify the bare circuit by attaching to the ends of the array the miniature resistors (see the electric diagram of this device, which we call R-pump, in Fig. 1.1). Their total resistance R exceeds the resistance quantum $R_Q \equiv h/4e^2 \cong 6.5$ kΩ so that the dimensionless parameter

$$z \equiv \frac{R}{R_Q} \gg 1. \tag{1.1}$$

The self-capacitance of these resistors should not be much larger than the junction capacitance C. In this paper we will analyze how such resistors affect tunneling and co-tunneling of Cooper pairs in the three-junction array and present preliminary experimental data.

PECULIARITIES IN OPERATING THE R-PUMP

In general terms, the principle of operation of the R-pump remains the same as that of the pump without resistors (for details of the Cooper pair pump operation see Ref. [4]). Two periodic signals $V_1(t)$ and $V_2(t)$ are applied to the gates to form an elliptic trajectory in the parameter plane $V_1 - V_2$. At zero voltage V across the pump, this trajectory encircles in the clockwise (counter-

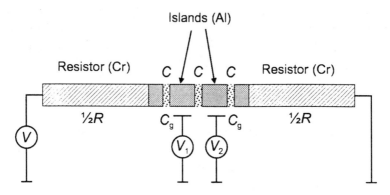

Figure 1.1 Schematic of the R-pump for single Cooper pairs. Three superconducting junctions of type Al/AlO$_x$/Al form two small islands. Miniature normal metal (Cr) resistors accomplish the structure. The device is driven by two harmonic voltages $V_{1,2} = V_{01,02} + A \cos(2\pi ft \pm \frac{1}{2}\theta)$, where the dc offset $V_{01,02}$ determines the centre of the cycling trajectory in the $V_1 - V_2$ plane; the phase shift $\theta \geq \frac{\pi}{2}$.

clockwise) direction the triple point of the boundaries between the stability domains for adjacent charge configurations with an excess Cooper pair. A complete cycle results in the charge $q = 2e$ ($q = -2e$) being sequentially transferred across all three junctions of the array.

The Cooper pair transitions in a circuit without dissipation (quasiparticle tunneling is neglected) occur due to Josephson tunneling (supercurrent). The strength of the Josephson coupling $E_J \equiv \frac{\hbar}{2e}I_c$ (here I_c is the critical current) is assumed to be $\ll E_c \equiv e^2/2C$, the charging energy, i.e. parameter $\lambda \ll 1$. When the domain boundary is crossed, the resonance condition is obeyed and Josephson coupling mixes up the two states, each corresponding to the position of the pair on either side of the junction. (Note that, at the same point, the resonance condition for co-tunneling in the opposite direction across two other junctions is met.) Slow passage across the boundary results in transition of the pair in the desired direction. This process is mapped on the adiabatic level-crossing dynamics (see, for example, Ref. [6]), and the probability of missing the transition is given by the Landau-Zener expression:

$$p_{LZ} = \exp\left(-\frac{\pi E_J^2}{2\hbar \dot{E}}\right), \tag{1.2}$$

where $\dot{E} \propto f$ is the velocity of variation of the energy difference E between the two states.

In contrast to the elastic (and, hence, reversible) tunneling of pairs in the pump without resistors, tunneling in the R-pump is accompanied by dissipation of energy. The strength of the dissipation effect depends on the active part

$\mathrm{Re}Z(\omega)$ of an electromagnetic impedance seen by the tunneling pair, namely by the dimensionless parameter $z' = \mathrm{Re}Z(0)/R_Q$. If the condition of weak Josephson coupling $E_J \ll E_c/\sqrt{z'}$ is satisfied, i.e. $\lambda\sqrt{z'} \ll 1$, the rate of Cooper pair tunneling is expressed as [7]

$$\Gamma(E) = \frac{\pi}{2\hbar}E_J^2 P(E), \tag{1.3}$$

where E is the energy gain associated with this transition and $P(E)$ is the function describing the property of the environment to absorb the energy released in a tunneling event. (The discussion of properties as well as numerous examples of $P(E)$ for different types of electromagnetic environment are given in the review paper by Ingold and Nazarov [8].) The probability of a missed transition is then given by the expression:

$$p_{\mathrm{m}} = \exp\left\{-\int_{-\infty}^{\infty}\Gamma[E(t)]dt\right\} = \exp\left(-\frac{\pi E_J^2}{2\hbar\dot{E}}\right) = p_{\mathrm{LZ}}. \tag{1.4}$$

Here we used the normalization condition $\int_{-\infty}^{\infty}P(E)dE = 1$, and the assumption of a value $\dot{E}$ constant in time was exploited. Thus, in the case of weak coupling, dissipation does not affect the total probability of tunneling $p = 1 - p_{\mathrm{m}}$. This conclusion is in accordance with the general result obtained by Ao and Rammer [6] for quantum dynamics of a two-level system in the presence of substantial dissipation.

Using the expansion of P in the region of small $E \ll E_c$ [7], [8] one arrives at the expression

$$\Gamma(E) \propto \lambda^2 E^{2z'-1}. \tag{1.5}$$

For $z' < \frac{1}{2}$ function $\Gamma(E)$ is peaked at $E = 0$, while in the case $(z' > \frac{1}{2})$ the rate is zero at $E = 0$ and the maximum is reached at finite energy $\tilde{E} \leq 4E_c$, or, in other terms, at finite positive voltage V across the junction. This property of $\Gamma(E)$ in circuits with substantial dissipation plays the crucial role in the operation of the three-junction R-pump.

For one junction in the three-junction array with gate capacitances $C_g \ll C$ the effective impedance $\mathrm{Re}Z(0) \equiv R' = \frac{1}{9}R$, i.e. the parameter $z' = \frac{1}{9}z$. The factor $\frac{1}{9}$, which considerably attenuates the effect of resistors, stems from the square of the ratio of the total capacitance of the series array $(= \frac{1}{3}C)$ to the junction capacitance C (see a similar network analysis in Ref. [8]).

Accordingly, the equivalent damping for two junctions of the three-junction array is larger and determined by $R' = \frac{4}{9}R$ and, hence, $z' = \frac{4}{9}z$. Therefore, the rate of Cooper pair transition simultaneously across two junctions, i.e. the rate of the co-tunneling, at small E is

$$\Gamma_{\mathrm{cot}}(E) \propto \lambda^4 E^{\frac{8}{9}z-3}. \tag{1.6}$$

This expression has been obtained by considering the environmental effect in a similar fashion as was done by Odintsov *et al.* [9] for the co-tunneling of normal electrons in a single-electron transistor. Accordingly, co-tunneling across all three junctions decays even stronger ($\propto \lambda^6 E^{2z-5}$), because the whole array experiences the full resistance R.

The comparison of Eq. (6) with Eq. (5) taken for the value $z' = \frac{1}{9}z$ shows that the rate of co-tunneling is drastically depressed if z is sufficiently large. For example, for $z = 9$ (or $R \approx 58.5$ kΩ) the rate of Cooper pair co-tunneling across two junctions $\Gamma_{\text{cot}} \propto E^5$ (i.e. it is similarly small as single-electron co-tunneling across a three-junction normal array [7]). On the other hand, the direct tunneling rate $\Gamma \propto E$, is similar to that of ordinary electron tunneling. The effective junction resistance in the latter case is $R_{\text{eff}} \approx 0.03\lambda^{-2}R_Q$. For the practical value $\lambda \sim 0.03$ this formula yields $R_{\text{eff}} \sim 200$ kΩ, i.e. the typical value of a normal single electron junction. Since the junction resistance determines the characteristic time constant RC, it can be expected that the frequency characteristics of the Cooper pair R-pump are at least not worse than those of the normal electron counterpart without resistors.

EXPERIMENT

The Al/AlO$_x$/Al tunnel junctions and Cr resistors were fabricated *in situ* by the three-angle shadow evaporation technique through a trilayer mask patterned by e-beam lithography and reactive-ion etching. The tunnel junction parameters of the best suited available sample were found to be: capacitance $C \approx 250$ aF and normal-state resistance $R_j \approx 160$ kΩ. The former value yields $E_c \approx 320$ μeV. On the assumption that the Ambegaokar-Baratoff relation between the critical current I_c and the junction resistance is valid, the aforementioned value of R_j yields the value $E_J = \Delta_{\text{Al}}R_Q/2R_j \approx 4.1$ μeV, where $\Delta_{\text{Al}} \approx 200$ μeV is the superconducting energy gap of aluminum at low temperature. These parameters of the sample give the ratio of characteristic energies $\lambda \approx 0.013$.

The thin-film Cr resistors (thickness $d \approx 7$ nm) had lateral dimensions of about 80 nm by 10 μm. For the purpose of an independent characterization of the resistors and junctions, we attached four resistors in twos to either end of the junction array, resulting in the equivalent resistance $\frac{1}{2}R$ on each side as depicted in Fig. 1.1. (For the measurement of the I-V curves we used, however, the two-wire configuration.) Either of four resistors had $R \approx 60$ kΩ and reasonably low self-capacitance per unit length, ≈ 60 aF/μm. The resistors exhibited negligibly small (< 2 %) non-linearity of the I-V curve in the region of small voltages at the millikelvin temperature [10].

The samples were measured in the dilution refrigerator at the temperature T of about 10 mK, i.e. well below the critical temperature of aluminum, $T_c \approx 1.15$ K. The bias and gate lines were equipped with pieces of the thermocoax

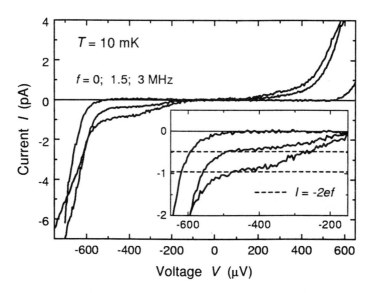

Figure 1.2 I-V characteristics of the R-pump without and with ac drive of two different frequencies. The inset shows the blow-up of the steps whose positions were expected to be strictly at $I = -2ef$ (shown by dashed lines), where the sign "-" is due to the phase relation chosen between two gate signals.

cable which was thermally anchored at the sample holder plate and served as a filter for frequencies above 1 GHz. The typical I-V curves with and without ac drive are presented in Fig. 1.2. The curves exhibit considerably smeared steps whose position ($I \approx -2ef$) clearly shows the linear current-versus-frequency dependence (although the quality of the steps is noticeably degraded with increasing frequency) [11]. Another feature of the steps is their position on the voltage axes: they appear at finite voltage V applied in the current direction, while at $V = 0$ neither co-tunneling nor pumping effect were observed. We attribute this effect to the damping effect of the resistors, which was too strong in this sample. Thus, the behavior of the Cooper pair R-pump differs significantly from that of the Cooper pair pump without resistors [4].

DISCUSSION

Although our first experiment evidenced the desirable effect of Cooper pair pumping, a further improvement is inevitably needed. First, the sample parameters should be optimized: Josephson coupling should be increased (possibly up to $\lambda = 0.1$) while the resistance value R should be somewhat reduced. These parameter values lead to an increase of the tunneling rate and, as a consequence, to Cooper pair pumping at zero voltage bias V. Secondly, the quasiparticle tunneling, resulting in sporadic translations of the cycle trajectory and, hence, to the pumping errors, should be reliably depressed: As the measurements showed, all our Al-Cr samples (as well as those with $E_c < \Delta_{Al}$) so far suffered from quasiparticle tunneling [12] and did not show the so-called parity effect [13], [14]. It is particularly remarkable that our Cr resistors do not prevent an uncontrolled poisoning of Josephson tunneling by non-equilibrium quasiparticles arriving at the islands from the external circuit. Such "buffer" effect of the normal electrodes was proposed and demonstrated in Al-Cu devices by Joyez *et al.* [15]. That is why the problem of the poisoning quasiparticle tunneling requires a more radical solution. Probably manufacture of small-capacitance niobium junctions ($\Delta_{Nb} \approx 1.4$ meV $\gg E_c$) could improve the situation by making the parity effect in these structures stable.

In conclusion, we proposed a simple superconductor-normal metal circuit enabling, in principle, the efficient pumping of Cooper pairs. Due to the energy dissipation in the resistors, the effect of Cooper pair co-tunneling is heavily damped. A possibility of pair pumping was demonstrated and improvements of the experiment are proposed.

Acknowledgments

This work is supported in part by the EU Project COUNT.

References

[1] D. V. Averin and K. K. Likharev, Single-electronics: correlated transfer of single electrons and Cooper pairs in small tunnel junctions. In B. L. Altshuler, P. A. Lee and R. A. Webb editors, *Mesoscopic Phenomena in Solids*, pages 173-271, Amsterdam, North-Holland: Elsevier, 1991.

[2] H. Pothier, P. Lafarge, P. F. Orfila, C. Urbina, D. Esteve and M. H. Devoret. *Physica B*, 169:573-574, 1991; *Europhys. Lett.*, 17:249-254, 1992.

[3] M. W. Keller, J. M. Martinis, N. M. Zimmerman and A. H. Steinbach. *Appl. Phys. Lett.*, 69:1804-1806, 1996.

[4] L. J. Geerligs, S. M. Verbrugh, P. Hadley, J. E. Mooij, H. Pothier, P. Lafarge, C. Urbina, D. Esteve and M. N. Devoret. Single Cooper pair pump. *Z. Phys. B: Condens. Matter*, 85:349-355, 1991.

[5] J. P. Pekola, J. J. Toppari, M. Aunola, M. T. Savolainen and D. V. Averin. Adiabatic transport of Cooper pairs in arrays of Josephson junctions. *Phys. Rev. B*, 60:R9931-R9934, 1999.

[6] P. Ao and J. Rammer. Quantum dynamics of a two-state system in a dissipative envinronment. *Phys. Rev. B*, 43:5397-5418, 1991.

[7] D. V. Averin and A. A. Odintsov. Macroscopic quantum tunneling of the electric charge in small tunnel junctions. *Phys. Lett. A*, 140:251-257, 1989.

[8] G. L. Ingold and Yu. V. Nazarov. Charge tunneling rates in ultrasmall junctions. In H. Grabert and M. H. Devoret, editors, *Single Charge Tunneling*, pages 21-107, New York, Plenum, 1992.

[9] A. A. Odintsov, V. Bubanja and G. Schön. Influence of electromagnetic fluctuations on electron co-tunneling. *Phys. Rev. B*, 46:6875-6881, 1992.

[10] A. B. Zorin, S. V. Lotkhov, H. Zangerle and J. Niemeyer. Coulomb blockade and co-tunneling in single electron circuits with on-chip resistors: towards the implementation of the R-pump. *J. Appl. Phys.*, 88:No.3, 2000.

[11] In the normal state maintained by application of perpendicular magnetic field $B = 1$ T, this sample exhibited excellent steps at $I = ef$ due to the single electron pumping, see S. V. Lotkhov, S. A. Bogoslovsky, A. B. Zorin and J. Niemeyer. The single electron R-pump: first experiment. *Preprint, available at* http://arXiv.org/abs/cond-mat/0001206.

[12] This was found from the current-versus-gate voltage dependencies whose behavior reflected the charging of the islands with the charge quantum $q = e$.

[13] D. V. Averin and Yu. V. Nazarov. Single-electron charging of a superconducting island. *Phys. Rev. Lett.*, 69:1993-1996, 1992.

[14] M. T. Tuominen, J. M. Hergenrother, T. S. Tighe and M. Tinkham. Experimental evidence for parity-based 2e periodicity in a superconducting single-electron transistor. *Phys. Rev. Lett.*, 69:1997-2000, 1992; A. Amar, D. Song, C. J. Lobb and F. C. Wellstood. 2e and e periodic pair currents in superconducting Coulomb-blockade electrometers. *Phys. Rev. Lett.*, 72:3234-3237, 1994.

[15] P. Joyez, P. Lafarge, A. Filipe, D. Esteve and M. H. Devoret. *Phys. Rev. Lett.*, 72:2458-2461, 1994.

NONLOCALITY IN SUPERCONDUCTING MICROSTRUCTURES

K. Yu. Arutyunov and J. P. Pekola

University of Jyväskylä, Department of Physics, P.B. 35, 40351 Jyväskylä, FINLAND

A. B. Pavolotski and D. A. Presnov

Moscow State University, Department of Physics, 119899, Moscow, RUSSIA

Keywords: superconductivity, microstructures, nonlocality

Abstract   We discuss experimental evidence of nonlocality in electron transport of small structures. It is shown that for superconductors reasonable agreement with experiment can be achieved by assuming exponential decay of the nonlocal interaction $\propto \exp(-L/\xi)$, where L is the distance between the interacting points and ξ is the correlation length. ξ is associated with the Ginzburg - Landau coherence length ξ_{GL}.

Nonlocality effects in electron transport appear when characteristic scale which determines transport properties becomes smaller than the corresponding scale of boundary conditions. A classical example is the skin effect. At high frequencies the electromagnetic field penetration depth δ_{skin} can be smaller than the mean free path ℓ. To treat correctly the effective force acting on a charge $eE(x)$ between successive collisions one should perform an integration over the skin layer δ_{skin}, whereby the effect is essentially nonlocal.

In typical thin film metallic microstructures the elastic mean free path ℓ is about ~ 20 nm even at the lowest temperatures and is much smaller than the sample dimension(s). From this 'classical' point of view one should not expect any nonlocality. However, in normal metals quantum coherence effects dominate transport properties at low temperatures ($T \leq 1$ K). In this limit the relevant physical scale is set not by the elastic mean free path ℓ, but by the phase breaking length ℓ_φ. The phase breaking length is material and temperature dependent and can reach few μm at sub-K temperatures, driving a sufficiently small system inno the mesoscopic limit: $\ell \ll L \leq \ell_\varphi$. Theoretically the subject

of nonlocality applied to mesoscopic structures has been addressed by
Buttiker [1] already in the middle 80ies. nonlocality of electron transport in normal metal nanostructures has been observed qualitatively in
a nice experiment [2]: attachment of a loop to a small bar apart from
classical current path led to the appearance of h/e Aharonov - Bohm
oscillations in the spectrum of magnetoconductance fluctuations of this
bar. It has been clearly shown that the relevant physical scale responsible for nonlocality of electron transport in mesoscopic normal metal
structures is the phase breaking length ℓ_φ [3].

An interesting issue of nonlocality arises in normal metal systems
proximity coupled to a superconductor. As superconducting correlations
are induced in the normal metal at the interface region within the normal
metal coherence length ξ_N , one might expect that this new scale governs
the nonlocal properties. From the other point of view, these structures
are formally normal metals and, hence, the 'competing' scale is the phase
breaking length ℓ_φ. For typical nanostructures $\xi_N \sim 80$ nm at 1 K, while
at this temperature limit ℓ_φ can be an order of magnitude longer. It has
been shown that independent of proximity induced superconductivity it
is still the phase breaking length ℓ_φ which scales the nonlocal interaction
in these systems [4].

In this Paper we concentrate on the nonlocal properties of superconductors. For superconductors the subject of nonlocality has been
addressed in early works of Pippard [5] proposing a nonlocal generalization of the London equation relating the supercurrent $j_S(\mathbf{r})$ at a
point $\mathbf{r}$ and the vector potential at some arbitrary point $\mathbf{r}'$. The kernel function of the corresponding integral equation is proportional to
$\propto \exp(-\mid \mathbf{r} - \mathbf{r}' \mid /\xi_0^{Pip})$ and the integration is performed over all values $\mathbf{r}'$. ξ_0^{Pip} is the temperature independent Pippard coherence length
$\xi_0^{Pip} = \hbar v_F / k T_c$, where v_F is the Fermi velocity and T_c is the critical temperature. For electrodynamics of type-I superconductors this
approach gives good agreement both with experiment and with microscopic theory.

Our interest is different from the nonlocal relation between supercurrent and the vector potential. If the superconducting order parameter
$\Psi = \mid \Psi \mid e^{i\varphi}$ is altered at a position (both its magnitude and phase),
then the corresponding response can be observed at a remote point.
This effect is essentially a nonlocal phenomenon. To avoid confusion,
it should be clarified that the label 'macroscopically quantum coherent'
is applicable to a superconductor with respect to the phase of the order parameter. In equilibrium the phase φ of the order parameter is
macroscopically locked. From this point of view the scale of nonlocality is infinite. However, it is not the case with the modulus $\mid \Psi \mid$. We

wish to understand which physical scale governs the healing of the $\mid \Psi \mid$ disturbance.

A convenient experimental method to vary the order parameter is to attach to the sample a small superconducting loop and apply a perpendicular magnetic field **B**. The magnetic field causes periodic modulation of the free energy in the loop and, hence, a variation of the critical temperature $T_c(B)$ [6]. Period of the oscillations is equal to Φ/ϕ_0 , where Φ is the magnetic flux through the loop and the magnetic flux quantum is $\phi_0 = h/2e = 2.07 \cdot 10^{-15}$ Wb. Preferably, the diameter of the loop should be neither too large to enable sufficient magnetic field resolution, nor too small to neglect monotonous suppression of superconductivity by magnetic field. For loops with area of few μm^2 made of conventional low temperature superconductors (Al, Sn, In) the latter criterion holds only for the very first periods of oscillations.

As the system under consideration is coherent, one might expect the periodic modulation of the free energy in the loop to cause a corresponding response in a remote point away from the loop. How far from the loop this effect will be observed depends on the spatial properties of the nonlocal interaction. Hereafter we assume the following functional dependence:

$$F(\mathbf{r}) = F(\mathbf{r}') \exp(-\mid \mathbf{r} - \mathbf{r}' \mid /\xi), \qquad (1.1)$$

where $F(\mathbf{r})$ stands for the amplitude of the oscillation at a point $\mathbf{r}$, and ξ is the relevant scale governing the decay of nonlocal interaction. Below we consider separately the cases when the oscillating parameter F is magnetoresistance, critical current or critical temperature.

The most straightforward way to study quantum oscillations in superconductors is to fix the temperature slightly below the critical point $T_c^0 \equiv T_c(B = 0)$ and to observe the dependence of the sample resistance on magnetic field, $R(B)$. This is the original way the Little - Parks effect has been observed [7]. Due to inevitable inhomogeneity of the structure, the monotonous destruction of superconductivity will take place in a finite range of magnetic fields with superimposed Little - Parks oscillations [8]. Theoretical generalization for the nonlocal configuration (loop with an attached bar) in a paraconductance regime has been proposed [9]. The experiment [10] did show the existence of the nonlocal interaction. It has been concluded [10] that the length scale for nonlocality is set by the fluctuation correlation length being associated with the Ginzburg - Landau coherence length $\xi = \xi_{GL}(T)$. Our own experiments in slightly different configuration (Fig. 1) allowed us to obtain similar conclusions.

For the temperature dependence of the Ginzburg - Landau coherence length one can use the familiar equation $\xi_{GL}(T) = \xi_{GL}(0)[1 -$

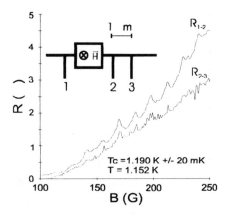

Figure 1 Magnetoresistance of the loop and the bar at a fixed temperature. Inset shows a schematic of the aluminum structure studied.

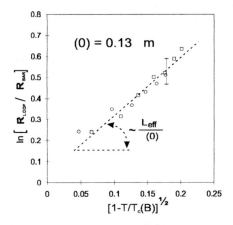

Figure 2 Analysis of the magnetoresistance oscillations at different temperatures for the configuration presented in the inset of Fig. 1. δR_{LOOP} and δR_{BAR}) are the magnitudes of the resistance oscillations in the loop and in the bar, respectively, with the monotonous component $R(B)$ being subtracted. Dashed line is a guide for eyes.

$T/T_c(B)]^{-1/2}$, where the critical temperature depends on magnetic field $T_c(B)$. Utilizing the postulated exponential decay of the nonlocal interaction, it is easy to show that the product $\ln(\delta R_{LOOP}/\delta R_{BAR})/[1 - T/T_c(B)]^{1/2}$ should be proportional to $L_{eff}/\xi_{GL}(0)$. Here L_{eff} is the effective distance between the loop and the bar, and δR stands for the magnitude of the magnetoresistance oscillations with the monotonous component being subtracted. Figure 2 illustrates reasonable agreement between experiment and the stated functional dependencies, equalizing L_{eff} with the distance between the loop and the middle point of the bar. One may conclude that the Ginzburg - Landau coherence length is a suitable candidate for the nonlocal correlation length ξ.

However, it should be noted that there are at least two drawbacks of such an analysis of the nonlocal interaction. The first one comes from the nature of the 'broad' resistive state of a superconductor enabling the observation of the non local Little - Parks effect. Theory [9] can treat finite resistance of a superconductor in a paraconductance limit $R(T) \approx R_N$ (at the top of the transition), while we have used for the analysis the whole transition region (Figs. 1 and 2). Second problem originates from the assumption that the critical temperature is a function of magnetic field only and not of the coordinate x. Unfortunately, this is not the case in experiment.

Due to spatial decay of the nonlocal interaction, the oscillating component of the critical temperature is essentially coordinate-dependent:

$$\delta T_c(B, x) = \delta T_c(B, 0) \exp(-x \mid /\xi), \tag{1.2}$$

where $\delta T_c(B, x) = T_c(B, x) - T_c^{env}(B, x)$, and $T_c^{env}(B, x)$ is the monotonous component of the critical temperature suppressed by magnetic field, and $x = L_{eff}$ is the distance between the loop and the point of observation (middle of the bar). To calculate the magnetic field dependence of $T_c^{env}(B, x)$ we consider our nanostructures as thin films in a parallel magnetic field:

$$T_c^{env}(B, x) \equiv T_c^{env}(B, 0) = T_c^{\parallel}(B) = T_c(0) \left[1 - \frac{\pi^2}{3} \left(\frac{d\xi_{GL} B}{\phi_0} \right)^2 \right], \tag{1.3}$$

where d is the film thickness. In the local limit $x = 0$:

$$T_c(B, 0) = T_c^{env}(B, 0) \left[1 - \left(\frac{\xi_{GL}(0)}{w} \right)^2 \left[n - \frac{w^2 B}{\phi_0} \right]^2 \right], \tag{1.4}$$

where w is the side of the square 'loop', and the quantum number n is selected to obtain the highest value of critical temperature for a given magnetic field [6]. Note that the magnitude of the critical temperature oscillations in the loop $max[\delta T_c(B, 0)]$ has a constant value independent of magnetic field. Corrections due to finite thickness of the loop 'walls' (line width) [8] were neglected giving marginal improvement to the fitting procedure. Recovering the postulated exponential decay (1.2) one can write:

$$\delta T_c(B,x) = \delta T_c(B,0)\exp\left[-\frac{x}{\xi(T^*,B,x)}\right]. \tag{1.5}$$

Following [9] let us associate the nonlocal correlation length ξ with the Ginzburg - Landau coherence length ξ_{GL}:

$$\xi(T^*,B,x) = \xi_{GL}(0)\left[1 - \frac{T^*}{T_c(B,x)}\right]^{-1/2}. \tag{1.6}$$

Substitution of (1.4) and (1.6) into (1.5) gives a transcendental equation for the critical temperature at a point x away from the loop. The only fitting parameter in calculations of the correlation length (1.6) is the effective temperature T^*. Experiments [11] and our own data clearly indicate that contrary to the local case (loop), the nonlocal oscillations die out with magnetic field: $max[\delta T_c(B,x > 0)] \neq const$. It confirms the intuitive expectations that the correlation length ξ has a finite value and decreases with magnetic field. At the same moment, the Ginzburg - Landau coherence length, with which the nonlocal correlation length is associated, diverges at the critical point. Being unable to solve completely the above contradiction, we can only expect that the best fit with experiment can be achieved with T^*close to, but not equal to the critical temperature $T_c(B)$. Experimental data and results of calculations are presented in Fig. 3.

Each experimental point in Fig. 3a has been obtained by recording simultaneously the resistance of the loop and that of the bar (Fig. 3a, inset) as a function of temperature T at a fixed perpendicular magnetic field B. Critical temperature is associated with the point $R(T_c, B) = R_N/2$, where R_N is the normal state resistance. It is clearly seen (Fig. 3b) that by solving equation (5) one can obtain a nice agreement with experimental $\delta T_c(B,x)$ values. The 'price' for the accurate fitting is the non-constant values of $T^*(B)$ and $\xi(B)$ (Fig. 3c). Probably, this fact indicates the limitations of the postulated simple exponential decay of nonlocal interaction (1.1), as well as the straightforward association of the correlation length ξ with the Ginzburg - Landau coherence length ξ_{GL}.

The presented simple formalism allows one to obtain quantitative information from the phenomena, related to the nonlocal Little - Parks effect. For example, the nonlocal oscillations of the critical current of the bar $I_c(B,x)$ can be calculated. To localize the point of observation, constrictions between the probes have been fabricated (Fig. 4). If the

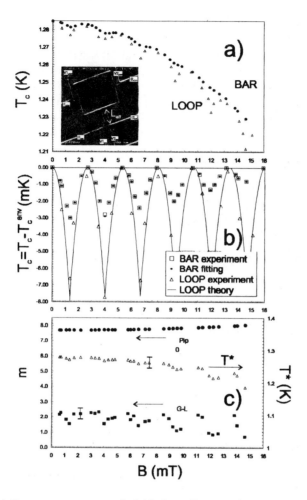

Figure 3 a) Temperature - magnetic field phase diagram of the aluminum nanostructure of the inset. b) Comparison of the experimental $\delta T_c(B, x)$ data with calculations. The monotonous component $T_c^{env}(B, 0)$ has been subtracted. c) Magnetic field dependence of the correlation length ξ and the effective temperature T^*. For details see text.

oscillating critical temperature $T_c(B, x)$ is known, then the critical current can be obtained by the well-known expression for the temperature dependence of the critical current of a one-dimensional wire:

$$I_c(T, B, x) = I_c(0, B, x) \left[1 - \frac{T}{T_c(B, x)} \right]^{3/2} \tag{1.7}$$

The validity of the above expression for our nanostructures has been verified by plotting the experimental values of critical current vs. temperature at zero magnetic field. The reason for the abrupt drop of the critical current after the first period of $I_c(T, B, x)$ oscillation (Fig. 4b) is not clear. We believe that this phenomenon is not related to nonlocality, but probably to the resistive state anomaly [12] and / or negative magnetoresistance [13], frequently observed in aluminum nanostructures.

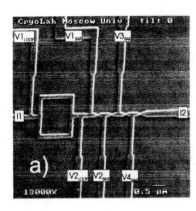
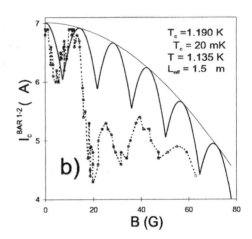

Figure 4 a) Scanning electron micrograph of the sample studied. Linear segments (bars) contain constrictions to localize the point of observation. b) Dependence of the critical current of the bar between contacts 1 and 2 on magnetic field. Solid lines represent calculations of $I_c^{env}(B, 0)$ (thin line) and $I_c(T, B, x)$ (thick line).

In summary we can say that the aluminum nanostructures studied do show nonlocal Little - Parks effect, i.e. oscillation of the critical temperature and the related phenomena (oscillations of the effective resistance and the critical current). Results can be interpreted in terms of a simple model, postulating exponential decay of the nonlocal interaction (1.1) with the correlation length ξ being associated with the Ginzburg - Landau coherence length ξ_{GL}.

Acknowledgments

This work has been supported by the Russian Foundation for Basic Research (Grant 98-02-16850) and the Academy of Finland under the Finnish Center of Excellence Programme 2000-2005 (Project No. 44875, Nuclear and Condensed Matter Programme at JYFL).

References

[1] M. Buttiker, Phys. Rev. B **32**, 1846 (1985)

[2] C. P. Umbach, P. Santhanam, C. Van Haesendonck, and R. A. Webb, Appl. Phys. Lett. **50**, 1289 (1987)

[3] A. Benoit, C. P. Umbach,R. B. Laibowitz, and R. A. Webb, Phys. Rev. Lett. **58**, 2343 (1987); W. J. Skocpol, P. M. Mankiewich, R. E. Howard, L. D. Jackel, and D. M. Tennant, Phys. Rev. Lett. **58**, 2347 (1987)

[4] V. T. Petrashov, R. Sh. Shaikhaidarov, I.A. Sosnin, JETP Lett. **64**, 839 (1996); V. N. Antonov, A. F. Volkov, H. Takayanagi, Phys. Rev. B **55**, 3836 (1997)

[5] A. B. Pippard, Proc. Roy. Soc. (London) A 216, 547 (1953)

[6] M. Tinkham, *Introduction to superconductivity*, McGraw-Hill, Inc., 1996

[7] W. A. Little and R. D. Parks, Phys. Rev. Lett. **9**, 9 (1962)

[8] R. P. Groff and R. D. Parks, Phys. Rev. **176**, 567 (1963)

[9] L. I. Glazman, F. W. J. Hekking, and A. Zyuzin, Phys. Rev. B **46**, 9074 (1992)

[10] N. E. Izraeloff, F. Yu, A. M. Goldman, and R. Boiko, Phys. Rev. Lett. **71**, 2130 (1993)

[11] C. Strunk, V. Bruyndoncx, V. V. Moshchalkov, C. Van Haesendonck, and I. Bruynseaede, Phys. Rev. B **54**, 12701 (1996)

[12] K. Yu. Arutyunov, D. A. Presnov, S. V. Lotkhov, A. B. Pavolotski, and L. Rinderer, Phys. Rev. B **59**, 6487 (1999)

[13] P. Santhanam, C. P. Umbach, and C. C. Chi, Phys. Rev. B **40**, 11392 (1989)

CHARACTERISATION OF COOPER PAIR BOXES FOR QUANTUM BITS

M.T. Savolainen, J.J. Toppari, L. Taskinen, N. Kim, K. Hansen and J.P. Pekola
Department of Physics, University of Jyväskylä. P.O. Box 35, FIN-40351 Jyväskylä, Finland
email: pekola@phys.jyu.fi

Key words: quantum coherence, quantum bit, Cooper pair box

Abstract: We have fabricated and measured single Cooper pair boxes (SCB) using
             superconducting single electron transistors (SET) as electrometers. The box
             storage performance for Cooper pairs was measured by observing the changes
             in the SCB island potential. We are also fabricating niobium structures, which
             are expected to have less problems with quasiparticle contamination than
             similar aluminium based devices because of the high critical temperature. The
             use of niobium may also reduce decoherence and thereby increase the time
             available for quantum logic operations.

1.   INTRODUCTION

1.1   Josephson junction qubits

In principle any quantum mechanical two level system can act as a qubit, but the realisation of a useful quantum computer has turned out to be highly non-trivial. Some of the problems are related to the difficulties in scaling the number of components of the devices. These and other problems may be eliminated with nanostructure solid state devices.

Averin has suggested the use of long arrays of Josephson junctions [1] and Schön et al. and Bouchiat et al. have proposed to use the Cooper pair box [2,3]. The two states that constitute the qubit in a Single Cooper pair

Box (SCB) are mixed charge eigenstates of the superconducting island. The mixing is provided by the Josephson coupling and under normal circumstances, i.e. when the Josephson energy is much smaller than the charging energy, at most two charge states mix (although there is of course a quasi infinite number corresponding to the possible number of Cooper pairs on the island). It is possible to induce quantum oscillations involving these states by applying fast gate pulses to the system as demonstrated by the experiments of Nakamura et al. [4]. These experiments also showed that the oscillations between the ground state and the excited state of the system can last at least a few and possibly several tens of nanoseconds or even longer [4]. The persistence of the oscillations is an aspect which is related to the question of decoherence, an issue of overwhelming importance for quantum computing.

A quantum computer designed along the lines suggested in [1-3] will ultimately rely on current or voltage measurements and it is therefore important also to identify and assess the inherent limitations of the suggested designs in this respect. We have analysed theoretically the Cooper pair pump and noticed that, unlike in the normal metal arrays in which the transfer of electrons can be made very accurate [5], the transfer of Cooper pairs is always inaccurate due to the presence of the Josephson coupling [6]. The inaccuracy in pumping can easily exceed ten percent in a typical superconducting array. Obviously this may affect the operation of Josephson junction array based quantum gates.

2.   SAMPLE FABRICATION

2.1   Single electron boxes

We fabricate our samples on silicon-oxide coated silicon using standard electron beam lithography and shadow evaporation methods. In addition we use a multilayer technique similar to the one described in [7] to improve the important coupling capacitance between the islands of the SET and the SCB.

The bottom layer of a sample contains a gold line which provides this coupling. The line is covered with a thin (300Å) insulating silicon monoxide layer and the SET and the SCB are deposited on top of these layers, aligned with the line. Figure 1 shows a SEM image of a sample and a schematic picture of the system. The large metallic structures are ground electrodes and the vertical line is the gold line which terminates at the SCB (upper part) and the SET (lower part). The three angle evaporation procedure is needed in order to place short normal metal parts of the electrodes close to the junctions. These normal metal parts, made out of copper, act as quasiparticle filters or -traps and are important to see a clear 2e-periodic gate modulation of the SET.

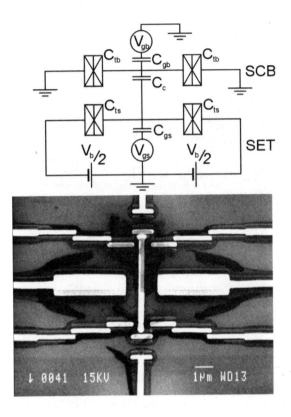

Figure 1. Schematic view and an SEM-image of one of the measured samples. The large structures in the vertical centre are ground electrodes.

2.2   Niobium structures

Due to the large superconducting gap of niobium, the systems based on this material should suffer less from quasiparticle contamination than similar structures of aluminium. In addition, the use of niobium might reduce decoherence and thereby increase the time available for quantum logic operations. We have fabricated single electron transistors made of niobium with conventional electron beam lithography and two angle shadow evaporation method (fig. 2). The results so far have been promising, with fairly high T_C (above 8 K), estimated from I-V curves. One of the main problems in the production of Nb structures is the high temperature needed for vaporisation of the material. This seems not to hinder the manufacture in our UHV system, probably due to the large distance from the hot region to the sample chip. It still remains a challenge to manufacture reliable and reproducible oxide layers for the junctions, however, and no experiments have yet been performed with Nb structures similar to the aluminum sample shown in fig.1.

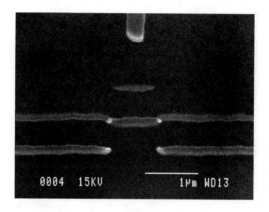

Figure 2. SEM picture of the all-Nb SET.

3.   EXPERIMENTS

As can be seen from fig.1 the samples are fairly symmetric between the SCB and the SET, the only difference being the gold line, which is in metallic contact with the SET. This allows us to interchange the role of the components so the nominal SCB can act as a SET and *vice versa*. This makes it possible to measure the important parameters of the system like gate and cross capacitances, charging energies and so on. Table 1 contains some parameters of the sample shown in the SEM image in fig. 1. The main difference between the box and the SET is the larger total capacitance of the latter due to the gold line. The charging energy of both the SET and the SCB is measured at the reference temperature of liquid helium (4.2 K) using the zero bias anomaly of the dynamic conductance [8] and applying the principle of Coulomb Blockade Thermometry. All other measurements are made at low temperature, around 100 mK, reached with a dilution refrigerator. The two gate capacitances $C_{g\,s/b}$ (*s* refers to the SET, *b* to the SCB) can be determined from the periodicity of the modulation of the current vs. the corresponding gate voltage $V_{g\,s/b}$.

There is an undesired cross capacitance C_{cross} between the box gate and the SET island and between the SET gate and the box island. We found this cross capacitance to be smaller than the gate capacitances by a factor of three. The effect is so big that we need to compensate it by applying a constant fraction of the gate voltage to the other gate with inverse polarity: $V_{gs}=-(C_{cross}/C_{gs})V_{gb}$.

The optimum operating point of the SET is found by I-V curves measured with different SET gate voltages. In the sample shown in fig. 3 the biasing point is about 1.6 mV where the current through the SET is most sensitive to external voltages (the 0.5 mV offset in the curve has been identified as a thermoelectric effect).

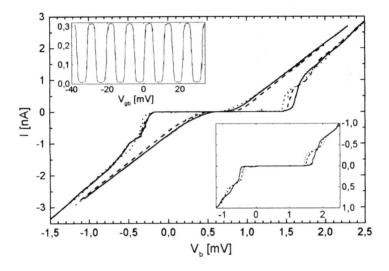

Figure 3. I-V characteristics measured from the SET-circuit with different gate voltages in both the superconducting and normal state. I-V curves of the SCB are depicted in the lower inset. Upper inset shows the current modulation in the SCB with the bias 1.6 mV. The modulation is not compensated and is thus due to the box gate - SET island cross capacitance. The period gives the cross capacitance directly.

After the basic characterisation of the sample we used the SET-electrometer to detect tunnelling events in the SCB. This is done by sweeping the gate voltage V_{gb} of the box while keeping the electrodes grounded and in addition applying the compensation procedure explained above in order to fix the SET operating point to the desired value, independent of V_{gb}. The combination of sweeping the gate voltage and tunnelling of single Cooper pairs into or out of the box generates a sawtooth-like modulation of the island potential. This potential change affects the SET-current via the capacitive coupling between the islands. The sawtooth-like current modulation and the compensation procedure are shown in fig. 4. The current through the SET, reflecting the potential of the box, is plotted as a function of V_{gb} with different levels of compensation. In the bottom figure the effect of the cross capacitance is clearly seen as a ca. 150 mV period in the amplitude of the tunnelling event modulations with period 25 mV. In the top figure this is fully compensated and only the effect due to tunnelling of Cooper pairs is seen.

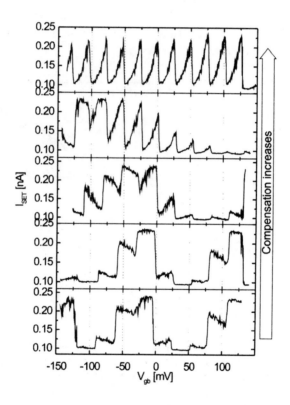

Figure 4. Measured SET current as a function of SCB gate voltage with different levels of compensation.

Table 1. The parameters of the sample of Figure 1

	SCB	SET
E_C	0.17 meV	0.11 meV
C_{island}	460 aF	700 aF
$C_{g\,b/s}$	23.5 aF	23.3 aF
C_{cross}	7.6 aF	7.6 aF (estimated)

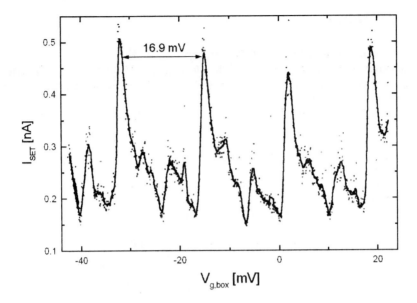

Figure 5. Measured SET current as a function of SCB gate voltage with full compensation. Small circles are the data points and the solid line is a ten point smoothing of these data. The long arrow indicates the period corresponding to Cooper pair tunnelling.

A more detailed plot of a sawtooth modulation is seen in fig.5 for the sample with the parameters given in table 1. The expected period of the 2e-tunneling can be estimated from the numbers in the table to 15 mV whereas the measured value is 17 mV. The small difference could be due to an error in the estimated cross capacitance. In addition to the major peaks the graph also shows smaller regularly spaced peaks.

ACKNOWLEDGMENTS

This work has been supported by the Academy of Finland under the Finnish Centre of Excellence Programme 2000-2005 (Project No. 44875, Nuclear and Condensed Matter Physics Programme at JYFL) and the EU (contract IST-1999-10673).

M.T. Savolainen *et al.*

REFERENCES

[1] D.V. Averin, Solid State Commun. **105**, 659 (1998).
[2] A. Shnirman, G. Schön, and Z. Hermon, Phys. Rev. Lett. **79**, 2371 (1997); A. Shnirman and G. Schön, Phys. Rev. B **57**, 15400 (1998); Yu. Makhlin, G. Schön, and A. Shnirman, Nature **398**, 305 (1999).
[3] V.Bouchiat, D.Vion, P.Joyez, D.Esteve and M.H.Devoret, Physica Scripta **T76**, 165 (1998).
[4] Y. Nakamura, Yu.A. Pashkin, and J.S. Tsai, Nature **398**, 786 (1999).
[5] M.W. Keller, J.M. Martinis, N.M. Zimmerman, and A.H. Steinbach, *Appl. Phys. Lett.* **69**, 1804 (1996).
[6] J.P. Pekola, J.J. Toppari, M. Aunola, M.T. Savolainen, and D.V. Averin, *Phys. Rev. B* **60**, R9931 (1999).
[7] E.H. Visscher, S.M. Verbrugh, J. Lindeman, P. Hadley, and J.E. Mooij, Appl. Phys. Lett. **66**, 306 (1995).
[8] Sh. Farhangfar, K.P. Hirvi, J.P. Kauppinen, J.P. Pekola, and J.J. Toppari, *J. Low Temp. Phys.* **108**, 191 (1997).

NOISE MEASUREMENTS OF A SUPERCONDUCTING SINGLE ELECTRON TRANSISTOR (SSET) AT T=0.3K

B. Buonomo,[1] P. Carelli,[2] M.G. Castellano,[1] F. Chiarello,[3] C. Cosmelli,[3] R. Leoni,[1] and G. Torrioli[1]

[1] Istituto Elettronica dello Stato Solido IESS-CNR, Via Cineto Romano 42, 00156 Roma, Italy

[2] INFM-Dipartimento di Ingegneria Elettrica Università dell'Aquila, Italy

[3] INFM-Dipartimento di Fisica Universita' "La Sapienza", P.le A. Moro, Roma, Italy

Key words: single electron tunneling; mesoscopic devices

Abstract: A superconducting single electron transistor, based on aluminum electrodes, was fabricated and studied. The characterization was made by measuring both the transport properties as a function of temperature and the noise performance. As a result, we observed modulation of the transfer characteristics up to 1.4K and we measured a charge sensitivity of $3.8 \cdot 10^{-4}$ e/$\sqrt{Hz}$ at 10 Hz and 0.3 K.

INTRODUCTION

The solid state implementation of the quantum computation needs a suitable detector to probe the quantum states of the system. In the superconductive implementation[1] the superconducting single-electron transistor (SSET) is the natural device for probing the charge states. The charge sensitivity of a superconducting single-electron transistor (SSET) is predicted to be considerably higher than its normal metal counterpart, at least when operated near the threshold for quasi-particle tunneling[2].

Operation with SSET has two main advantages. First, the current amplitude modulation close to the quasi-particle threshold (~4Δ/e, where Δ is the superconducting energy gap) is higher than for a normal metal SET. This allows SET operation at higher temperatures[3], where a normal metal SET would have exceedingly small modulation. This feature is directly due to the shift of the Coulomb blockade by the superconducting energy gap: temperature does smear the I-V curve at the onset of the Coulomb gap, but does not have comparable effect on the superconducting energy gap, even for temperatures of the order of Δ/k. The modulation in the I-V characteristics, then, is less rounded than for a normal metal SET.

We report on the fabrication and tests of SSET's, operated between T=0.3 K and T=1.4 K. The characterization was made by measuring the SSET transport properties and then the output noise spectral density at different working points (not systematically); a calibration signal of known amplitude allowed us to refer the noise at the input as equivalent charge noise.

SAMPLE FABRICATION

The SSET consists of two tunnel junctions, with capacitance C_1 and C_2 and tunnel resistance R_{T1} and R_{T2}, an Al island and a gate capacitor C_g (see Fig.1). The tunnel junctions are made by an Al base electrode, an insulator grown onto its surface (Al oxide) and an Al counter electrode on top of

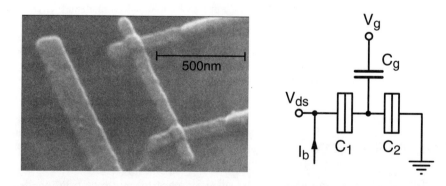

Fig.1 Left: Scanning electron micro-graph of a SET made by Al films (light gray) on an oxidized Si substrate (dark background). The two tunnel junction areas are obtained by the Al film overlap (measured dimension 0.08x0.12μm²). The island and the coplanar gate capacitor (about 1μm long) are also shown. Right: Electrical equivalent circuit for a current biased SET.

them. We fabricated Al/AlOx/Al sub-micron junctions using e-beam lithography and the shadow mask technique [4].

The Al evaporation is made at three different large angles θ [5,6], (typically, θ=+52°, -52° and 0°, where θ=0° is the direction normal to the substrate plane). Before the last Al evaporation at θ=0°, the insulating layer of the tunnel junction was grown by a controlled oxidation of the Al in pure oxygen at a pressure of about 1 mbar for 5 minutes, giving a specific tunnel resistance per junction of about $1 \div 2$ kΩ·μm^2. At the end the resist mask is lifted off in acetone producing in this way the structure shown in Fig.1.

RESULTS AND DISCUSSION

The measurements are carried out in a ^{3}He refrigerator operating in an unshielded environment. All the lines to the sample are heavily filtered at rf: this is obtained by feed-through capacitive filters and Thermocoax cables[7] about 0.4m long.

We measured three samples (A,B and C). We used both current (sample A,B and C) and voltage polarization (sample C) of the SSET. In the first scheme the two junctions (see Fig.1, right) are dc current biased at I_b through a resistance of 500MΩ (not shown) and the voltage across them, V_{DS} changes periodically, with a period equal to the charge quantum e, by changing the gate voltage V_G.

Sample A has total resistance $R_T=R_{T1}+R_{T2}\approx140$kΩ. From a fit of the transfer characteristics with the orthodox model, we found the values of the SET capacitances [8]: $C_1\approx C_2\approx C_\Sigma/2\approx0.6$fF and $C_g=e/\Delta V_G\approx0.05$fF. The voltage gain was $K_v=dV_{DS}/dV_G =C_g/C_1\approx0.1$. At 321mK the Josephson energy for each junction is $E_{J(1,2)}\approx (\Delta/8) (R_K/R_{T(1,2)})\approx9$μeV, where the quantum resistance $R_K=h/e^2\approx25.8$kΩ, and it is much smaller than the charging energy of a single junction $E_{c(1,2)} \approx e^2/C_\Sigma \approx140$μeV. The latter is four times larger than the thermal energy $k_BT\approx28$μeV.

In order to see the influence of the temperature, the transfer characteristics were recorded up to 1.45K (about the critical temperature T_c of our Al films): at this temperature the voltage modulation with the gate disappears. Fig.2 shows some of the source-drain voltage V_{DS} versus gate voltage V_G characteristics, recorded at four intermediate temperatures (base temperature-0.32K, 0.5, 0.8 and 1K), for our current biased SET. To show at the same time all the working points also the bias current I_b is changed quasi-statically (as a triangular wave function). The insert shows the transfer characteristics at 1.35K, which have a much smaller amplitude (about 20μV) than at 320mK (about 150μV). As already mentioned in the introduction, the

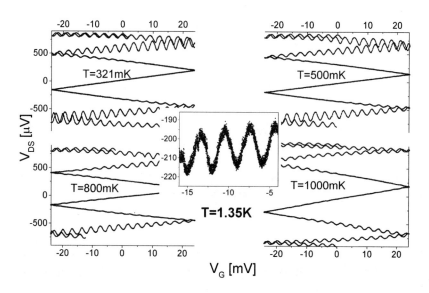

Fig. 2. Source-Drain voltage V_{DS} versus gate voltage V_G characteristics for current biased SSET (sample A) at four different temperatures. To show all the working points, the bias current I_b was changed quasi-statically (as a triangular wave function). Insert: Sample A transfer characteristics at constant I_b at T=1.35K.

I_b-V_{DS} remains sharp close to the voltage $4\Delta/e$ also for $k_B T \approx \Delta$. This fact allows the presence of modulation in the transfer characteristics up to T_c even if $k_B T > E_{c(1,2)}$.

Sample B has a total resistance $R_T = R_{T1} + R_{T2} \approx 230k\Omega$, $C_i \approx C_2 \approx C_\Sigma/2 \approx 0.29fF$, with $C_g = e/\Delta V_G \approx 0.03fF$. Its I-V characteristics, taken under current bias and for different values of the gate voltages, are shown in Fig.3 (left). The SET output voltage was sent to a commercial low-noise amplifier and then to a spectrum analyzer (HP3565A), to measure the noise spectral density. In order to refer the measured output noise to the input as charge noise, a calibration signal was used: an ac signal corresponding to 10^{-2} e was superimposed to the dc gate voltage and the SET response at that frequency was measured, the ratio between the two quantities giving directly the SET responsivity. Besides, by changing the ac frequency we measured the SSET signal bandwidth, typically flat up to $f_c = 400Hz$. The f_c roll-off value is basically determined by the capacitance C_L given by the rf filtering of the refrigerator.

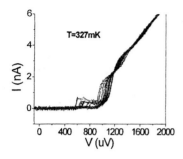 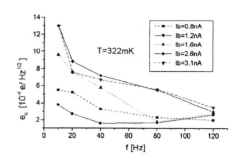

Fig.3 left: I-V characteristics of sample B taken with current polarization for different values of the gate voltage. Right: Charge noise of sample B taken at different values of the bias current and frequency.

Fig.3 right shows the square root of the charge spectral density e_n taken at 10, 20, 40, 80 and 120Hz for different values of the bias current I_b. The working point was adjusted (by changing the dc gate voltage within the limits corresponding to one charge quantum on the island) in such a way to get the maximum gain dV_{DS}/dQ_G (where $Q_G = C_g V_G$) for each current bias value. Sample B, at T=322mK, shows a charge noise $e_n = 1.6 \cdot 10^{-4}$ e/Hz$^{1/2}$ at f=40Hz. This was achieved biasing the SET above the superconducting gap $4\Delta/e$, in the region of sequential quasi-particle tunneling; at this working point the bias current was 1.2nA and the responsivity was $dV_{DS}/dQ_G \approx 1300\mu V/e$. The charge noise at f=10Hz was $e_n = 3.8 \cdot 10^{-4}$ e/Hz$^{1/2}$.

Sample C was measured using both current and voltage polarization of the SSET. In the second case, we used a transimpedance amplifier [9]. Its simplified scheme is shown in Fig.4. In our case, R_f=8.2MΩ is the feedback resistor, the op-amp is an OPA 627AP and V_b is the external bias voltage source. In this way, by measuring the current $I_{DS} = V_{out}/R_f$, rather than the voltage across the SET, the frequency bandwidth is increased. This because in principle we connect a low impedance amperometer across the SET electrodes, shunting in this way the capacitance C_L. Besides, the feedback loop forces the inverting input of the op-amp to the same potential as the non inverting input, voltage biasing the SSET at V_b. With this method we recorded the I_{DS}-V_G for various V_b at a temperature T=313mK.

The acquired characteristics allow us to build up the contour plot for the SET current [10], which shows the thresholds for the various tunnel processes (Fig. 5). The current I_{DS} is periodic in the gate voltage, with periodicity e/C_g. The top dark triangular shape with upper vertex at bias voltage $4\Delta/e + e/C_\Sigma$ and lower vertex at $4\Delta/e$ forms the threshold for the quasi-particle tunneling through the SSET. In this process, a single quasi-particle tunnels onto the island through one junction and then another quasi-

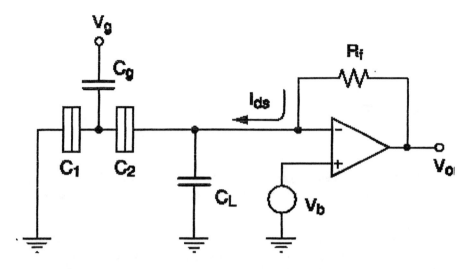

Fig. 4 A simplified scheme of the SSET and the trans-impedance amplifier. R_f is the feedback resistor, C_L the line capacitance and V_b is the external bias voltage source.

particle tunnels out through the other one. This allow us to evaluate $E_{c(1,2)}$ $\approx e^2/C_\Sigma \approx 320\mu eV$ and $C_1 \approx C_2 \approx C_\Sigma/2 \approx 0.25fF$. The total resistance is $R_T = R_{T1} + R_{T2} \approx 240k\Omega$.

Also visible are the intersecting lines for the Josephson quasi-particle cycle (JQP) corresponding to the tunneling of a Cooper pair onto the island and then the sequential tunneling of two quasi-particles through the other junction (negative slope) and vice versa (positive slope). These are the X-shaped lines intersecting at $2e/C_\Sigma \approx 640\mu V$: we clearly see the upper part of the X's. It is also possible to see, with some difficulties, the isolated current peaks located at e/C_Σ, placed on the extensions of the JQP lines, which are related to more complicated tunneling processes [10].

We then measured the noise of sample C under voltage and current bias configuration. The spectra were obtained by merging 4 different spectra acquired with overlapping frequency bandwidth in order to increase the resolution; the noise was then referred to the input as equivalent charge noise using a calibration signal, as discussed before.

In the measurements shown in fig. 6a, the SSET is voltage-biased at the same $V_b = 1.1mV$ and two different gate voltages, with gain dI_{DS}/dQ_G of 0.9 and 8 nA/e respectively. The device bandwidth, measured by changing the frequency of the calibration signal, was flat up to $f_c = 3000Hz$. In this case, the f_c roll-off value is inversely proportional to the product $C_L R_f$ and proportional to the op-amp gain[9]. The low frequency part of the spectral noise density has a slope $1/f^\alpha$ with $\alpha \approx 1.5$. The equivalent charge noise at the

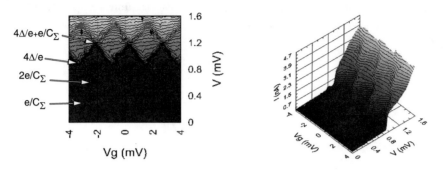

Fig. 5 Stability contour map (2D left, 3D right) of sample C obtained by the characteristics I_{DS}-V_G for various values of V_b at T=313mK.

input is $1\ 10^{-2}$ e/Hz$^{1/2}$ at 1 Hz and $1.4\ 10^{-3}$ at 10 Hz, comparable (within a factor two) to what is found in the literature for similar devices [ref.11]. We remark that our measurements are taken at a higher temperature (300 mK) than most reported results. The two curves of fig. 6a almost overlap for most of the frequency range, irrespective of the different gain of the SET in the two conditions: this an indication that the noise is due to charge input noise [ref.11].

The situation changes when the SSET is biased at constant current. We measured both sample B and sample C, getting similar results. Fig.6b shows the curves relative to sample C, again at two different gate voltages (with gain dV_{DS}/dQ_G of 100 and 370 µV/e) and same I_b=1.2nA. In contrast with the voltage-bias measurements, the power of the $1/f^{\alpha}$ (low frequency) spectrum is $\alpha\approx0.7$, and the charge noise level is lower ($2\ 10^{-3}$ e/Hz$^{1/2}$ at 1 Hz). Also the dependence on the responsivity is different from the voltage-bias case: we observe that the curve corresponding to lower SET gain has a higher equivalent charge noise referred to the input, which means that the noise is not due to an input source. We checked that the amplifier contribution to the SET output noise is negligible and cannot be responsible for the observed behavior. The charge noise corresponding to the point with higher responsivity is quite low, being $7\ 10^{-4}$ e/Hz$^{1/2}$ at 10 Hz. We also notice that in Fig. 3, relative to sample B measured under current bias, the responsivity was higher ($dV_{DS}/dQ_G\approx1300$µV/e instead of 760 µV/e for the best curve of sample C) and the noise was still lower ($e_n=3.8\ 10^{-4}$ e/Hz$^{1/2}$ for sample B compared to $7\ 10^{-4}$ e/Hz$^{1/2}$ for sample C).

We think that the main difference in the two setups is the different intrinsic dynamics. As a fact, the linearity range in the SET response is much smaller for the voltage biased case, hence a low frequency input noise has the additional effect of moving the working point along the current-gate

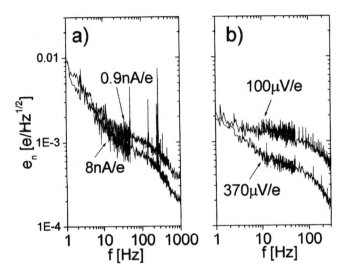

Fig. 6 Square root of the charge noise spectral densities for sample C, with voltage bias a) and current bias b). The gate responsivity for a given bias point is given in the figure.

voltage characteristics: this manifests as additional low frequency noise, while its cause is just that the SET working point has changed. By operating the SSET with a charge feedback, in order to keep the operation point fixed, the excess noise should disappear.

CONCLUSION

We have performed measurements on SSET's fabricated in our lab in order to characterize their properties and compare their performance with the literature. We found that our devices operated properly, with modulated transfer characteristics, in a temperature range between the base temperature of a ^{3}He refrigerator and 1.4K. The noise measurements performed on our samples show a different SSET behavior depending on whether the device was biased at constant current or constant voltage.

This work has been partially supported by INFM under the project PRA on "Solid State Quantum Information".

REFERENCES

[1] A. Shnirman, G.Shon, Z.Hermon, Quantum manipulation of small Josephson junctions, *Phys. Rev. Lett* **79**, 2371 (1997)

[2] A. N.Korotkov, Charge sensitivity of superconducting single-electron transistor, *Appl. Phys. Lett.* **69**, 2593-2595 (1996)

[3] D.Song, A.Amar, C.J.Lobb, F.C.Wellstood, Advantages of superconductive coulomb blockade electrometers, *IEEE Trans. on Appl. Supercond.* **5**, 3085 (1995)

[4] T.A.Fulton and G.J.Dolan, Observation of single electron charging effects in small tunnel junctions, *Phys. Rev. Lett.* **59**, 109 (1987)

[5] A.J. Manninen, J.Ksuoknuuti, M.M.Leivo, J.Pekola, Cooling a superconductor by quasi particle tunneling, *Appl. Phys. Lett.* **74**, 3020 (1999)

[6] R.Leoni, P.Carelli, M.G. Castellano, F. Melchiorri, C. Pasqui, G. Torrioli, Electron cooling by small tunnel junctions for a self-refrigerating bolometer, *Proc. LT22 Conf.* Helsinki 1999, *Physica B* (to be published)

[7] B. Zorin, The thermocoax cable as the microwave frequency filter for single electron circuits, *Rev. Sci. Instrum.* **66**, 4296 (1995)

[8] D.V.Averin and K.K.Likharev, Single electronics: a correlated transfer of single electrons and Cooper pairs in systems of small tunnel junctions, in *Mesoscopic Phenomena in Solids*, edited by B.L.Altshuler P.A.Lee and R.A.Webb (Elsevier Science Publisher B.V., 1991), p. 173-271

[9] B.Starmark, P.Delsing, D.B.Haviland, T.Claeson, Noise measurements of a single electron transistor using a transimpedance amplifier, in *Extended abstracts of the Sixth International Superconductive Electronics Conference* (PTB, Berlin, 1997), pp. 391-393

[10] P.Hardley, E.Delvigne, E.H.Visscher, S.Lahtenmaki, J.E.Mooij, 3-e tunneling processes in a superconducting single-electron tunneling transistor, *Phys Rev,* **B 58**, 15317-15320 (1998)

[11] B.Starmark, T.Henning, T.Claeson, P.Delsing, A.N. Korotkov, Gain dependence of the noise in the single electron transistor, *J. Appl. Phys.* **86**, 2132 (1999)

QUANTUM COHERENCE AND DECOHERENCE IN MAGNETIC NANOSTRUCTURES

Eugene M. Chudnovsky
CUNY Lehman College, Bronx, NY 10468-1589
www.lehman.cuny.edu/departments/faculty/chudnovsky.html

Abstract   The prospect of developing magnetic qubits is discussed. The first part of the article makes suggestions on how to achieve the coherent quantum superposition of spin states in small ferromagnetic clusters, weakly uncompensated antiferromagnetic clusters, and magnetic molecules. The second part of the article deals with mechanisms of decoherence expected in magnetic systems. Main decohering effects are coming from nuclear spins and magnetic fields. They can be reduced by isotopic purification and superconducting shielding. In that case the time reversal symmetry of spin Hamiltonians makes spin-phonon coupling ineffective in destroying quantum coherence.

Keywords: nanomagnetism, tunneling, decoherence

1.   INTRODUCTION

Is quantum computing physics or engineering? Some time ago I asked this question to Murray Gell-Mann who replied that he had a formal proof that it is physics: articles on quantum computing are published by the Physical Review Letters. In the spirit of his answer I will concentrate on the physics aspects of the problem, leaving out the engineering aspects. The purpose of this article is to discuss the prospect of using small magnetic clusters as qubits.

Qubits based upon quantum superposition of the $|\uparrow>$ and $|\downarrow>$ states of individual electrons and nuclei have been discussed during all years of quantum computermania [1]. In parallel, but with little overlap, there has been intensive theoretical and experimental research on spin tunneling between the $| \uparrow >$ and $| \downarrow >$ states of molecular magnets and in ferromagnetic and antiferromagnetic nanoparticles [2]. These are composite objects with total spin S ranging from $S = 10$ for Fe-8 and Mn-12 molecular clusters to S of a few thousand in nanoparticles. Experimental study of the magnetization reversal in individual nanoparticles gives evidence that very small particles are uniformly magnetized [3]. At low enough temperature they, like molecular magnets, possess a large fixed-length spin formed by the strong exchange interaction.

Quantum tunneling between the $| \uparrow >$ and $| \downarrow >$ states of spin-10 molecular nanomagnets has been unambiguously established in experiment [4, 5]. Early measurements of low temperature magnetic relaxation [2] and more recent measurements of individual nanoparticles [6] also provided evidence of quantum tunneling of spin. There is experimental evidence of quantum coherent oscillations between the $| \uparrow >$ and $| \downarrow >$ states in nanoparticles [7] and molecular magnets [8].

On one hand, individual nanoparticles and high-spin molecules, because of their large magnetic moments, must be easier to operate as qubits than individual electron and nuclear spins. In fact, the techniques of measuring individual particles of $S \sim 10^2 - 10^3$ already exist [9]. On the other hand, as the size of the system increases, its interaction with the dissipative environment also increases and it is not obvious whether the decoherence in large spin systems can be made low enough to allow their application as qubits. To achieve this goal the decoherence rate γ must be made small compared to the frequency $\Delta/\hbar$ of coherent oscillations between the $| \uparrow >$ and $| \downarrow >$ states. For a large spin system it requires effort for two reasons. Firstly, the interaction with the environment is proportional to the size of the system. Secondly, $\log(\Delta)$ scales linearly with $-S$ so that, except for some special cases, Δ becomes immeasurably small for $S \geq 30$.

In this article we will make suggestions on how to obtain large Δ for a relatively large spin and how to achieve the condition $\gamma << \Delta/\hbar$. It is our believe that this is doable in spin systems, in part due to their special properties with respect to time reversal. Spin Hamiltonians and tunneling rates will be considered in Section 2. Mechanisms of decoherence will be discussed in Section 3.

2. COHERENCE

2.1 FERROMAGNETIC CLUSTERS

The basic Hamiltonian for a ferromagnetic cluster of spin S that exhibits quantum coherence is

$$H = -AS_z^2 + V \ , \tag{1.1}$$

where A is a positive constant and V is an operator that does not commute with S_z but is invariant under the transformation $S_z \to -S_z$. The first term in Eq.(1) is typically produced by a crystal field or by the shape anisotropy of the cluster. At $V = 0$ the Hamiltonian (1) has a double degenerate ground state corresponding to two opposite orientations, $|\uparrow>$ and $|\downarrow>$, of $\mathbf{S}$ along the Z-axis. In terms of the magnetic quantum number, $S_z|m> = m|m>$, these ground states are $|S>$ and $|-S>$. Note that this degeneracy is independent of any geometry of the problem including the shape of the magnetic cluster and is solely due to the odd symmetry of $\mathbf{S}$ with respect to time reversal.

Our case of interest will be an integer large S. In that case the problem is almost classical so that even a large perturbation $V \neq 0$ generates only a small probability of tunneling between the $|\uparrow>$ and $|\downarrow>$ states. The degeneracy of the ground state is then removed and the new ground state can be approximated by

$$|0> = \frac{1}{\sqrt{2}}(|\uparrow> + |\downarrow>) \ . \tag{1.2}$$

It is separated from the first excited state,

$$|1> = \frac{1}{\sqrt{2}}(|\uparrow> - |\downarrow>) \ , \tag{1.3}$$

by the energy gap Δ which is determined by the strength and the nature of V and is small compared to the scale A of the energy levels of $-AS_z^2$. The probability of finding $\mathbf{S}$ looking up (down) then oscillates in time according to $\cos(\Delta \cdot t/\hbar)$, which is quantum coherence.

One of the possible forms of V is $V = -g\mu_B H_x S_x$ induced by the external field applied along the X-axis. If the field is small, the tunneling splitting can be obtained by the perturbation theory [10]

$$\Delta = \frac{4AS}{(2S-1)!} \left(\frac{g\mu_B|H_x|}{2A} \right)^{2S} \ . \tag{1.4}$$

From practical point of view this case is not very promising since the inevitably present weak misorientation of the field, resulting in $H_z \neq 0$, will destroy the coherence.

A more promising case, which to a good approximation corresponds to the Fe-8 spin-10 molecular nanomagnet, is $V = BS_x^2$. At $|B| << A$ the perturbation theory gives [10]

$$\Delta = 8A \frac{(2S)!}{[(S-1)!]^2} \left(\frac{|B|}{16A} \right)^S . \tag{1.5}$$

In Fe-8 $\Delta/\hbar$ is of order 10^4s^{-1}.

For large S and arbitrary B the tunneling splitting has been computed by the instanton method [11, 12] and by mapping the spin problem onto a particle problem [13, 14]

$$\Delta = 16\pi^{-1/2} S^{3/2} |B| \frac{\left[\frac{A}{|B|} \left(1 + \frac{A}{|B|} \right) \right]^{3/4}}{\left[\left(1 + \frac{A}{|B|} \right)^{1/2} + \left(\frac{A}{|B|} \right)^{1/2} \right]^{(2S+1)}} . \tag{1.6}$$

From practical point of view, in ferromagnetic nanoparticles with $S >> 1$ the case of interest is $|B| >> A$. In that case Eq.(6) gives

$$\Delta = 16\pi^{-1/2} S^{3/2} A^{3/4} |B|^{1/4} \exp \left[-2S \left(\frac{A}{|B|} \right)^{1/2} \right] , \tag{1.7}$$

and one can see that the effect of large S in the exponent is suppressed by a a small factor $(A/|B|)^{1/2}$. There are two ways to achieve this suppression and to increase Δ. The first is to use magnetic clusters with very strong easy plane anisotropy and relatively weak easy axis anisotropy in that plane. Particles of Tb and Dy may satisfy this condition. The second way is to place the particle above the surface of a superconductor [15]. In that case the magnetic dipole interaction of $\mathbf{S}$ with its mirror image inside the superconductor effectively reduces the uniaxial anisotropy A. The particle and the superconductor should be selected such that the magnetic field induced by the particle, $H \sim 4\pi\mu_B S$, does not exceed the first critical field of the superconductor, H_{c1}. Manipulating the distance between the particle and the superconductor, one can achieve the condition $A << |B|$. This can be done by, e.g., controlling the distance with a ferroelectric buffer in the electric field. Note that in the absence of the external magnetic field the odd symmetry of $\mathbf{S}$ with respect to time reversal preserves the coherence of such a setup independently of the shape of the superconducting surface and electric fields in the problem.

The tunneling of $\mathbf{S}$ also can be induced by its hyperfine interaction with nuclear spins, $V = B\mathbf{S}\cdot\mathbf{I}$, where $\mathbf{I} = \sum \mathbf{I}_i$ is the total nuclear spin of the cluster obtained by summing over spins of individual nuclei. This problem is rather involved. It has been studied in Ref.[16] and is relevant to tunneling in Mn-12. The total Hamiltonian conserves the magnitude of I and the Z-projection of $\mathbf{S} + \mathbf{I}$. In the millikelvin range nuclear spins must order, developing $I_{max} = \sum |I_i|$. It is easy to see that the problem is the one of quantum coherence only if $I_{max} = S$. In that case the classical ground states correspond to $\mathbf{S}$ and $\mathbf{I}$ of equal length looking opposite to each other along the Z-axis, $|S > | - I_{max} >$ and $| - S > |I_{max} >$. Tunneling removes the degeneracy of the ground state. The corresponding splitting can be obtained by the perturbation theory for $B << A$:

$$\Delta = 8(A + B)S^2 \left[\frac{B}{2(A + B)} \right]^{2S} .$$

(1.8)

In Mn-12 $S = 10$ while $I_{max} = 30$, so that the coherence of the above type is impossible. It is not out of the question, however, that chemists will produce a molecular cluster with $S = I$ in the future.

2.2 ANTIFERROMAGNETIC CLUSTERS

Tunneling in antiferromagnetic clusters [17] turns out to be much stronger than in ferromagnetic clusters, making them promising candidates for quantum coherence. Consider an anisotropic antiferromagnetic cluster with two compensated sublattices of spin $\mathbf{S}_1$ and $\mathbf{S}_2$, described by the Hamiltonian

$$H = -A(S_{1z}^2 + S_{2z}^2) + B\mathbf{S}_1 \cdot \mathbf{S}_2$$

(1.9)

with positive A and B satisfying $A << B$.

Let us show that this model can be mapped onto the model with strong transverse anisotropy. The Lagrangian corresponding to Eq.(1.9) is [2]

$$
\begin{aligned}
L &= S(\dot{\phi}_1 \cos\theta_1 + \dot{\phi}_2 \cos\theta_2) - S(\dot{\phi}_1 + \dot{\phi}_2) + AS^2(\cos^2\theta_1 + \cos^2\theta_2) \\
&- BS^2\cos\theta_1\cos\theta_2 - BS^2\sin\theta_1\sin\theta_2\cos(\phi_1 - \phi_2) ,
\end{aligned}
$$

(1.10)

where ϕ_1, θ_1, ϕ_2, θ_2 are spherical coordinates of vectors $\mathbf{S}_1$ and $\mathbf{S}_2$ of fixed length S. The path integral over these angles is dominated by $\cos\theta_1 = -\cos\theta_2 \equiv \cos\theta$, $\sin\theta_1 = \sin\theta_2 \equiv \sin\theta$. Introducing $\phi = \phi_1 - \phi_2$, one obtains, up to a phase term, the effective Lagrangian

$$L_{eff} = S\dot{\phi}\cos\theta + (2A + B)S^2\cos^2\theta - BS^2\sin^2\theta\cos\phi .$$

(1.11)

With the notations $\phi/2 = \Phi$ and $2S = \sigma$, it can be transformed into

$$L_{eff} = \sigma\dot{\Phi}\cos\theta + \frac{A}{2}\sigma^2\cos^2\theta - \frac{B}{2}\sigma^2\sin^2\theta\cos^2\Phi ,$$

(1.12)

which is equivalent to the Hamiltonian

$$H = -\frac{A}{2}\sigma_z^2 + \frac{B}{2}\sigma_x^2 \ .$$

(1.13)

One can then use the known result, Eq.(7), to obtain the tunneling splitting,

$$\Delta = 32(2\pi)^{-1/2}S^{3/2}A^{3/4}B^{1/4}\exp\left[-4S\left(\frac{A}{B}\right)^{1/2}\right] \ .$$

(1.14)

Here B is the exchange constant which is typically $10^4 - 10^6$ the anisotropy constant A. Consequently, antiferromagnetic particles consisting of a few thousand magnetic atoms can exhibit a significant tunneling rate between the $|\uparrow\downarrow>$ and $|\downarrow\uparrow>$ states.

One should notice, that, in order to manipulate these states by the magnetic field, some magnetic non-compensation of the sublattices is needed. In this case tunneling is possible only due to the presence in the Hamiltonian of the transverse field or the transverse anisotropy, e.g., $b(S_{1x}^2 + S_{2x}^2)$. Let the non-compensated spin be s. It has been demonstrated that the antiferromagnetic tunneling with $\Delta \propto \exp[-4S(A/B)^{1/2}]$ holds upto $s \sim (b/B)^{1/2}S << S$. At greater s it switches to the ferromagnetic tunneling with $\Delta \propto \exp[-2s(A/b)^{1/2}]$ [18] . Strongly non-compensated ferrimagnetic clusters, like, e.g., Mn-12, are always in the ferromagnetic tunneling regime, while weakly non-compensated ferritin particles [7, 2] can be in the antiferromagnetic tunneling regime.

3.   DECOHERENCE

A magnetic cluster of the type described above will always be imbedded in a non-magnetic solid dissipative environment. The potential for low decoherence arises from a number of reasons. The first of them is that strong electrostatic interactions are involved, through exchange couplings, only in the formation of the single spin $\mathbf{S}$ of the cluster. All other interactions of $\mathbf{S}$ have relativistic smallness of order $(v/c)^2$ to a some power. Due to this fact the ferromagnetic resonance in some materials has a quality factor of one million. The second reason is selection rules for spins due to the time reversal symmetry discussed below.

In quantum computation one is interested in creating an arbitrary superposition of the $|\uparrow>$ and $|\downarrow>$ states,

$$|\Psi> = C_1|\uparrow> + C_2|\downarrow> \ .$$

(1.15)

Using equations (2) and (3) this state can be re-written in terms of $|0>$ and $|1>$:

$$|\Psi> = C_1'|0> + C_2'|1> \ ,$$

(1.16)

where $C_1' = (C_1 + C_2)/\sqrt{2}$ and $C_2' = (C_1 - C_2)/\sqrt{2}$. It is clear, therefore, that the spontaneous decay of the excited state $|1>$ into the ground state $|0>$, accompanied by the emission of the energy quantum Δ, should be a major concern for preserving quantum coherence.

Note that, in principle, there may be decohering processes involving other excited levels of the spin Hamiltonian. For large spin, due to a small tunneling rate, such levels are separated from $|0>$ and $|1>$ by the energy gap (say A) that is large compared to Δ. An example of such a process would be an Orbach two-phonon process corresponding to the transition $|1> \to |A>$ caused by the absorption of a phonon, followed by the spontaneous decay $|A> \to |0>$ with the emission of a phonon. Also, there may be processes involving the excited states of the environment. An example of such a process would be a two-phonon Raman process that corresponds to the emission and absorption of two real phonons satisfying $\hbar(\omega_1 - \omega_2) = \Delta$. All such processes are strongly temperature-dependent. Their strength is measured by $\exp(-A/k_BT)$ or by some high power of T/Θ where $\Theta \sim 10^2 K$ is the Debye temperature [19]. In the millikelvin range the rate of such processes is negligible. Consequently, they are of little concern for the decoherence. On the contrary, processes of the spontaneous decay $|1> \to |0>$ can exist even at T=0. One should therefore concentrate on such processes.

As follows from the previous section, cases of interest for quantum coherence are described by Hamiltonians which contain even powers of spin operators. Due to the time reversal symmetry, this will always be the case in the absence of the external magnetic field. Consequently, the states $|0>$ and $|1>$ most generally can be written as

$$|0> = \sum_{m=-S}^{S} \alpha_m |m> \tag{1.17}$$

$$|1> = \sum_{m=-S}^{S} \beta_m |m> ,$$

where $\alpha_m = \alpha_{-m}$ and $\beta_m = -\beta_{-m}$. These equations are exact, as compared to the approximate equations (2) and (3). They simply reflect the fact that $|m>$ is a complete set of vectors in the Gilbert space of the spin Hamiltonian and that $|0>$ and $|1>$ have different symmetry with respect to time reversal.

Let K be the antilinear antiunitary operator of time reversal. The spin operator is odd with respect to time reversal, $KSK^\dagger = -S$. On the contrary, the spin Hamiltonian that only contains even powers of components of S is even with respect to time reversal, $KHK^\dagger = H$. Consider now a decohering operator D. D can be due to the interaction of $\mathbf{S}$ with phonons, electromagnetic fields, nuclear spins, etc. These interactions have different symmetry with respect to time reversal. Let D_o and D_e be time-odd and time-even operators respectively. Since the state $|O>$ is even with respect to time reversal and $|1>$ is odd, the following general statement is true

$$< 0|D_e|1 >= 0 \quad . \tag{1.18}$$

The spin-phonon interaction is of the form

$$H_{sp} = a_{iklm}S_iS_k\frac{\partial u_l}{\partial r_m} + h.c. \quad , \tag{1.19}$$

where $\mathbf{u}$ is the lattice displacement and a_{iklm} is a tensor reflecting the symmetry of the lattice. In terms of the operators of creation and annihilation of phonons

$$\mathbf{u} = \frac{i}{(2MN)^{1/2}} \sum_{\mathbf{k},\lambda} \frac{\mathbf{e}_{\mathbf{k},\lambda}e^{i\mathbf{k}\cdot\mathbf{r}}}{(\omega_{k\lambda})^{1/2}}(a_{k\lambda} - a_{k\lambda}^\dagger) \quad , \tag{1.20}$$

where M is the unit cell mass, N is the number of cells in the lattice, $\mathbf{e}_{\mathbf{k},\lambda}$ is the phonon polarization vector, $\lambda = t, t, l$, and $\omega_{k\lambda} = v_\lambda k$ is the phonon frequency (v_λ being the speed of sound). The spin-phonon interaction given by Eq.(19) describes transitions between different spin states accompanied by the emission and absorption of phonons. For instance, in the absence of tunneling the excited state $|S-1>$ of the Hamiltonian $H = -AS_z^2$, corresponding to the ferromagnetic resonance, can relax to the ground state $|S>$ at $T = 0$ by spontaneously emitting a phonon at a rate [19]

$$\gamma = C\frac{A^2S^2\Delta^3}{\hbar^4\rho v^5} \quad , \tag{1.21}$$

where ρ is the mass density of the lattice and C is a constant of order unity. For, e.g., $A \sim 1K$, $S \sim 10^3$, $(\Delta/\hbar) \sim 10^9 s^{-1}$, $\rho \sim 1g/cm^3$, and $v \sim 10^5 cm/s$, Eq.(21) gives $\gamma \sim 10^3 s^{-1}$.

We should now notice that $KH_{sp}K^\dagger = H_{sp}$, i.e., the spin-phonon operator is even with respect to time reversal. Consequently, $< 0|H_{sp}|1 >= 0$. This also can be checked by the direct calculation of matrix elements of H_{sp} using expressions (17). Thus, $|1>$ cannot spontaneously decay into $|0>$ with an emission of a phonon and we conclude that in a millikelvin range the spontaneous emission of phonons cannot decohere the coherent superposition of the $|\uparrow>$ and $|\downarrow>$ spin states. This is a strong statement which requires some clarification. Indeed, the spin-phonon interaction originates from the spin-orbit coupling of the form $H_{so} \propto \mathbf{L}\cdot\mathbf{S}$

where $\mathbf{L}$ is the orbital momentum. Its density can be presented as $\rho \epsilon_{ikl} r_k \dot{u}_l$. This operator seems to have non-zero matrix elements between $|0>$ and $|1>$ given by equations (17). However, this is only because the formulation of the spin tunneling problem presented above does not account for the conservation of the total angular momentum. In fact, the coherence is possible only if $\mathbf{L} + \mathbf{S} = 0$ [20]. Consequently, as $\mathbf{S}$ tunnels between $|\uparrow>$ and $|\downarrow>$, so does the mechanical angular momentum of the body. This is an analogue of the Mössbauer effect for spin tunneling. The energy associated with the mechanical rotation is $\hbar^2 S^2 / 2I_{in}$ where I_{in} is the moment of inertia of the solid matrix containing the magnetic cluster. This energy should not exceed Δ, otherwise it would be energetically favorable for $\mathbf{S}$ to localize in the $|\uparrow>$ or $|\downarrow>$ state. Since I_{in} scales as the fifth power of the size of the matrix, it is easy to see that such localization may occur in a free particle of size less then 10 nm, while for bigger systems the conservation of the total angular momentum is rather formal than practical question. With the account of the momentum conservation, the ground state and the first excited state are $\frac{1}{\sqrt{2}}(|S> | -L> \pm | -S> |L>)$, where the absolute values of $\mathbf{S}$ and $\mathbf{L}$ are equal. Now, again, the $|0>$ state is even with respect to time reversal while the $|1>$ state is odd. The spin-orbit operator is even, $KH_{so}K^\dagger = H_{so}$, because $\mathbf{L}$ is odd, $KLK^\dagger = -\mathbf{L}$. Thus, as before, $<0|H_{so}|1> = 0$.

As has been shown in the previous section (see also Refs. 21,22), nuclear spins always destroy the coherence at $H = 0$ unless tunneling is induced by the hyperfine interaction with $I = S$. Except for that exotic possibility, nuclear spins must be always eliminated from magnetic qubits by the isotopic purification. Similarly, the presence of free non-superconducting electrons in the sample will decohere tunneling through the spin scattering of electrons, $D_e \propto \mathbf{s} \cdot \mathbf{S}$. Although, this operator is time-even, free electrons incidentally passing through the magnetic cluster, perturb $|0>$ and $|1>$, breaking their properties with respect to time reversal. Thus, strongly insulating materials should be chosen for magnetic qubits. The effect of incidental phonons due to, e.g., relaxation of elastic stresses in the matrix (the $1/f$ noise), should be similar to the effect of incidental electrons in perturbing $|\Psi>$. Thus, the perfection of the lattice should be given a serious thought when manufacturing magnetic or any other qubits. One should then worry about the decohering effect of spin interactions which are odd with respect to time reversal. These are Zeeman terms due to magnetic fields. For example, $D_o = -g\mu_B S_z H_z(t)$ has a non-zero matrix element between $|0>$ and $|1>$. The effort should be made, therefore, to shield the magnetic qubit from the magnetic fields during the process of quantum computation.

This can be done by placing the magnetic cluster inside a nanoscopic superconducting ring. Such a ring may be used to control and measure the states of the qubit. Connecting such rings by superconducting lines may be the way to make a miltiqubit system.

Acknowledgements

I am grateful to Joe Birman, Lev Bulaevski, and Dima Garanin for discussions. This work has been supported by the U.S. National Science Foundation Grant No. DMR-9978882.

References

[1] H.-K. Lo, S. Popescu, and T. P. Spiller, *Introduction to Quantum Computation and Information* (World Scientific, 1998).

[2] E. M. Chudnovsky and J. Tejada, *Macroscopic Quantum Tunneling of the Magnetic Moment* (Cambridge University Press, 1998).

[3] W. Wernsdorfer, E. Bonet Orozco, K. Hasselbach, A. Benoit B. Barbara, N. Demoncy, A. Loiseau, H. Pascard, and D. Mailly, Phys. Rev. Lett. **78**, 1791 (1997); E. Bonet, W. Wernsdorfer, B. Barbara, A. Benoit, D. Mailly, and A. Thiaville, Phys. Rev. Lett. **83**, 4188 (1999).

[4] J. R. Friedman, M. P. Sarachik, J. Tejada, and R. Ziolo, Phys. Rev. Lett. **76**, 3830 (1996).

[5] W. Wernsdorfer and R. Sessoli, Science **284**, 133 (1999).

[6] W. Wernsdorfer, E. Bonet Orozco, K. Hasselbach, A. Benoit, D. Mailly, O. Kubo, H. Nakano, and B. Barbara, Phys. Rev. Lett. **79**, 4014 (1997).

[7] D. D. Awschalom, J. F. Smyth, G. Grinstein, D. P. DiVincenzo, and D. Loss, Phys. Rev. Lett. **68**, 3092 (1992).

[8] E. del Barco, N. Vernier, J. M. Hernandez, J. Tejada, E. M. Chudnovsky, E. Molins, and G. Bellessa, Europhys. Lett. **47**, 722 (1999).

[9] W. Wernsdorfer, private communication.

[10] D. A. Garanin, J. Phys. A: Math. Gen. **24**, L61 (1991).

[11] M. Enz and R. Schilling, J. Phys. **C19**, 1765 (1986).

[12] E. M. Chudnovsky and L. Gunther, Phys. Rev. Lett. **60**, 661 (1988).

[13] J. L. van Hemmen and A. Sütö, Physica **B141**, 37 (1986).

[14] O. B. Zaslavsky, Phys. Lett. **A145**, 471 (1990).

[15] E. M. Chudnovsky and J. Friedman, Phys. Rev. Lett., to appear.

[16] D. A. Garanin, E. M. Chudnovsky, and R. Schilling, Phys. Rev. **B61**, 12204 (2000).

[17] B. Barbara and E. M. Chudnovsky, Phys. Lett. **A145**, 205 (1990).

[18] E. M. Chudnovsky, J. Magn. Magn. Mat. **140-144**, 1821 (1995).

[19] A. Abragam and B. Bleaney, *Electron Paramagnetic Resonance of Transition Ions* (Clarendon-Oxford, 1970).

[20] E. M. Chudnovsky, Phys. Rev. Lett. **72**, 3433 (1994).

[21] Anupam Garg, Phys. Rev. Lett. **70**, 1541 (1993); J. Appl. Phys. **76**, 6168 (1994).

[22] N. V. Prokof'ev and P. C. E. Stamp, J. Phys.: Condens. Matter **5**, L663 (1993); Rep. Prog. Phys. **63**, 669 (2000).

MQT OF MAGNETIC PARTICLES

Wolfgang Wernsdorfer

Lab. L. Néel - CNRS, BP166,

38042 Grenoble Cedex 9, France

wernsdor@labs.polycnrs-gre.fr

Keywords: Magnetism, nanoparticle, uniform rotation, SQUID, Néel–Brown model, macroscopic quantum tunnelling.

Abstract   This brief review focuses on quantum tunnelling of magnetisation studied in individual nanoparticles or nanowires where the complications due to distributions of particle size, shape etc. are avoided. We discuss briefly various measuring techniques and review the basic concepts of a new micro-SQUID technique which allows us to study single nanometer-sized magnetic particles at low temperature. In the case of sufficiently small particles, the magnetisation reversal occurs by uniform rotation and the measurement of the angular dependence of the magnetisation reversal yields the effective magnetic anisotropy. The influence of the temperature on the magnetisation reversal is discussed. Probabilities of switching, switching field distributions, and telegraph noise measurements are proposed to check the predictions of the Néel Brown theory describing thermal activated magnetisation reversal. At very low temperature, we will see how quantum effects can be revealed.

Introduction

Macroscopic quantum tunnelling (MQT) represents one of the most fascinating phenomena in condensed matter physics. MQT means the tunnelling of a macroscopic variable through a barrier of an effective potential of a macroscopic system. It has been predicted that MQT can be observed in magnetic systems with low dissipation. In our case, it is the tunnelling of the magnetisation vector of a single-domain particle through its anisotropy energy barrier or the tunnelling of a domain wall through its pinning energy.

In order to realise why magnetic nanostructure are interesting for MQT studies, let us consider Fig. 1 which presents a scale of size ranging from macroscopic down to nanoscopic sizes. The unit of this scale is the

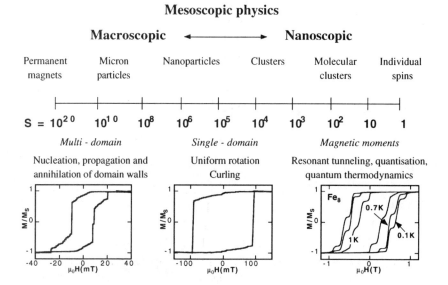

Figure 1 Scale of size which goes from macroscopic down to nanoscopic sizes. The unit of this scale is the number of magnetic moments in a magnetic system (roughly corresponding to the number of atoms). The hysteresis loops are typical examples for a magnetisation reversal via nucleation, propagation and annihilation of domain walls (left), via uniform rotation (middle), and quantum tunnelling (right)

number of magnetic moments in a magnetic system. At the macroscopic level, magnetism is governed by domains and domain walls. Magnetisation reversal occurs via nucleation, propagation and annihilation of domain walls (see the hysteresis loop on the left in Fig. 1 which was measured on an elliptic CoZr particle of 1 μm $\times$ 0.8 μm, and a thickness of 50 nm). Size and width of domain walls depend on the material of the magnetic system, on its size, shape and surface, and on its temperature [1].

When the system size is small enough, the formation of domain walls requires too much energy, i.e., the magnetisation remains in the so-called single domain state. Hence, the magnetisation has to reverse mainly by uniform rotation (see hysteresis loop in the middle of Fig. 1). In this review, we discuss mainly this size range where the physics are rather simple. For even smaller system sizes, one must take into account the band structure of spins which are complicated by band structure modifications at the particle's boundaries [2].

At the smallest size (below which one must consider individual atoms and spins) there are either free clusters made of several atoms [3] or molecular clusters which are molecules with a central complex containing magnetic atoms. In the last case, measurements on the Mn_{12} acetate and Fe_8 molecular clusters showed that the physics can be described by a collective moment of spin $S = 10$. By means of simple hysteresis loop measurements, the quantum character of these molecules showed up in well defined steps which are due to resonance quantum tunnelling between energy levels (see hysteresis loop on the right in Fig. 1). In addition, the Fe_8 molecular cluster is particularly interesting because of its biaxial anisotropy which allowed us to observe for the first time [4] the existence of what in a semi-classical description is the quantum spin phase or Berry phase associated with the magnetic spin of the cluster. This could be done by measuring the tunnelling matrix element Δ as a function of a magnetic field applied along the hard anisotropy axis. Oscillations of Δ were revealed which are due to constructive or destructive interference between two spin paths [5]. Hence, molecular chemistry had a large impact in the research of quantum tunnelling of magnetisation at molecular scales.

In this contribution, we discuss quantum effects on larger systems of 10^3 to 10^6 magnetic moments which we call magnetic nanoparticles. We show that these systems are large enough to behave classically during most of the time they are being observed. However, they may undergo quantum transitions under certain conditions. The advantage of magnetic systems lays mainly in the fact that they have several parameters which can be varied easily. Among them, external parameters like the angle of the applied field or the value of an applied transverse field can tune the tunnelling probability which allows a detailed comparison between measurement and theory.

1. MEASUREMENT TECHNIQUES

The dream of measuring the magnetisation reversal of an individual magnetic particle goes back to the pioneering work of Néel [6]. The first realisation was published by Morrish and Yu in 1956 [7] who employed a quartz–fiber torsion balance. More recently, insights into the magnetic properties of individual and isolated particles were obtained with the help of electron holography , vibrating reed magnetometry, Lorentz microscopy, and magnetic force microscopy (for a review, see [8]).

The first magnetisation measurements of individual single-domain nanoparticles and nanowires at very low temperatures were presented by Wernsdorfer *et al.* [9]. The detector, a Nb micro-bridge-Superconducting

Quantum Interference Device (SQUID), and the studied particles were
fabricated using electron-beam lithography. By measuring the electrical
resistance of isolated Ni wires with diameters between 20 and 40 nm,
Giordano and Hong studied the motion of magnetic domain walls [10].
Other low temperature techniques which may be adapted to single particle measurements are Hall probe magnetometry [11], magnetometry
based on the giant magnetoresistance or spin-dependent tunnelling with
Coulomb blockade [12]. At the time of writing, the micro-SQUID technique allowed the most detailed study of the magnetisation reversal of
nanometer-sized particles [8]. In the following, we review the basic ideas
of the micro-SQUID technique.

1.1.   MICRO-SQUID MAGNETOMETRY

The Superconducting Quantum Interference Device (SQUID) has been
used very successfully for magnetometry and voltage or current measurements in the fields of medicine, metrology and science. SQUIDs
are mostly fabricated from a $Nb-AlO_x-Nb$ trilayer, several hundreds
of nanometers thick. The two Josephson junctions are planar tunnel
junctions with an area of at least 0.5 μm^2. In order to avoid flux pinning in the superconducting film the SQUID is placed in a magnetically
shielded environment. The sample's flux is transferred via a superconducting pick up coil to the input coil of the SQUID. Such a device is
widely used as the signal can be measured by simple lock-in techniques.
However, this kind of SQUID is not well suited to measuring the magnetisation of single submicron-sized samples as the separation of SQUID
and pickup coil leads to a relatively small coupling factor. A much better coupling factor can be achieved by coupling the sample directly with
the SQUID loop. In this arrangement, the main difficulty arises from
the fact that the magnetic field applied to the sample is also applied to
the SQUID. The lack of sensitivity to a high field applied in the SQUID
plane, and the desired low temperature range led to the development
of the micro-bridge-DC-SQUID technique [8] which allows us to apply
several teslas in the plane of the SQUID without dramatically reducing
the SQUID's sensitivity.

The planar Nb micro-bridge-DC-SQUID can be constructed by using standard electron beam lithography, and the magnetic particle is
directly placed on the SQUID loop (Fig. 2) [8]. The SQUID detects the
flux through its loop produced by the sample magnetisation. For hysteresis loop measurements, the external field is applied in the plane of
the SQUID, so that the SQUID is only sensitive to the flux induced by
the stray field of the sample magnetisation. Due to the close proximity

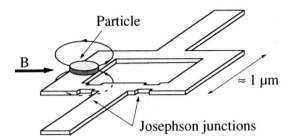

Figure 2 Drawing of a planar Nb micro-bridge-DC-SQUID on which a ferromagnetic particle is placed. The SQUID detects the flux through its loop produced by the sample magnetisation. Due to the close proximity between sample and SQUID a very efficient and direct flux coupling is achieved

between sample and SQUID, magnetisation reversals corresponding to 10^3 μ_B can be detected, i.e., the magnetic moment of a Co nanoparticle with a diameter of $2-3$ nm.

2. MAGNETISATION REVERSAL

For a sufficiently small magnetic sample it is energetically unfavourable for a domain wall to be formed. The specimen then behaves as a single domain. The simplest classical model describing magnetisation reversal is the model of uniform rotation of magnetisation, developed by Stoner and Wohlfarth [13], and Néel [14]. One supposes a particle of an ideal material where exchange energy holds all spins tightly parallel to each other, and the magnetisation does not depend on space. The original study by Stoner and Wohlfarth assumed only uniaxial shape anisotropy which is the anisotropy of the magnetostatic energy of the sample induced by its non-spherical shape. Later on, Thiaville has generalised this model for an arbitrary effective anisotropy which includes any magnetocrystalline anisotropy and even surface anisotropy [15]. The main interest in Thiaville's calculation is that measuring the critical surface of the switching field allows one to find the effective anisotropy of the nanoparticle. Knowledge of the latter is important for temperature dependent studies and quantum tunnelling investigations.

A typical example are BaFeO nanoparticles which have a dominant uniaxial magnetocrystalline anisotropy. Figure 3 shows the three dimensional angular dependence of the switching field measured on a BaFeO particle of about 20 nm. Although this particle has a simple critical surface being close to the original Stoner–Wohlfarth astroid, it cannot be generated by the rotation of a 2D astroid. However, taking into account

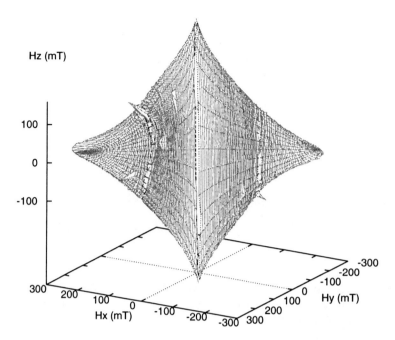

Figure 3 Three-dimensional angular dependence of the switching field of for a BaFeCoTiO particle [17] with a diameter of about 20 nm. The anisotropy field is $H_a \sim 0.4$ T

the shape anisotropy and hexagonal crystalline anisotropy of BaFeO, good agreement with the Stoner–Wohlfarth model is found [16].

The thermal fluctuations of the magnetic moment of a single-domain ferromagnetic particle and its decay towards thermal equilibrium were introduced by Néel [6] and further developed by Brown [18] and Coffey et al. [19]. In Néel and Brown's model of thermally activated magnetisation reversal, a single domain magnetic particle has two equivalent ground states of opposite magnetisation separated by an energy barrier which is due to shape and crystalline anisotropy. The system can escape from one state to the other by thermal activation over the barrier. Just as in the Stoner–Wohlfarth model, they assumed uniform magnetisation and uniaxial anisotropy in order to derive a single relaxation time.

The Néel–Brown model is widely used in magnetism, particularly in order to describe the time dependence of the magnetisation of collections of particles, thin films and bulk materials. However until recently, all the reported measurements, performed on individual particles, were not consistent with the Néel–Brown theory. This disagreement was attributed to the fact that real samples contain defects, ends and surfaces which could play an important, if not dominant, role in the physics of magnetisation reversal [20, 21]. A few years later, micro-SQUID measurements on Co nanoparticles showed for the first time a very good agreement with the Néel–Brown model by using waiting time, switching field and telegraph noise measurements [22, 23, 24, 25]. It was also found that sample defects, especially sample oxidation, play a crucial role in the physics of magnetisation reversal.

3.   TUNNELLING OF MAGNETISATION

This section focuses on Magnetic Quantum Tunnelling (MQT) studied in individual nanoparticles or nanowires where the complications due to distributions of particle size, shape, and so on are avoided. We concentrate on the necessary experimental conditions for MQT and review some experimental results.

On the theoretical side, it has been shown that in small magnetic particles, a large number of spins coupled by strong exchange interaction, can tunnel through the energy barrier created by magnetic anisotropy. It has been proposed that there is a characteristic crossover temperature T_c below which the escape of the magnetisation from a metastable state is dominated by quantum barrier transitions, rather than by thermal over barrier activation. Above T_c the escape rate is given by thermal over barrier activation.

In order to compare experiment with theory, predictions of the crossover temperature T_c and the escape rate Γ_{QT} in the quantum regime should be expressed as a function of parameters which can be changed experimentally. Typical parameters are the number of spins S, effective anisotropy constants, applied field strength and direction, etc. Many theoretical papers have been published during the last few years [26]. We discuss here a result specially adapted for single particle measurements, which concerns the field dependence of the crossover temperature T_c.

The crossover temperature T_c can be defined as the temperature where the quantum switching rate equals the thermal one. The case of a magnetic particle, as a function of the applied field direction, has been considered by several authors [27, 28, 29]. We have chosen the result for a particle with biaxial anisotropy as the effective anisotropy of most particles can be approximately described by strong uniaxial and weak transverse anisotropy. The result due to Kim can be written in the following form [29]:

$$T_{\mathrm{c}}(\theta) \quad \sim \quad \mu_0 H_{\parallel} \varepsilon^{1/4} \sqrt{1 + a \left(1 + \mid \cos\theta \mid^{2/3}\right)} \frac{\mid \cos\theta \mid^{1/6}}{1 + \mid \cos\theta \mid^{2/3}}, \quad (1.1)$$

where $\mu_0 H_{\parallel} = K_{\parallel}/M_S$ and $\mu_0 H_{\perp} = K_{\perp}/M_S$ are the parallel and transverse anisotropy fields given in tesla, $K_{\parallel}$ and $K_{\perp}$ are the parallel and transverse anisotropy constants of the biaxial anisotropy, θ is the angle between the easy axis of magnetisation and the direction of the applied field, and $\varepsilon = (1 - H/H_{sw}^0)$. Equation (1.1) is valid for any ratio $a = H_{\perp}/H_{\parallel}$. The proportionality coefficient of (1.1) is of the order of unity (T_c is in units of kelvin) and depends on the approach used for calculation [29].

The most interesting feature which may be drawn from (1.1) is that the crossover temperature is tuneable using the external field strength and direction because the tunnelling probability is increased by the transverse component of the applied field. Although at high transverse fields, T_c decreases again due to a broadening of the anisotropy barrier. Therefore, quantum tunnelling experiments should always include studies of angular dependencies. When the effective magnetic anisotropy of the particle is known, MQT theories give clear predictions with no fitting parameters.

The first magnetisation measurements of individual single-domain nanoparticles at low temperature (0.1 K < T < 6 K) were presented by Wernsdorfer et al. [9]. The detector (a Nb micro-bridge-DC-SQUID [8]) and the particles studied (ellipses with axes between 50 and 1000 nm and thickness between 5 and 50 nm) were fabricated using electron beam lithography. Electrodeposited wires (with diameters ranging from 40 to 100 nm and lengths up to 5000 nm) were also studied [22]. Waiting time and switching field measurements showed that the magnetisation reversal of these particles and wires results from a single thermally activated domain wall nucleation, followed by a fast wall propagation reversing the particle's magnetisation. For nanocrystalline Co particles of about 50 nm and below 1 K, a flattening of the temperature dependence of the mean switching field was observed which could not be explained by thermal activation. These results were discussed in the context of MQT. However, the width of the switching field distribution and the probability of switching are in disagreement with such a model because nucleation is very sensitive to factors like surface defects, surface oxidation and perhaps nuclear spins. The fine structure of pre-reversal magnetisation states is then governed by a multivalley energy landscape (in a few cases

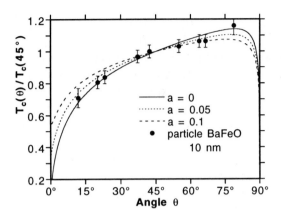

Figure 4 Angular dependence of the crossover temperature T_c for a 10 nm BaFe$_{10.4}$Co$_{0.8}$Ti$_{0.8}$O$_{19}$ particle with $S \approx 10^5$. The lines are given by (1.1) for different values of the ratio $a = H_\perp/H_\parallel$. The experimental data are normalised by $T_c(45°) = 0.31$ K

distinct magnetisation reversal paths were effectively observed [9, 8]) and the dynamics of reversal occurs via a complex path in configuration space.

Later, we published results obtained on nanoparticles synthesised by arc discharge, with dimensions between 10 and 30 nm [23]. These particles were single crystalline, and the surface roughness was about two atomic layers. Their measurements showed for the first time that the magnetisation reversal of a ferromagnetic nanoparticle of good quality can be described by thermal activation over a single-energy barrier as proposed by Néel and Brown [6, 18]. The activation volume, which is the volume of magnetisation overcoming the barrier, was very close to the particle volume, predicted for magnetisation reversal by uniform rotation.

A quantitative agreement with the Néel–Brown model of magnetisation reversal was also found on BaFe$_{12-2x}$Co$_x$Ti$_x$O$_{19}$ nanoparticles($0 < x < 1$) [24], short BaFeO, in the size range of $10 - 20$ nm. However, strong deviations from this model were evidenced for the smallest particles containing about $10^5 \mu_B$ and for temperatures below 0.4 K. These deviations are in good agreement with the theory of macroscopic quantum tunnelling of magnetisation. The measured angular dependence of $T_c(\theta)$ is in excellent agreement with the prediction given by (1.1) (Fig. 4). The normalisation value $T_c(45°) = 0.31$ K compares well with the theoretical value of about 0.2 K.

Although the above measurements are in good agreement with MQT theory, we should not forget that MQT is based on several strong assumptions. Among them, there is the assumption of a giant spin, i.e., all magnetic moments in the particle are rigidly coupled together by strong exchange interaction. This approximation might be good in the temperature range where thermal activation is dominant but is it not yet clear if this can be made for very low energy barriers. Future measurements will tell us the answer.

The proof for MQT in a magnetic nanoparticle could be the observation of level quantisation of its collective spin state. This was recently evidenced in molecular Mn_{12} and Fe_8 clusters having a collective spin state $S = 10$. Also the quantum spin phase or Berry phase associated with the magnetic spin $S = 10$ of the Fe_8 molecular cluster was evidenced [4]. In the case of BaFeCoTiO particles with $S = 10^5$, the field separation associated with level quantisation is rather small: $\Delta H = H_a/2S \sim 0.002$ mT. Future measurements should focus on the level quantisation of collective spin states of $S = 10^2$ to 10^4 and their quantum spin phases.

References

[1] A. Aharoni (1996) An Introduction to the Theory of Ferromagnetism, Oxford University, London

[2] A.J. Freemann and R. Wu, J. Magn. Magn. Mat., **100**, 497 (1991).

[3] I.M.L. Billas, A. Châtelain, W.A. de Heer, J. Magn. Magn. Mat., **168**, 64 (1997).

[4] W.Wernsdorfer and R. Sessoli, Science, **284**, 133 (1999).

[5] A. Garg, Europhys. Lett., **22**, 205 (1993).

[6] Néel L., Ann. Geophys., **5**, 99 (1949) and C. R. Acad. Science, Paris, **228**, 664 (1949).

[7] A. H. Morrish and S. P. Yu, Phys. Rev., **102**, 670 (1956).

[8] W. Wernsdorfer, A. Benoit, D. Mailly, J. Appl. Phys., **87**, 5094 (2000); W. Wernsdorfer, Ph.D. thesis, Joseph Fourier University, Grenoble, (1996).

[9] W. Wernsdorfer , K. Hasselbach, D. Mailly, B. Barbara, A. Benoit, L. Thomas and G. Suran, J. Magn. Magn. Mat., **145**, 33 (1995).

[10] K. Hong and N. Giordano, J. Magn. Magn. Mat., **151**, 396 (1995) and Europhys. Lett., **36**, 147 (1996).

[11] A.D. Kent, S. von Molnar, S. Gider, D.D. Awschalom, J. Appl. Phys., **76**, 6656 (1994).

[12] S. Guéron, M. M. Deshmukh, E. B. Myers, and D. C. Ralph, Phys. Rev. Lett., **83**, 4148 (1999).

[13] E. C. Stoner and E. P. Wohlfarth, Philos. Trans. London Ser. A, **240**, 599 (1948).

[14] L. Néel, C. R. Acad. Science, Paris, **224**, 1550 (1947).

[15] A. Thiaville, Phys. Rev. B, **61**, 12221 (2000).

[16] E. Bonet Orozco, W. Wernsdorfer, B. Barbara, A. Benoit, D. Mailly, and A. Thiaville, Phys. Rev. Lett., **83**, 4188 (1999).

[17] O. Kubo, T. Ido and H. Yokoyama, IEEE Trans. Magn., **MAG-23**, 3140 (1987).

[18] W. F. Brown, J. Appl. Phys., **30**, 130S (1959); J. Appl. Phys., **34**, 1319 (1963).

[19] L.J. Geoghegan, W.T. Coffey and B. Mulligan, Adv. Chem. Phys. **100**, 475 (1997), and Adv. Chem. Phys. **103**, 259 (1998).

[20] M. Ledermann, S. Schultz and M. Ozaki, Phys. Rev. Lett., **73** 1986 (1994).

[21] W. Wernsdorfer, K. Hasselbach, A. Benoit, G. Cernicchiaro, D. Mailly, B. Barbara, L. Thomas, J. Magn. Magn. Mat., **151**, 38 (1995).

[22] W. Wernsdorfer, B. Doudin, D. Mailly, K. Hasselbach, A. Benoit, J. Meier, J.-Ph. Ansermet, B. Barbara, Phys. Rev. Lett., **77**, 1873 (1996).

[23] W. Wernsdorfer, E. Bonet Orozco, K. Hasselbach, A. Benoit B. Barbara, N. Demoncy, A. Loiseau, D. Boivin, H. Pascard and D. Mailly, Phys. Rev. Lett., **78**, 1791 (1997).

[24] W. Wernsdorfer, E. Bonet Orozco, K. Hasselbach, A. Benoit, D. Mailly, O. Kubo, H. Nakano and B. Barbara, Phys. Rev. Lett., **79**, 4014 (1997).

[25] W.T. Coffey , D.S.F. Crothers, J.L. Dormann, Yu. P. Kalmykov, E.C. Kennedy and W. Wernsdorfer, Phys. Rev. Lett., **80**, 5655 (1998).

[26] Quantum Tunneling of Magnetization - QTM'94, NATO ASI, Serie E: Applied Sciences, Vol. 301, ed. L. Gunther and B. Barbara, (1995).

[27] O.B. Zaslavskii, Phys. Rev. B, **42**, 992 (1990).

[28] M.C. Miguel and E.M. Chudnovsky, Phys. Rev. B, **54**, 388 (1996).

[29] Gwang-Hee Kim and Dae Sung Hwang, Phys. Rev. B, **55**, 8918 (1997).

ABRUPT TRANSITION BETWEEN THERMALLY-ASSISTED AND PURE QUANTUM TUNNELING IN MN$_{12}$

K. M. Mertes, Yicheng Zhong, and M. P. Sarachik
Physics Department
City College of New York, New York, NY 10031, USA

Y. Paltiel, H. Shtrikman, and E. Zeldov
Department of Condensed Mater Physics
The Weizmann Institute of Science, Rehovot 76100, Israel

Evan Rumberger and D. N. Hendrickson
Department of Chemistry and Biochemistry
University of California at San Diego, La Jolla, CA 92093, USA

Single-molecule magnets are organic materials which contain a large (Avogadro's) number of identical magnetic molecules; Mn$_{12}$ acetate is a particularly simple and much-studied example of this class. The Mn$_{12}$ clusters are each composed of twelve Mn atoms (see Fig. 1) coupled by superexchange through oxygen bridges to give a sizable spin magnetic moment, S=10. These magnetic molecules are regularly arranged on a tetragonal crystal lattice with spacings between them sufficiently large that inter-cluster magnetic interactions are weak. As illustrated by the double well potential of Fig. 2, strong uniaxial anisotropy (of the order of 60 K) yields doubly degenerate ground states in zero field and a set of excited levels corresponding to different projections $m_s = \pm 10, \pm 9, \ldots, 0$ of the total spin along the easy c-axis of the crystal. Measurements[1, 2] below the blocking temperature of 3 K reveal a series of steps in the curves of M versus H at roughly equal intervals of magnetic field, as shown in Fig. 3, due to enhanced relaxation of the magnetization whenever levels on opposite sides of the anisotropy barrier coincide in energy. Strong temperature dependence was found which indicates that thermal processes play a central role. The steps in the magnetization curves have thus been attributed to thermally-assisted quantum tunneling of the spin magnetization.

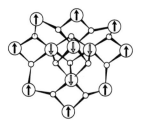

Figure 1 Schematic of the Mn_{12} molecule composed of four inner spin ($S = -3/2$) Mn^{4+} ions and eight outer spin ($S = +2$) Mn^{3+} ions with oxygen bridges, yielding a total spin $S = 10$ ground state at low temperatures.

As indicated in Fig. 2, the magnetization is thermally activated to a level near the top of the metastable well from which it tunnels across the barrier. Thermal activation becomes exponentially more difficult as one proceeds up the ladder to higher energy levels; on the other hand, the barrier is lower and more penetrable, so that the tunneling process becomes exponentially easier. Which level (or group of adjacent levels) dominates the tunneling is determined by competition between the two effects. As the temperature is reduced and thermal activation becomes more difficult, the states that are active in the tunneling move gradually to lower energies and deeper into the potential well.

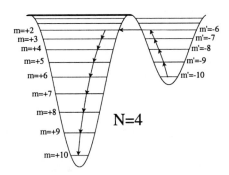

Figure 2 Double-well potential in the presence of a longitudinal magnetic field applied along the easy c-axis; the thermally-assisted tunneling process is indicated by arrows.

Chudnovsky and Garanin[3] have recently proposed that as the temperature is reduced, the levels that dominate the tunneling can shift to lower energies either continuously ("second order") or abruptly ("first order" transition), depending on the form of the potential. It may therefore be possible to observe an abrupt transition from thermally assisted tunneling, perhaps near the top of the well, directly to pure quantum mechanical tunneling from the lowest state of the metastable well as the temperature is reduced. We have reported earlier experiments that indicated an abrupt change to tunneling from the lowest state in Mn$_{12}$ [4]. In the present paper we present detailed measurements at closely spaced temperatures that provide strong evidence for an abrupt "first order" transition from thermally-assisted to pure quantum mechanical tunneling from the ground-state.

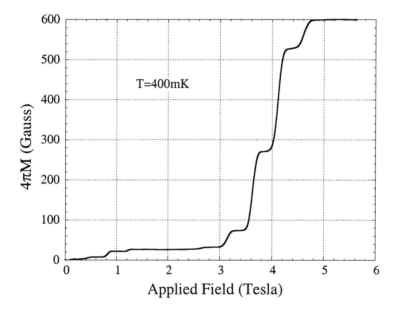

Figure 3 Magnetization versus longitudinal magnetic field for a Mn$_{12}$ sample cooled in the absence of field; note the steps at specific values of magnetic field corresponding to faster magnetic relaxation.

The spin Hamiltonian for Mn$_{12}$ is given by:

$$\mathcal{H} = -DS_z^2 - g_z\mu_B H_z S_z - AS_z^4 + \ldots\ldots$$

where $D \approx 60K$ is the anisotropy, the second term is the Zeeman energy, and the third on the right-hand side represents the next higher-order term in longitudinal anisotropy; additional contributions (transverse internal magnetic fields, transverse anisotropy) are not explicitly shown. Tunneling occurs from level m' in the metastable well to level m in the

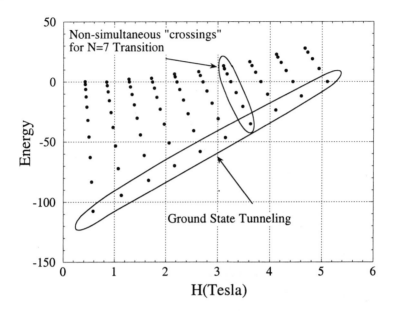

Figure 4   Energy level diagram obtained for the spin Hamiltonian for Mn$_{12}$. The dots indicate magnetic fields where pairs of energy levels on opposite sides of the barrier cross, giving rise to enhanced magnetic relaxation via tunneling.

stable potential well for magnetic fields:

$$H_z = N \frac{D}{g_z \mu_B} \left[1 + \frac{A}{D} \left(m^2 + m'^2 \right) \right],$$

where $N = |m + m'|$ is the step number. The second term in brackets is small compared to 1. Thus, steps N_i occur at approximately equally spaced intervals of magnetic field, $D/(g_z \mu_B) \approx 0.42$ Tesla; for a given step all pairs of levels cross at roughly the same magnetic field. However, careful measurements show that there is structure within each step due to the presence of the term in AS_z^4; as shown diagrammatically in Fig. 4, the levels do not cross simultaneously, an effect that is more pronounced for levels that are deeper in the well. Values of $A = 1.173(4)$ mK/mol as well as g_z and D have been determined by EPR[5] as well as g_z and D have been determined by EPR[5] and neutron scattering [6] experiments. Comparison of the measured magnetic fields with those calculated from Eq. (2) therefore provides an experimental tool that allows identification of the states that are predominantly responsible for the tunneling.

The magnetization of small single crystals of Mn$_{12}$ was determined from measurements of the local magnetic induction at the sample surface using 10×10 μm^2 Hall sensors composed of a two-dimensional electron gas (2DEG) in a GaAs/AlGaAs heterostructure. The 2DEG was aligned

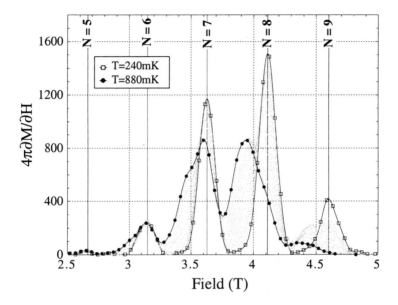

Figure 5 For a set of closely spaced temperatures, the first derivative of the magnetization with respect to magnetic field is plotted as a function of magnetic field. The amplitude is a measure of the rate of magnetic relaxation. There is substructure within each of the four maxima corresponding to steps $N = |m' + m| = 6, 7, 8$, and 9.

parallel to the external magnetic field, and the Hall bar was used to detect the perpendicular component (only) of the magnetic field arising from the sample magnetization.

Our results are shown in the next few figures. For different temperatures between 0.88 and 0.24 K, Fig. 5 shows the first derivative, $\partial M/\partial H$, of the magnetization M with respect to the externally applied magnetic field H[7]. The maxima occur at magnetic fields corresponding to faster magnetic relaxation due to level crossings on opposite sides of the anisotropy barrier. In the temperature range of these measurements, maxima are observed for $N = |m + m'| = 5$ through 9. Considerable structure associated with different pairs m, m' is clearly seen within each step N, with a transfer of "spectral weight" to higher values of m' as the temperature is reduced. The issue is whether this transfer occurs gradually or abruptly.

In order to address this question, we examine one of the steps in greater detail; Fig. 6 shows the data for $N = 7$ on an expanded scale. As the temperature is reduced, the maximum gradually moves to higher field and its amplitude changes. There is structure at certain temper-

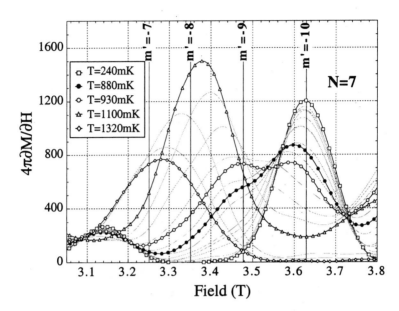

Figure 6 The first derivative of the magnetization with respect to magnetic field versus magnetic field shown on an expanded scale for $N = |m' + m| = 7$. The vertical lines denote the magnetic fields corresponding to tunneling between different pairs of levels (m',m) on opposite sides of the potential barrier: (-7,0), (-8,1), (-9,2), and (-10,3). For clarity several intermediate temperatures were omitted from the legend.

atures that clearly incidates the presence of more than one maximum, implying that more than one pair of levels is active; where a single maximum appears, it is probably the convolution of two or three maxima. It is noteworthy that the contribution from $m' = -9$ is minimal, or quite small, compared with other levels. In contrast, the contribution of $m' = -10$ becomes increasingly dominant as the temperature is lowered. Although not clearly discontinuous, there does appear to be an abrupt transfer of weight to tunneling from the lowest (ground) state of the metastable well.

This is shown more explicity in Fig. 7. Here, the position of the maxima are plotted as a function of temperature for all measured N. Within each step, the magnetic fields corresponding to tunneling from levels m' in the metastable well are indicated by horizontal lines, as labeled. For each step, the position of the maximum shifts gradually and continuously to higher magnetic field as the temperature is decreased, and then switches abruptly to $m' = -10$ at some temperature below which the field of the maximum remains constant. Although no levels are

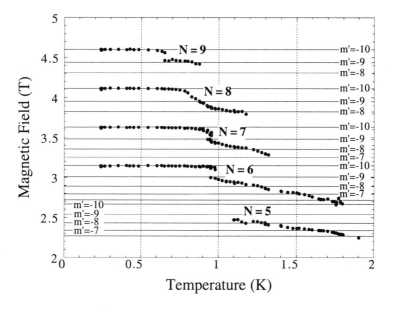

Figure 7 The magnetic field of the maxima in $\partial M/\partial H$ corresponding to enhanced magnetic relaxation plotted as a function of temperature. The fields corresponding to tunneling from different levels m' within each step N are indicated by horizontal lines. With decreasing temperature, the maxima initially shift gradually upward in field, then exhibit an abrupt shift to tunneling from $m' = -10$ within a narrow temperature range, below which the maximum remains constant.

skipped entirely for the conditions of our experiments, there is a sudden shift to a magnetic field that is independent of the temperature, and the value of this magnetic field agrees quantitatively with the calculated position for tunneling from $m' = -10$.

To summarize, magnetization measurements in Mn$_{12}$ acetate taken at closely spaced intervals of temperature exhibit an abrupt shift over a narrow temperature range to rapid magnetic relaxation at a "resonant" magnetic field corresponding to tunneling from the lowest state of the metastable potential well; this resonant field is then independent of temperature as the temperature is reduced further. Our data provide strong evidence for an abrupt transition between thermally-assisted tunneling and pure quantum tunneling from the ground state.

Acknowledgments

Work at City College was supported by NSF grant DMR-9704309 and at the University of California, San Diego by NSF grant DMR-9729339. EZ acknowledges the support of the German-Israeli Foundation for Scientific Research and Development.

References

[1] J. R. Friedman, M. P. Sarachik, J. Tejada, and R. Ziolo, Phys. Rev. Lett. **76**, 3830 (1996).

[2] L. Thomas, F. Lionti, R. Ballou, R. Sessoli, D. Gatteschi, and B. Barbara, Nature (London) **383**, 145 (1996).

[3] E. M. Chudnovsky and D. A. Garanin, Phys. Rev. Lett. **79**, 4469 (1997); D. A. Garanin and E. M. Chudnovsky, Phys. Rev. B **56**, 11102 (1997); *ibid.* **59**, 3671 (1999).

[4] A. D. Kent, Y. Zhong, L. Bokacheva, D. Ruiz, D. N. Hendrickson, and M. P. Sarachik, Europhys. Lett. **49** 521 (2000).

[5] A. L. Barra, D. Gatteschi, and R. Sessoli, Phys. Rev. B **56**, 8192 (1997).

[6] Yicheng Zhong, M. P. Sarachik, Jonathan R. Friedman, R. A. Robinson, T. M. Kelley, H. Nakotte, A. C. Christianson, F. Trouw, S. M. J. Aubin, and D. N. Hendrickson, J. Appl. Phys. **85**, 5636 (1999); I. Mirebeau, M. Hennion, H. Casalta, H. Andres, H. U. Güdel, A. V. Irodova, and A. Caneschi, Phys. Rev. Lett. **83**, 628 (1999).

[7] Throughout this paper, we have used H instead of the total field $B = H + \alpha(4\pi M)$; the field due to the sample magnetization is on the order of 300 Oersted.

QUANTUM EFFECTS IN THE DYNAMICS OF THE MAGNETIZATION IN SINGLE MOLECULE MAGNETS

D. Gatteschi,[1] R. Sessoli,[1] and W. Wernsdorfer[2]

[1] *Department of Chemistry, University of Florence, Via Maragliano 75/77, 50144 Firenze, Italy*

[2] *Laboratoire L. Neél, C.N.R.S. , 38042 Grenoble, France*

Key words: clusters, magnetization, relaxation, tunneling

Abstract: Molecular clusters of interacting paramagnetic metal ions show a slow dynamics of the magnetization at low temperature, which is characterized by quantum effects. Oscillation of the tunnel splitting is observed when the field is applied along the hard axis due to the topological interference of the tunnel pathways. An unsual isotope effect, based on the magnetic moment of the nuclei, is observed in the dynamics of the magnetization in the quantum regime.

INTRODUCTION

The development of quantum computing requires a new type of hardware in which data can be stored in a network of quantum mechanical two-level systems. Several different systems have been suggested as possible candidates, including spin 1/2 particles and two level atoms. Molecules are natural candidates, and one of the best explored routes so far, is that of using NMR techniques to manipulate quantum information in classical fluids [1]. In this case the spin 1/2 is provided by nuclei, for instance ^{13}C in $CHCl_3$.

This nucleus is magnetically coupled to spin 1/2 ^1H, and the system can behave as a two qu-bits ensemble. Experiments which realize a controlled NOT gate have been accurately described. Although these systems are fascinating, the possibility of making large arrays of qubits is not guaranteed, because on increasing the size of the molecules the interaction between the distant nuclei decrease. So other molecular two level systems should be investigated with the goal of finding the best suited quantum computing hardware.

A possible alternative is provided by molecular magnetic clusters, which sometimes are called single molecule magnets in order to indicate that at low temperature their magnetization relaxes as slowly as in a bulk magnet. Every molecular cluster may act as an individual and identifiable qubit. In fact the clusters are characterized by a large ground spin state and an Ising type anisotropy, which leaves a pair of quasi degenerate levels lying lowest. Evidence for tunneling between these levels has been achieved, and recent experiments have provided further insight in the mechanism of tunneling in these systems. A brief overview of these novel findings is presented here.

FE8: A QUANTUM NANOMAGNET

The octanuclear iron(III) cluster of formula $[Fe_8O_2(OH)_{12}(tacn)_6]Br_8.9H_2O[2]$, and whose structure is reported in Figure 1, is very suited to investigate the quantum effects in the dynamics of the magnetization.

Antiferromagnetic interactions between the localized spin S=5/2 of the high spin iron(III) ions are mediated by the bridging oxygen atoms. Due to the complex spin topology the ground state is S=10, as depicted by the arrows in Figure 1. This cluster can therefore be seen as a nanoscopic ferrimagnet. The spin structure has been suggested by the results of the modelling of the thermodinamical properties, as the temperature dependence of the magnetic susceptibility, which have also suggested a gap of ca. 30 K between the ground S=10 state and the first excited state which has S=9. Recently polarized neutron diffraction experiments have confirmed the spin structure reported with the arrows in Figure 1 [3].

The 21 manifold of the S=10 spin state is splitted in zero field. This splitting, which originates a sizeable magnetic anisotropy, has been investigated trough Electron Paramagnetic Resonance[4] as well as through Inelastic Neutron Scattering[5]. Both techniques suggest a biaxial magnetic anisotropy which can be described by

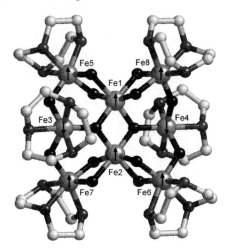

Fig.1 View of the structure of the Fe8 cluster. The big spheres are iron atoms, while oxygen, nitrogen and carbon are black, grey and pale grey respectively. The arrows represent the spin structure of the ground state.

the following spin hamiltonian, where also four order terms have been included.

$$H = \mu_B \mathbf{S \cdot g \cdot B} + D\, \mathbf{S}_z^2 + E\,(\mathbf{S}_x^2 - \mathbf{S}_y^2) + B_4^0\, \mathbf{O}_4^0 + B_4^2\, \mathbf{O}_4^2 + B_4^4\, \mathbf{O}_4^4 \quad (1)$$

where:
$$\mathbf{O}_4^0 = 35\, \mathbf{S}_z^4 - 30\, S(S+1)\mathbf{S}_z^2 + 25\, \mathbf{S}_z^2,$$
$$\mathbf{O}_4^2 = \{(7\, \mathbf{S}_z^2 + S(S+1)-5)\,(\mathbf{S}_+^2 + \mathbf{S}_-^2) + (\mathbf{S}_+^2 + \mathbf{S}_-^2)\,(7\, \mathbf{S}_z^2 + S(S+1)-5)\}/4,$$
$$\mathbf{O}_4^4 = (\mathbf{S}_+^4 + \mathbf{S}_-^4)/2$$

The parameters determined from neutron scattering are D=-0.29 K, E=0.046. $B_4^0 = 1 \times 10^{-6}$ K, $B_4^2 = 1.1 \times 10^{-7}$ K $B_4^4 = 8.5 \times 10^{-6}$ K.

The 21 levels are splitted as shown in Figure 2. A double well potential description is only accurate for the levelling which are mainly described by large |m| components, being m the eigenvalues of the $\mathbf{S}_z$ operator. The almost complete degeneracy of the two lowest spin states in the two potential wells separated by the anisotropy barrier may be lifted by applying a large magnetic field parallel to the easy axis of the anisotropy. Consequently, the system can be prepared in a chosen spin state. If the field is removed the equilibrium states, which corresponds to zero magnetization, is reached very slowly at low temperature due to the presence of an energy barrier. This gives rise to a magnetic hysteresis which is fully molecular in origin.

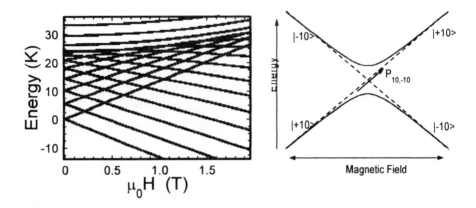

Fig.2. Energy splitting of the S=10 manifold when the magnetic field is applied along the easy axis (left). Avoided crossing of the m=10 and m'=-10 states (right).

The hysteresis loops recorded along the easy axis show almost equally separated steps at $H_z \approx n0.22$ T (n= 1, 2, 3,...). These steps are due to a faster relaxation of magnetization at particular field values where energy coincidence between states on the opposite site of the barrier is established. A similar behavior was first observed in Mn12 acetate clusters where equally separated steps were observed at $H_z \approx n0.45$ T [6,7]. The main difference between the clusters is that the hysteresis loops of Fe8 become temperature independent below 0.36 K. This corresponds to a mechanism of tunneling between the degenerate ground levels. In zero longitudinal field the tunneling occurs between the m=+10 and m=-10 states and the tunnel splitting between these two states is calculated from (1) to be of the order of 10^{-8} K. The tunnel splitting can be directly measured by using the Landau Zener method[8]. By sweeping the longitudinal filed around a resonance field, for instance $H_z=0$, as shown in the inset of Figure 2. The probability for the system to cross from the state m to the state m' is given by:

$$P_{m,m'} = 1 - \exp\left[-\frac{\pi\Delta^2_{m,m'}}{4\eta g\mu_B \, |m - m'| \, \mu_0 \, (dHz/dt)}\right]$$

(2)

where m and m' are the quantum number of the levels involved in the avoided-crossing that gives rise to the tunnel, dH/dt is the constant sweeping rate, the g is taken equal to 2 and μ_B is the Bohr magneton.

If P<<1 the tunnel probability is proportional to the relative variation of the magnetization after each field sweep. By measuring the decay of the magnetization it is therefore possible to estimate $\Delta_{m,m'}$. In order to apply the Landau-Zener model we first saturate the magnetization in a negative field of −1.4 T yielding Min=-Ms. We have also checked the predicted Landau-Zener dependence of P on the sweeping rate and found it verified in the range 0.001-1.0 T/s. For smaller sweeping rate the relaxation is slower due to the hole digging mechanism[9], which we will not discuss here.

TRANSVERSE FIELD DEPENDENCE OF THE TUNNEL SPLITTING

Since 1993 it was predicted that in a system described as a large spin exhibiting a biaxial magnetic anisotropy the application of a field along the hard axis originates oscillations of the tunnel splitting Δ which is quenched for critical values of the field[10].

If in first approximation we consider only the 2nd order terms of the magnetic anisotropy in eq. (1) the quench of the tunneling splitting is observed for H=(j+0.5)ΔH⊥ (with j=0,1,2..), where ΔH⊥ =2kB/gμB(2E(E+D))$^{\varsigma}$. Maxima are observed for H=jΔH⊥(with j=0,1,2). These oscillations can be explained in a semiclassical picture as due to the constructive-destructive interference of two topological paths for the reversal of the magnetic moment. In Figure 3 we report the unit sphere where the hard, intermediate, and easy axes are indicated as X,Y, and Z. By applying a field in the XY plane the unit vector can point towards the two degenerate minima A or B. The two minima can be connected by two paths. These give rise to destructive interference whenever the shaded area between the two path is kπ/S. This phenomenon is similar to the Aharanov-Bohm oscillations

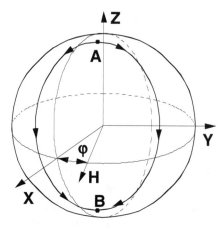

Fig.3 Unit sphere showing the tunneling pathway connecting the two minima A and B. X, Y, and Z are the hard, intermediate, and easy axis, respectively.

of the conductivity of mesoscopic rings and is generally known as the Berry phase[11].

The investigation of traditional nanoscale magnetic particles has not allowed to observe such a phenomenon, which was however observed in Fe8 crystals by using a micro SQUID magnetometer[12]. In Figure 5 we report the variation of the tunnel splitting as a function of H_X measured at 40 mK, *i.e.*, in the pure quantum tunneling regime[13].

The behavior can be quantitatively reproduced by a full diagonalization procedure using of the hamiltonian matrix of eq. (1). The parameters of the neutron scattering experiments have been used but to correctly reproduce the periodicity the B_4^4 has been modified to the value of -5.7×10^{-5} K to obtain the curves reported in Figure 4. This is in fact the parameter that most influence the tunnel splitting Δ of the ground states. A smaller correction is instead required if the parameters of high field EPR spectra experiments are used.

Another important result, shown in Figure 4, is the fact that the tunnel splitting can be measured not only for the ± 10 doublet but also for other pairs of levels if a static magnetic field along Z brings them in coincidence. For instance when a field is applied to have M=-10 and M=(10-n) in energy coincidence a parity effect is observed depending on n. In fact for n odd a quench of the tunnel splitting is now observed for $H_X=0$ and for those value of H_Y at which maxima are observed when n is even. The experimental behavior is nicely reproduced by the calculations as shown on the right of Figure 5. The parity effect observed for n odd is analogue to that expected for half integer spins for which the tunnel splitting is expected to be zero in zero field (Kramers' degeneracy). In fact in both cases the two states are connected by an odd numbers of steps.

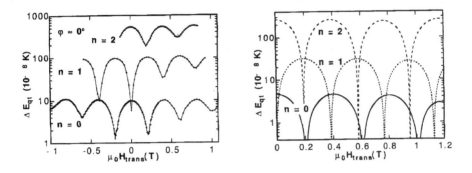

Fig.4 Observed (left) and calculated transverse field dependence of the tunnel splitting observed in Fe8 at 40 mK. The calculated curves are obtained using the parameter of eq. (2) according to ref. 17: D=-0.29 K, E=0.046. $B_4^0 = 1 \times 10^{-6}$ K, $B_4^2 = 1.1 \times 10^{-7}$ K $B_4^4 = -5.7 \times 10^{-5}$ K.

TUNNEL SCENARIO AND HYPERFINE FIELDS

The tunnel splitting measured and calculated as shown in the previous section is extremely small, $\Delta < 10^{-7}$ K. In order to observe the tunneling the local longitudinal field must be smaller than 10^{-8} T, otherwise the two levels are no more in coincidence within the tunnel splitting. The dipolar field generated by the surrounding clusters depends on the magnetic history but it is usually distributed over 10 mT. Most of the cluster therefore experience a field which hamper the tunneling. In order to solve this dilemma Prokof'ev and Stamp have taken into account the magnetic field generated by the magnetic nuclei[14]. As the nuclei are not polarized at the temperature at which the pure quantum tunneling is observed (T≤0.4 K) they can still fluctuate and give rise to a broadening of the levels which re-establishes the tunnel condition. The broadening is of the order of 1 mT and it may restore the tunneling conditions for the molecules for which the dipolar field is smaller than the width of the level. In order to test this hypothesis we have prepared an characterized two isotopically enriched Fe8 cluster compounds[15]. A ^{57}Fe, I=1/2, enriched (95%) starting material has been used. On the other side a partial deuteration of the cluster has been performed, thus replacing the hydrogen nuclei with the deuterium ones, which, even with a larger I, generate a smaller hyperfine field. The comparison of the time needed to relax 1% of the saturation magnetization, reported in Figure 5, shows a marked difference between the three isotopic samples. The largest is the hyperfine field generated by the nuclei the smaller is the time. It is important to notice that while isotopic effects are well known in physics they are usually originated by the different mass of the nuclei. In this case the mass is increased in both isotopically modified samples while the effect is reversed.. The hyperfine field influences the relaxation essentially at low temperature T<0.4 K in the tunneling regime. This is a further confirmation of the role that the magnetic nuclei have in promoting the tunneling as previously suggested.

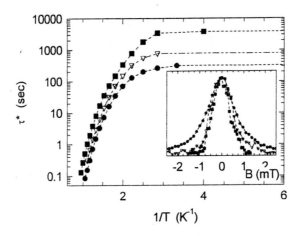

Fig.5. Time needed to relax 1% of the magnetization for natural Fe8 (∇), ^{57}Fe ($\bullet$) and ^{2}H ($\blacksquare$) enriched samples. In the inset the intrinsic line-width for the three samples.

We have also developed a technique to measure the intrinsic line-width of the quantum resonance getting rid of the dipolar broadening using the hole digging method[9]. We have therefore measured the intrinsic line width of the resonance of the three isotopic samples. Again a sizeable difference in the observed line width (see the inset of Figure 5) is observed among the three isotopic samples. The largest line width is observed for the ^{57}Fe enriched sample which experiences the largest hyperfine field. We have also calculated the line width taking into account the hyperfine interaction of each iron electronic spin with the nuclei and also considering the projection on the total spin state S=10 of the cluster. The results are in good agreement with the experimental values[15].

CONCLUSIONS

Despite the fact that quantum coherence has not been observed in single molecule magnets, these have shown a very rich behavior with many quantum effects in the dynamics of the macroscopic magnetization never observed before in traditional nanomagnets. The tunneling scenario, as well as the sources of decoherence, like the dipolar interactions and the nuclear magnetic field, have also been object of investigation. The strong transverse field dependence of the tunnel splitting could be exploited to control the hysteresis loop and in particularly the coercitivity of the material. The hyperfine field effect opens the perspectives of controlling the dynamics of the magnetization through the control of the nuclear magnetic moment.

REFERENCES

[1] J. A. Jones, R. H. Hansen, and M. Mosca, *J. Magn. Reson.* **135**, 353 (1998).
[2] K. Wieghardt, K. Pohl, I. Jibril, and G. Huttner, *Angew. Chem., Int. Ed. Engl.* **23**, 77 (1984).
[3] Y. Pontillon *et al.*, *J. Am. Chem. Soc.* **121**, 5342 (1999).
[4] A. L. Barra, D. Gatteschi, and R. Sessoli, *Chemistry-a European Journal* **6**, 1608 (2000).
[5] R. Caciuffo *et al.*, *Phys. Rev. Lett.* **81**, 4744 (1998).
[6] J. R. Friedman, M. P. Sarachik, J. Tejada, and R. Ziolo, *Phys. Rev. Lett.* **76**, 3830 (1996).
[7] L. Thomas *et al.*, *Nature (London)* **383**, 145 (1996).
[8] L. Gunther, *Europhys. Lett.* **39**, 1 (1997).
[9] W. Wernsdorfer *et al.*, *Phys. Rev. Lett.* **82**, 3903 (1999).
[10] A. Garg, *Europhys. Lett.* **22**, 205 (1993).
[11] M. V. Berry, Proc. Roy. Soc. London A **392**, 45 (1984).
[12] W. Wernsdorfer *et al.*, *J. Appl. Phys.* **78**, 7192 (1995).
[13] W. Wernsdorfer and R. Sessoli, *Science* **284**, 133 (1999).
[14] N. V. Prokof'ev and P. C. E. Stamp, *Phys. Rev. Lett.* **80**, 5794 (1998).
[15] W. Wernsdorfer *et al.*, *Phys. Rev. Lett.* **84**, 2965 (2000).

QUANTUM COHERENCE AND VERY LOW TEMPERATURE MAGNETIC EXPERIMENTS IN MESOSCOPIC MAGNETS

J. Tejada
J. M. Hernandez
E. del Barco
Universitat de Barcelona — Facultat de Física
Diagonal, 645, E-08028 Barcelona
Spain

N. Biskup
J. Brooks
NHFML, CM/T Group
1800 E. P. Dirac Dr., Tallahassee, Florida 32310
USA

M. D. Zyzler
Comisión Nacional de Energía Atómica — Centro Nacional de Bariloche
8400 S.C. de Bariloche
Argentina

Keywords: quantum coherence, mesoscopic magnets, quantum computation

Abstract The question addressed in this paper is that of the circumstances under which it is possible to observe quantum coherence phenomena in mesoscopic magnets such as single domain particles and molecular clusters. As these phenomena are only detectable at very low temperatures we also discuss the low temperature characterisation of these magnetic systems.

At a given temperature any magnetic system has a certain state that corresponds to the absolute minimum of its free energy. All magnetic systems that exhibit hysteresis are expected to relax toward that minimum. The magnetic hysteresis results, therefore, from the existence of metastable states which correspond to the different spin configurations. Single domain particles, both mesoscopic particles and molecular magnets, can b in metastable states due to energy barriers produced by the magnetic anisotropy. The rate of relaxation associated to the thermal fluctuations is $\Gamma \propto \exp(U/T)$ where U is the energy barrier. The occurrence of magnetic relaxation at temperatures where the thermal fluctuations die out has been explained in terms of quantum tunneling [1].

For certain mesoscopic magnetic particles and magnetic clusters the value of both the total net spin, $S < 1000$, and anisotropy barrier height, $U < 1000K$, are such that these materials may be considered as good candidates to enhance the recording density if the operation is performed at low temperature. Moreover, the discrete structure of the spin levels corresponding to the different classical spin orientations, suggest that these materials are good candidates to explore the border between quantum and classical mechanics and for a role in quantum information processing.

Within the last few years magnetic mesoscopic systems have emerged as a truly interdisciplinary field involving the contribution of chemists and both theorists and experimental physicists working in magnetism and quantum solid physics. The first seminal discovery on quantum magnetic hysteresis [1, 2, 3] has induced a lot of work in this field resulting in similar discoveries in mesoscopic antiferromagnetic particles [4], new theories [5, 6, 7] and discoveries such as those related to quantum interference effects [8, 9], spin phonon avalanches [10] and quantum coherence [11] and the nature of the transition between the quantum and classical spin relaxation [12, 13]. The effect of nuclear spins on the tunneling transitions has also been worked out theoretically and detected experimentally [14, 15].

To a first approximation the spin Hamiltonian of both nanometre scale magnetic particles and magnetic clusters in a magnetic field H is written

$$\mathcal{H} = -DS_z^2 + \mathcal{H}' - g\mu_B \mathbf{S} \cdot \mathbf{H} + \mathcal{H}_{\text{dis}} \tag{1.1}$$

where D is the magnetic anisotropy constant, S is the spin of the particle/molecule, $\mathcal{H}'$ stands for other anisotropy terms and $\mathcal{H}_{\text{dis}}$ represents the interaction of the spin system with the environment, z refers to the easy axis direction. The first term of Hamiltonian of Eq.1.1 generates spin levels S_z inside each well separated by an energy $D(2S_z - 1)$ and in zero magnetic field the spin levels in the two wells (which are separated by an energy barrier $U = DS^2$) are degenerate, see Figs. 1 and 2. The symmetry-violating terms in the Hamiltonian inducing tunneling are those associated to the transverse component of both the magnetic field and magnetic anisotropy.

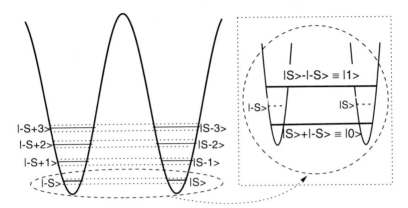

Figure 1 Structure of the energy spectrum of the spin system. The inset shows the energy of the ground-state splitted levels (symmetric and antisymmetric).

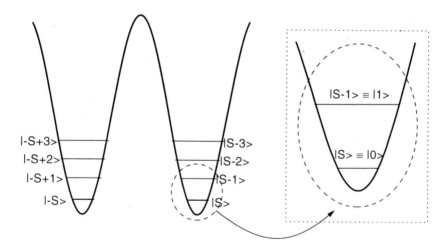

Figure 2 Structure of the energy spectrum of the spin system. The inset shows the energy of the lowest level localized in a well ($S_z = S$) and the first localized level above it ($S_z = S - 1$).

Although many of the mesoscopic magnets have very large spin values and/or small magnetic anisotropy being unlikely, therefore, to detect the discreet structure of the spin projection on the easy axis direction, the case of antiferromagnetic mesoscopic particles and molecular clusters is different as the gap between the spin levels corresponding to the different

S_z values, $-S < S_z < S$ may be of the order of 10 K in temperature units.

When applying a magnetic field, it can be chosen to be either parallel, $(H_\parallel)$ or perpendicular $(H_\perp)$ to the easy axis. The application of $H_\parallel$ reduces the barrier height between the two possible orientations of the spin, while the effect of $H_\perp$ is also that of increasing the spin tunneling rate between the degenerated Sz levels. In other words, the values of T, $H_\parallel$ and $H_\perp$ determine the levels between which occurs the resonant spin tunneling. At sufficient low temperature the populations of the excited levels subside to exponentially small values and for zero longitudinal field, $H_\parallel = 0$, the only possible tunneling process would occur between the two lowest states $S_z = S$ and $S_z = -S$ which may be denoted as the the states "yes" and "not" respectively. By increasing the transversal component of the field, the rate of tunneling between the "yes" and "not" states increases as a consequence of their mixing giving rise to the symmetric and antisymmetric combinations, see Fig. 1. At zero transversal field the tunneling process between the two degenerate lowest states is suppressed and the two levels entering the discussion for the resonance experiments are those corresponding to $S_z = S$ and $S_z = S-1$ lying in the same potential well, see Fig. 2.

In other words, the attractiveness of working with mesoscopic magnets and very low temperature is that we have two different sets of quantum states with an energy gap between them of 100 mK, case of the quantum splitting, to 10 K, case of the uniform precession of the magnetic moment, respectively which may be used to study quantum coherence phenomena.

The magnetic measurements were carried out by using a top loading He3-He4 dilution refrigerator (Oxford Kelvinox) which has incorporated a 5 T superconductor magnet. The sample is inside the liquid mixture and its temperature may be varied between 50 mK and 4 K. The temperature control is done by changing the power dissipation of a resistor and the temperature is measured by using a ruthenium-oxide resistor. Above 50 mK the thermometer has been calibrated against a superconducting fixed point device, a CMN thermometer and two calibrated germanium resistors. The magnetic moment of the sample is registered using a superconductor gradiometer by the extraction method. This gradiometer is coupled through out a superconducting transformer to a Quantum Design dc SQUID which is placed near the 1 K pot. The superconductor transformer has allowed us to enhance the signal to noise ratio without losing to much sensitivity. The temperature of the dc SQUID is keep constant as it is thermally linked to the 1 K pot. We have also shielded the dc SQUID from the magnetic field created by the superconductor magnet. The dc SQUID magnetometer has been calibrated by using pure paramagnetic samples and extrapolating the low temperature data in the mK regime to higher temperatures at which we have them scaled up with the data obtained by measuring the same samples by a Quantum Design rf SQUID.

The resonance experiments in the MHz range were performed using a split-ring resonator [11] while the high frequency experiments up to 110 GHz were carried out using the AB millimeter wave vector network analyzer (MVNA) [16]. In both cases, the dc applied field is parallel to the cavity axis and the sample is oriented such that its easy axis is mostly perpendicular to the magnetic field and the operation temperature was varied between 20 mK and 4 K.

The materials used in our experiments are; a pure Fe_8 single crystal of 2 mm length and a very diluted sample, 0.1% in weight, of Fe_2O_3 particles of 5 nm mean diameter dispersed in a polymer non magnetic matrix.

The parameters entering in the spin Hamiltonian of both samples may be obtained from EPR and magnetic measurements. The value of the barrier height, $U = DS^2$, may be estimated from the blocking temperature, $25T_b = DS^2$, and the anisotropy field, H_{an}, defined as the field where the demagnetisation curve branches from the initial magnetisation curve, is $H_{an} = 2DS$. That is, by performing zero field (ZFC) and field cooled (FC) magnetisation measurements and isothermal magnetisation vs field measurements at very low temperature, it is possible to estimate both the spin and the anisotropy D term of the spin Hamiltonian. In Table 1 we give the values of T_b, H_{an}, D and S for the two materials. For the case of Fe_8 the EPR data also suggest the existence of a second anisotropy term in the spin Hamiltonian ES_x^2 with $E = 0.092$ K.

We have performed relaxation experiments and isothermal magnetisation measurements for the two samples below the blocking temperature in order to verify the nature of the spin dynamic. The $M(H_{\parallel})$ data of Fe_8 show periodic steps with a period of 0.24 T corresponding to the enhanced relaxation via thermally assisted resonant tunneling between matching spin levels. In both cases the relaxation was studied after switching off the magnetic field from saturation. The relaxation law for the Fe_8 sample follows very well an stretch exponential while in the case of the Fe_2O_3 particles the relaxation follows the logarithmic law. The most important result is that in both cases we get the regime, below 0.4 for Fe_8 and below 1 K for the particles, in which the relaxation does not depend on temperature, suggesting that at enough low temperature the spin tunneling process occurs between the two lowest-lying states $S_z = \pm S$.

Another important aspect which may play an important role in the quantum decoherence phenomena is related to the magnetic interaction between the spin of the particles/molecules. We have performed low field magnetisation measurements until very low temperature. The field cooled $M(T)$ data, at $H = 70$ Oe, for the Fe_2O_3 increases when decreasing temperature until the lowest temperature reached in our experiments, $T = 50$ mK, in agreement with the fact below 1 K the tunneling occurs from the ground state and the particles behave, therefore, superparamagnetically. The situation in Fe_8 clusters is very different, the magnetisation increases slowly with temperature, presents a maximum around 200 mK and decreases for lower temperatures. The appearance of the cusp in the magnetisation disappear for applied magnetic fields bigger than 400 Oe. The lifetime of the superposition of the quantum spin oscillations should be affected, therefore, by these facts, as they may contribute to the dissipation causing decoherence.

Let us discuss now the results of the resonant experiments on the Fe8 sample; we will show data for both the single crystal above mentioned and for an oriented powder sample of Fe_8 microcrystals with their easy axes perpendicular to the applied field. In Fig. 3 we show the dependence of the absorption resonant peaks on the applied magnetic field ($H_\perp$) for both the single crystal and the oriented powder. The position of all the observed resonances for both single crystal and oriented powder has been fitted by the eigenstates of the spin Hamiltonian resonances experimentally $\mathcal{H} = -DS_z^2 + ES_x^2 - \mathbf{H} \cdot \mathbf{S}$.

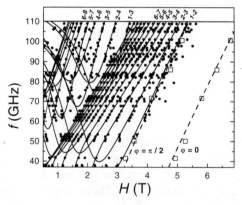

Figure 3 Absorption resonant peaks for oriented powder (open squares) and single crystal (solid circles). The lines represent the theoretical calculations from the diagonalization of the Hamiltonian refeq:ham for $\varphi = \pi/2$ in the case of single crystal (solid lines), and for $\varphi = 0$ and $\varphi = \pi/2$ for the oriented powder (dashed lines).

The quantum splitting, Δ, governing the tunneling probability of all Sz levels, depend on both the magnitude of the applied field and its angle, ϕ, with the hard axis [8], see Fig. 4. The evolution of the spin levels in the two wells as a function of the intensity of the transverse field, as deduced from.the diagonalisation of the spin Hamiltonian, is shown in Fig. reffig:levels; the levels on the left side of line 1 correspond to the quantum splittings of the different Sz levels due to tunneling and the levels on the right of line 2 are the spin levels resulting from the strong mixing of the Sz levels because the magnetic field is larger than the anisotropy field. Between these two lines, there is an intermediate region in which for a given field value there are both type of levels.

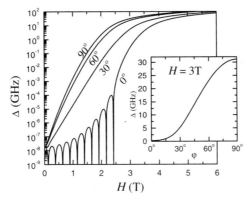

Figure 4 Dependence of the ground-state splitting on the transverse field in Fe8 for different angles φ. The inset shows the angular dependence of the splitting at a fixed value of the field $H = 3$ T.

The intensity of the resonance in the absence of dissipation and for a given value of the applied magnetic field is proportional to

$$\int_0^\pi g(\phi)\delta(\omega - \Delta[\phi, H])\mathrm{d}\phi , \qquad (1.2)$$

where $g(\phi)$ is the distribution of Fe8 crystallites over the angle ϕ. As we are mostly interested in the computation of the position of the maximum of the resonance as a function of H and ϕ and no preferred orientation on the angle ϕ is expected we may rewrite Eq. 1.2 as

$$\chi'' \propto \int_0^\infty \delta\left(\omega - \frac{\Delta}{\hbar}\right)\left(\frac{\mathrm{d}\Delta(\phi, H)}{\mathrm{d}\phi}\right)^{-1}\mathrm{d}\Delta = \left(\frac{\mathrm{d}\Delta}{\mathrm{d}\phi}\right)^{-1}\Bigg|_{\Delta=\hbar\omega} . \qquad (1.3)$$

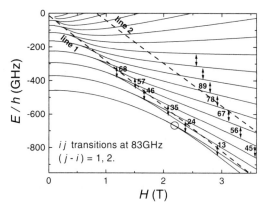

Figure 5 Evolution of the spin levels in the two wells on the intensity of the transverse field for $\varphi = \frac{\pi}{2}$. Lines 1 and 2 limit the transition zone between the quantum splittings of the $\hat{I}_z$ levels (low frequency) and the strongly mixed $\hat{I}_z$ states (high frequency). Open circle at 2.25 T in the lowest level indicates the absorption peak corresponding to a frequency of 680 MHz extracted from ref. [11].

It is clear, therefore, from both Eq. 1.3 and the dependence of the quantum splitting on the angle ϕ showed in Fig. 4 that two maxima in the resonance at the angles $\phi = 0$ and $\phi = \pi/2$ are expected. These two maxima of resonance absorption correspond, consequently, to two different field values, H_1 and H_2, which verify $\Delta(\pi/2, H_1) = \hbar\omega$ and $\Delta(0, H_2) = \hbar\omega$ respectively where $\omega = 2\pi f$ and f is the frequency of the ac field. These two resonances are expected in the case of the oriented powder because although the magnetic field is applied perpendicular to the easy axis, its orientation in the hard plane is at random. In the case of single crystals, as all molecules have the same orientation it is only expected one resonance. For both samples we have considered the occurrence of all possible transitions between spin levels i and j such that $j - i = 1$ and $j - i = 2$. The results of our fitting procedure are shown in Fig. 3, solid lines correspond to the single crystal and the dashed lines correspond to the oriented powder.

The two peaks at 680 MHz (40 GHz) observed at $H_1 = 2.25 \pm 0.05$ T and $H_2 = 3.60 \pm 0.05$ T for the oriented powder correspond to the quantum splitting of the ground state $Sz = \pm 10$ states for the two field orientations above mentioned. The classical two states for which we observe the coherence are two symmetric S states at some angle with the applied dc field. The quantum counterparts of these classical states are written as a superposition of the eigenstates of S_z

$$|g\rangle = \sum_{m=-10}^{m=10} A_m |m\rangle , \quad |e\rangle = \sum_{m=-10}^{m=10} B_m |m\rangle , \qquad (1.4)$$

where $A_m = A_{-m}$ and $B_m = B_{-m}$. Fig. 6 shows $|A_m|^2$ and $|B_m|^2$ for the first resonance $H = 2.25$ T, as a function of S_z.

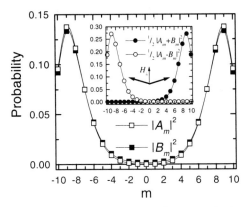

Figure 6 Probability distribution over m in the ground state and the first excited state, $|A_m|^2$ and $|B_m|^2$, respectively. The inset shows a double-degenerate classical ground state.

We have also been able to fit all resonances observed in the single crystal for 40 different frequencies, see Fig. 3. As it is shown in that figure, all the transition correspond to either $i \rightarrow j$ transitions with $j-i = 1$ or $j-i = 2$. The transitions $ij =12, 34, 56, 78$, etc observed for field values lower than the barrier height off each level, correspond to the quantum splitting of different S_z levels. The resonances corresponding to $j - i = 2$ are transitions between non consecutive S_z levels.

To conclude we have observed coherent quantum oscillations for different orientations of the $S = 10$ of Fe$_8$ molecular clusters. The decoherence time estimated from our resonant experiments is of about 10^{-7} s suggesting that these materials may be tested for quantum computing hardware.

Acknowledgments

The authors would like to acknowledge support by the European Commission through the FEDER project 2FD1997-1264(MAT).

References

[1] J. R. Friedman, M. P. Sarachik, J. Tejada, and R. Ziolo, Phys. Rev. Lett. **76**, 3830 (1996).

[2] J. M. Hernandez, X. X. Zhang, F. Luis, J. Bartolomé, J. Tejada, and R. Ziolo, Europhys. Lett. **35**, 301 (1996).

[3] L. Thomas, F. Lionti, R. Ballou, D. Gatteshi, R. Sessoli, and B. Barbara, Nature **383**, 145 (1996).

[4] J. Tejada, X. X. Zhang, E. del Barco, J. M. Hernandez, and E. M. Chudnovsky, Phys. Rev. Lett. **79**, 1754 (1997).

[5] E. M. Chudnovsky and D. A. Garanin, Phys. Rev. Lett. **56**, 11102 (1997).

[6] J. Villain, F. Hartmann-Boutron, R.Sessoli, and A. Rettori, Europhys. Lett. **27**, 159 (1994).

[7] J. F. Fernández, F. Luis, and J. Bartolomé, Phys. Rev. Lett. **80**, 5659 (1998).

[8] A. Garg., Europhys. Lett. **22**, 205 (1993).

[9] W. Wernsdorfer and R. Sessoli, Science **284**, 133 (1999).

[10] E. del Barco, J. M. Hernandez, M. Sales, J. Tejada, H. Rakoto, J. M. Broto, and E. M. Chudnovsky, Phys. Rev. B **60**, 11898 (1999).

[11] E. del Barco, N. Vernier, J. M. Hernandez, J. Tejada, E. M. Chudnovsky, E. Molins, and G. Bellessa, Europhys. Lett. **47**, 722 (1999).

[12] E. M. Chudnovsky and D. A. Garanin, Phys. Rev. Lett. **79**, 4469 (1997); D. A. Garanin, X. Martínez-Hidalgo, and E. M. Chudnovksy, Phys. Rev. B **57**, 13639 (1998).

[13] A. D. Kent, Y. Zhong, L. Bokacheva, D. Ruiz, D. N. Hendrickson, and M. P. Sarachik, Europhys. Lett. **49**, 521 (2000).

[14] D. A. Garanin, E. M. Chudnovsky, and R. Schilling, Phys. Rev. B **61**, 12204 (2000).

[15] W. Wernsdorfer, A. Caneschi, R. Sessoli, D. Gatteschi, A. Cornia, V. Villar, and C. Paulsen, Phys. Rev. Lett. **84**, 2965.

[16] E. del Barco, J. M. Hernandez, J. Tejada, N. Biskup, R. Achey, I. Rutel, N. Dalal, and J. Brooks, to appear in Phys. Rev. B (2000).

QUANTUM GATES AND NETWORKS WITH CAVITY QED SYSTEMS

D. Vitali, V. Giovannetti, and P. Tombesi

Dipartimento di Matematica e Fisica, Università di Camerino,

INFM, Unità di Camerino, via Madonna delle Carceri 62032, Camerino, Italy

vitali@campus.unicam.it

Abstract   We show how to implement quantum gates and networks using high-Q cavities in which the qubits are represented by single cavity modes restricted in the space spanned by the two lowest Fock states. The operations can be efficiently implemented using atoms sequentially crossing the cavities.

1.   INTRODUCTION

In the last years, numerous physical systems have been proposed as possible candidates for the implementation of a quantum computer. The desirable conditions which have to be satisfied are a reliable and easy way to prepare and detect the quantum states of the qubits, the possibility to engineer highly entangled states, the scalability to large number of qubits and a very low decoherence rate [1]. Up to now, experimental implementations have involved linear ion traps [2], liquid state NMR [3], and cavity QED systems [4, 5].

Here, elaborating on the suggestions of Ref. [6], we propose to use the Fock's states $|0\rangle$ and $|1\rangle$ of a high-Q cavity mode as the two logical states of a qubit. A quantum register of N qubits is therefore a collection of N identical cavities in which the state of an appropriately chosen cavity mode is within the space spanned by the vacuum and the one photon state. The register transformations are achieved sending off-resonant two-level atoms through the cavities and making them mutually interacting by means of suitable classical fields. With this respect, the present proposal is similar to that of Refs. [7, 8]; the important difference is that, in these papers, the logical qubits are represented by two circular Rydberg levels of the atoms. In our proposal, the role of atoms and cavity modes are exchanged. In this way, the present scheme becomes scalable in principle. In practice, its scalability can be limited by the spontaneous emission from the Rydberg levels or by other technical limitations, but the present proposal has the advantage that the needed technology is essentially already available to realize some proof-of-principle demonstrations of quantum computation with few qubits. In fact, in the present paper we shall specialize to the case of microwave cavities, for which a high level of quantum state control and engineering has been already experimentally shown [5, 9]. It is clear however that, in principle, the method can be applied to optical cavities too, in which one can have a miniaturization of the scheme and therefore a faster operation.

The system.   The interaction of a two-level atom quasi-resonant with a high-Q cavity mode is well described by the time dependent Hamiltonian [10]:

$$\mathcal{H}(t) = \hbar\omega b^\dagger b + \frac{\hbar\omega_{eg}}{2}[|e\rangle\langle e| - |g\rangle\langle g|] + \hbar\Omega(t)[|e\rangle\langle g|b + |g\rangle\langle e|b^\dagger], \quad (1.1)$$

in which b and ω are the annihilation operator and the angular frequency of the cavity mode respectively; $|e\rangle$ and $|g\rangle$ are the excited and lower circular Rydberg state and $\hbar\omega_{eg}$ is their energy difference. Finally $\Omega(t)$ is the atom-field interaction Rabi frequency which is time dependent because of the atomic motion through the cavity. In particular for a Fabry-Pérot-type cavity, with Gaussian transverse beam profile, we have $\Omega(t) = \Omega_0 e^{-(\frac{t}{\tau})^2}$, where 2τ is the atomic transit time, which depends of course on the inverse of the atomic velocity. For $|t| \gg \tau$, i.e. when the atom is outside the cavity, the energy eigenvectors of the system are $|g\rangle \otimes |n\rangle \equiv |g, n\rangle$ and $|e, n\rangle$, with $|n\rangle$ the generic Fock state of the cavity mode. Apart for the ground state $|g, 0\rangle$ which remains unchanged, in the presence of the time-dependent interaction, these terms are coupled by photon emission or absorption, and the istantaneous energy eigenstates at fixed time t are the dressed states $|\mathcal{V}_\pm^{(n)}(t)\rangle \propto (\delta/2 \pm \sqrt{(\delta/2)^2 + \Omega^2(t)(n+1)})|e, n\rangle + \Omega(t)\sqrt{n+1}|g, n+1\rangle$, with eigenvalues $E_\pm^{(n)}(t) = \hbar\omega(n+1) \pm \hbar\left((\delta/2)^2 + \Omega^2(t)(n+1)\right)^{1/2}$, where $\delta = \omega_{eg} - \omega$ is the atom-cavity detuning. Now, if the atom velocity is slow enough, and the system for $t \ll -\tau$ is prepared in a generic energy eigenstate, then in its time evolution it will adiabatically follow this eigenstate, with negligible transitions toward other states [11]. In the following we shall always work in this adiabatic regime.

2.   THE C-NOT GATE

Domokos *et al.* have shown in Ref. [8] that, using induced transitions between the dressed states, it is possible to implement a C-NOT gate in which a cavity containing at most one photon is the control qubit and the atom is the target qubit. This idea is the starting point for the implementation of the C-NOT between two cavities we propose here. In Ref. [8], when the atom enters the cavity, a classical field S of frequency ω_S equal to the energy difference between the dressed states $|\mathcal{V}_+^{(1)}(t = 0)\rangle$ (originating from $|e, 1\rangle$) and $|\mathcal{V}_-^{(0)}(t = 0)\rangle$ (originating from $|g, 1\rangle$) is switched on for a time interval $2\tau_S$, so that the following driving Hamiltonian is added to $\mathcal{H}(t)$ of Eq. (1.1)

$$\mathcal{H}_S(t) = -\hbar \Xi_0 \cos(\omega_S t + \varphi_S) e^{-(\frac{t}{\tau_S})^2} [|e\rangle\langle g| + |g\rangle\langle e|], \qquad (1.2)$$

where φ_S is the phase of the classical field S and Ξ_0 is the coupling costant. Appropriately choosing the value of τ_S, it is now possible to selectively couple S with these dressed states, leaving the other components of the vector state essentially unperturbed. Moreover, with a suitable choice of the intensity S, it is possible to apply a Rabi π pulse between the two states. In this way, when the atom exits the cavity, the resulting state has the $|e, 1\rangle$ and $|g, 1\rangle$ components exchanged. In this way one has realized a C-NOT gate in which, when the cavity (the control qubit) has one photon, the atom undergoes a NOT operation, while when the cavity contains no photons, the atomic state remains unchanged. We shall refer to this gate as the C-NOT(cavity $\rightarrow$ atom).

In a similar manner, we can build also a C-NOT gate in which the roles of atom and cavity are exchanged. Let us assume in fact to tune the frequency ω_S to the transition between the dressed state $|\mathcal{V}_-^{(0)}(t = 0)\rangle$ and the state $|g, 0\rangle$, and apply again a π pulse inside the cavity as before. Now, when the atom leaves the cavity, the $|g, 0\rangle$ and $|g, 1\rangle$ components in the state of the system are mutually exchanged. The $|e, 0\rangle$ and $|e, 1\rangle$ components are instead not affected by the interaction with the classical source S. This means having realized a C-NOT gate in which, when the atom is in the ground state, the cavity states $|0\rangle$ and $|1\rangle$ flip, while nothing happens to the cavity state for the atom in the excited state. In analogy with the previous case, we refer to this new gate as C-NOT(atom $\rightarrow$ cavity).

We now have all the elements to realize the C-NOT gate between two distinct but identical cavities, A and B, with the first one acting as the control qubit and the second one as the target qubit (see Fig. 1). Suppose that the initial states of the two cavities are respectively $|\phi\rangle_A = \alpha_A|0\rangle_A + \beta_A|1\rangle_A$ and $|\psi\rangle_B = \alpha_B|0\rangle_B + \beta_B|1\rangle_B$. *i)* A first atom, a_1, prepared in the ground state $|g\rangle$, enters cavity A, where it undergoes the C-NOT(cavity $\rightarrow$ atom) transformation realized with the classical field source S_A, and described above. *ii)* Then a_1 leaves A and enters cavity B: here the classical field S_B is switched on in order to obtain a C-NOT(atom $\rightarrow$ cavity) transformation. In the interaction picture and neglecting all the parasitic but controllable phase terms, the state of the total system at this stage is then:

$$\alpha_A|0\rangle_A \otimes \overline{|\psi\rangle}_B \otimes |g\rangle + \beta_A|1\rangle_A \otimes |\psi\rangle_B \otimes |e\rangle, \tag{1.3}$$

where $\overline{|\psi\rangle}_B$ is the NOT-conjugate vector of $|\psi\rangle_B$, that is, $\beta_B|0\rangle_B + \alpha_B|1\rangle_B$.   *iii)* The atom enters again A, where it undergoes the C-NOT(cavity $\rightarrow$ atom) transformation, so that the state of the system becomes

$$\left\{\alpha_A|0\rangle_A \otimes \overline{|\psi\rangle}_B + \beta_A|1\rangle_A \otimes |\psi\rangle_B\right\} \otimes |g\rangle, \tag{1.4}$$

which just describes the desired gate operation.

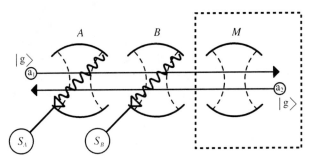

Figure 1   Schematical description of the C-NOT gate in which cavity A is the control qubit and cavity B the target qubit. M is the auxiliary cavity, transferring the entanglement with the cavities from the first atom (a_1) to the second atom (a_2).

The practical realization of step *iii)*, i.e. the return of the atom in the first cavity, is not trivial. The inversion of the motion of atom a_1, could be realized in principle with an atomic fountain configuration. However this implies having free-fall velocities, which are too slow in order to have the necessary interaction times within the cavities. For this reason we propose to transfer the quantum information from this atom onto a second one of the same type, but travelling in the opposite direction. With this respect, the scheme adopts the "quantum memory" scheme experimentally verified in Ref. [12]. This quantum information transfer is implemented introducing a third cavity, the auxiliary cavity M of Fig. 1, which, differently from A and B, is resonant with the $e \rightarrow g$ transition. If M is prepared in the vacuum state $|0\rangle_M$, and the transit time of a_1 is appropriately chosen, then the atomic state component $|e\rangle$ releases one photon in M through a resonant π Rabi oscillation. After that, the state of the total system will be

$$\alpha_A|0\rangle_A \otimes \overline{|\psi\rangle}_B \otimes |g\rangle \otimes |0\rangle_M + \beta_A|1\rangle_A \otimes |\psi\rangle_B \otimes |g\rangle \otimes |1\rangle_M. \tag{1.5}$$

The entanglement of a_1 with A and B is now transferred to the auxiliary cavity M; atom a_1 is factorized and it can be neglected from now on. At this stage, a second atom a_2 is prepared in the ground state $|g\rangle$ and injected into the apparatus with the same absolute value of velocity of a_1, but with the opposite direction. Entering M, it absorbes the photon left by the first atom through a similar π Rabi oscillation, and the entanglement with the cavities A and B is transferred from M to a_2:

$$\alpha_A |0\rangle_A \otimes \overline{|\psi\rangle}_B \otimes |g\rangle \otimes |0\rangle_M + \beta_A |1\rangle_A \otimes |\psi\rangle_B \otimes |e\rangle \otimes |0\rangle_M. \qquad (1.6)$$

At this stage also the state of the cavity M is factorized and the state (1.6) is equivalent to that of Eq. (1.3). In practice, one has realized an *atomic mirror*. Notice that a_2 has to cross cavity B without interacting with it, before reaching cavity A. This result is achieved simply by switching off the classical source S_B; the adiabatic regime prevents that, apart for adjustable dynamical phase factors, the state changes during the transit within B.

Using standard parameter values [8, 10], the whole gate operation takes place in about 1 msec, which has to be compared with the typical decoherence timescales, that is, the atomic radiative lifetimes and the cavity relaxation times. For circular Rydberg atoms with $n \simeq 50$, the atomic radiative lifetime is of the order of 30 msec, and therefore it does not represent a serious problem. The cavity damping times currently realized for microwaves have instead the same order of magnitude. However relaxation times of the order of 10 msec will be hopefully achieved in the near future, and in this case, one would have a perfectly working C-NOT gate between two cavities. It is clear therefore that, for the present implementation of quantum information processing, the main source of decoherence in the microwave domain is just the cavity leakage. If optical cavities are instead considered, also atomic spontaneous emission may represent an important source of decoherence.

One qubit operations. One qubit operations are straightforward to implement on qubits represented by two internal atomic states because it amounts to apply suitable Rabi pulses. This task is less trivial for bosonic degrees of freedom as our cavity modes, because the two lowest Fock states are coupled to the more excited ones. The most practical solution is to implement one-qubit operations on the two lowest Fock states sending again atoms through the cavity. To be more specific, one has to send an atom prepared in the ground state $|g\rangle$ through the cavity, with the classical field S tuned at the frequency corresponding to the transition between the states $|\mathcal{V}_-^{(0)}(t=0)\rangle$ and $|g,0\rangle$. If one sets the time duration and the intensity of the classical source S as in the case of the C-NOT(cavity $\rightarrow$ atom), i.e. such to realize a π pulse between the selected levels, one implements a "not-phase" gate, $|0\rangle \rightarrow e^{i\theta}|1\rangle$ $|1\rangle \rightarrow e^{-i\theta}|0\rangle$, where θ depends linearly on the phase φ_S of the classical field of Eq. (1.2) and it is therefore easily controllable. If otherwise the atom inside the cavity undergoes a $\pi/2$ instead of a π pulse, one realizes the "Hadamard-phase" gate $|0\rangle \rightarrow \left(|0\rangle + e^{i\theta'}|1\rangle\right)/\sqrt{2}$, $|1\rangle \rightarrow \left(e^{-i\theta'}|0\rangle - |1\rangle\right)/\sqrt{2}$, where also θ' depends linearly on the classical field phase φ_S. These two gates can be used to build the more general one-qubit operation and therefore, together with the C-NOT gate, form an universal set of gates.

3.   QUANTUM PHASE GATE

It is possible to realize another universal two-qubit gate, the quantum phase gate (QPG) [4], between the two cavities, slightly elaborating on the quantum phase gate operating on qubits carried by the Rydberg atom and the two lowest Fock states of a cavity mode, recently demonstrated experimentally [5]. Of course, since both the C-NOT and the QPG are universal quantum gates, it is always possible to implement one of them, by simply supplementing the other one with appropriate one-qubit operations. However, the experimental result of Ref. [5] suggests an alternative physical implementation of quantum logic operations between cavity qubits, which does not involve induced transitions between dressed states, and extends the scheme of [5] to a directly scalable model.

The QPG transformation reads $|a, b\rangle \rightarrow \exp\left(i\phi\delta_{a,1}\delta_{b,1}\right)|a, b\rangle$, where $|a\rangle$ and $|b\rangle$ describe the basis states $|0\rangle$ and $|1\rangle$ of two generic qubits. This means that the QPG leaves the initial state unchanged unless when both qubits are in state $|1\rangle$. The QPG of Ref. [5] did not involve levels g and e, but i and g, where i is a lower circular Rydberg level, which is uncoupled with the high-Q cavity. In this way, the QPG gate in the case $\phi = \pi$ can be realized by setting the atomic transition $g \rightarrow e$ perfectly at resonance with the relevant cavity mode (by appropriately Stark-shifting the atomic levels inside the cavity), and by selecting the atomic velocity so that the atom undergoes a complete 2π Rabi pulse when crossing the cavity. In fact, at resonance, such a pulse transforms the state $|g, 1\rangle$ into $e^{i\pi}|g, 1\rangle$, while nothing happens if the atoms is in i or the cavity is in the vacuum state.

We now show that this atom-cavity QPG can be used to realize a QPG between two cavity modes, by considering an arrangement very similar to that of the C-NOT gate shown in Fig. 1. The cavities A and B are again the two qubits, while M is again the auxiliary cavity needed to "reflect" the atom and disentangle it. The two classical sources inside the cavities S_A and S_B are no more needed, while we consider the possibility to apply Stark shift electric fields inside the cavities, in order to tune the $g \rightarrow e$ transition in and out of resonance from the cavity mode. The scheme of the QPG implementation is shown in Fig. 2 and involves only two atom crossings, as in the C-NOT gate of the preceding section, and three $\pi/2$ pulses between the i and g levels (the Hadamard gates H of Fig. 2), which can be realized with resonant classical microwave sources applied between the high-Q cavities.

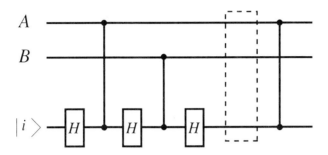

Figure 2 Logical scheme of the quantum phase gate (QPG) between two cavities. The vertical lines denote the QPG, while H denotes the Hadamard gates (i.e., $\pi/2$ pulses) applied on the atom. The dashed box denotes the "atomic mirror" (see Sec. II).

Let us assume a generic state of the two cavity qubits

$$|\psi\rangle = a_0|00\rangle + a_1|01\rangle + a_2|10\rangle + a_3|11\rangle \qquad (1.7)$$

and that a first atom, initially prepared in state i, is subject to a $\pi/2$ pulse, so to enter cavity A in state $(|i\rangle + |g\rangle)/\sqrt{2}$. The cavity mode is perfectly resonant with the $g \rightarrow e$ transition and the atom velocity is selected so that the atom undergoes a 2π Rabi pulse if it is in state g and the cavity contains one photon (the QPG of Ref. [5]). The resulting state at the exit of cavity A is

$$\frac{|i\rangle \otimes |\psi\rangle}{\sqrt{2}} + \frac{|g\rangle}{\sqrt{2}} \otimes (a_0|00\rangle + a_1|01\rangle - a_2|10\rangle - a_3|11\rangle). \qquad (1.8)$$

Then the atom undergoes another resonant $\pi/2$ pulse on the $i \rightarrow g$ transition and the state of the system becomes

$$|i\rangle \otimes (a_0|00\rangle + a_1|01\rangle) + |g\rangle \otimes (a_2|10\rangle + a_3|11\rangle). \qquad (1.9)$$

Then the atom crosses cavity B, where it is subjected again to the atom-cavity QPG as in cavity A, so that the state of the system becomes

$$|i\rangle \otimes (a_0|00\rangle + a_1|01\rangle) + |g\rangle \otimes (a_2|10\rangle - a_3|11\rangle). \qquad (1.10)$$

At this point, as in the C-NOT case of the preceding section, in order to realize the final transformation and disentangle the atom from the cavities, one has to "reflect" it. This is again achieved with the "atomic mirror" scheme of Sec. II (see Eqs. (1.5) and (1.6)). The second atom is then subjected to a $\pi/2$ pulse before entering the cavities and the state of Eq. (1.10) becomes

$$\frac{|i\rangle}{\sqrt{2}} \otimes (a_0|00\rangle + a_1|01\rangle + a_2|10\rangle - a_3|11\rangle) \qquad (1.11)$$

$$+ \frac{|g\rangle}{\sqrt{2}} \otimes (a_0|00\rangle + a_1|01\rangle - a_2|10\rangle + a_3|11\rangle).$$

The last step is the QPG between the atom and cavity A, i.e., the atom has to cross cavity B undisturbed (this is achieved by strongly detuning the $g \to e$ transition with a Stark shift field) and then has to undergo another full Rabi cycle in cavity A. The final state is

$$\frac{|i\rangle + |g\rangle}{\sqrt{2}} \otimes (a_0|00\rangle + a_1|01\rangle + a_2|10\rangle - a_3|11\rangle), \qquad (1.12)$$

which is desired result, corresponding to a QPG between cavities A and B with conditional phase shift $\phi = \pi$, and with a disentangled atom.

4.   THE TOFFOLI GATE

In principle, the most general quantum operation involving n qubits can be realized in terms of the one and two-qubit operations described above. This decomposition however implies a degree of network complexity, and a number of resources and steps which is rapidly increasing with the number of qubits n. One of the main advantages of the present proposal is that it is particularly suited for the efficient implementation of many-qubit quantum gates, which, in many cases, can be realized with the same number of steps of the two-qubit C-NOT gate of Sec. II.

A particularly clear example of the possibilities of the proposed scheme is provided by the Toffoli gate

$$|x\rangle_A|y\rangle_B|z\rangle_C \to |x\rangle_A|y\rangle_B \left|\left[z + (x \wedge y)\right]_{mod2}\right\rangle_C, \qquad (1.13)$$

in which the target qubit C is controlled by the first two, A and B. The effect of the Toffoli gate on a generic three qubit state is to exchange the components $|110\rangle$ and $|111\rangle$. The implementation of this gate needs the same arrangement of aligned cavities crossed by Rydberg atoms used for the C-NOT gate of Fig. 1, except that now one has *three* cavities (with the corresponding classical sources S_A, S_B and S_C) instead of two. The auxiliary cavity M is again needed for the atomic mirror scheme used to disentangle the atom. The atom is initially prepared in state $|g\rangle$ and when it is in the first cavity A, is subject to a π pulse between the dressed state $|\mathcal{V}_+^{(0)}(0)\rangle$ and $|g, 0\rangle$. This pulse creates atom-cavity entanglement and the generic state of the total system becomes

$$
\Big[\alpha_1|000\rangle + \alpha_2|001\rangle + \alpha_3|010\rangle + \alpha_4|011\rangle\Big] \otimes |e\rangle
$$
$$
+ \Big[\alpha_5|100\rangle + \alpha_6|101\rangle + \alpha_7|110\rangle + \alpha_8|111\rangle\Big] \otimes |g\rangle. \tag{1.14}
$$

Then, when the atom reaches the second cavity B, it undergoes another π pulse, at the new frequency ω_2 corresponding to the transition between $|g, 0\rangle$ and $|\mathcal{V}_+^{(2)}(0)\rangle$, so that the transformation

$$
|0\rangle_B \otimes |g\rangle \rightarrow |2\rangle_B \otimes |e\rangle, \tag{1.15}
$$

is realized. This means temporarily leaving the logical subspace, even though this allows us to realize a significant simplification of the scheme. The state after this second step is therefore

$$
\Big[\alpha_1|000\rangle + \alpha_2|001\rangle + \alpha_3|010\rangle + \alpha_4|011\rangle + \alpha_5|120\rangle + \alpha_6|121\rangle\Big] \otimes |e\rangle
$$
$$
+ \Big[\alpha_7|110\rangle + \alpha_8|111\rangle\Big] \otimes |g\rangle. \tag{1.16}
$$

When the atom enters in C, the classical field S_C is applied so to realize the C-NOT(atom $\rightarrow$ cavity C) of Sec. II and the state of Eq. (1.16) becomes

$$
\Big[\alpha_1|000\rangle + \alpha_2|001\rangle + \alpha_3|010\rangle + \alpha_4|011\rangle + \alpha_5|120\rangle + \alpha_6|121\rangle\Big] \otimes |e\rangle
$$
$$
+ \Big[\alpha_7|111\rangle + \alpha_8|110\rangle\Big] \otimes |g\rangle. \tag{1.17}
$$

At this point one has to disentangle the atom from the three cavities and also to adjust the state components in which the cavity C contains two photons. Both problems can be solved using again the auxiliary cavity M and a second atom crossing the apparatus in the opposite direction as in the "atomic mirror" configuration of Sec. II. The cavity M transfers the entanglement with the cavities from the first to the second atom, which is not subject to any classical pulse in C. Then the second atom enters B, where it undergoes a π pulse at the frequency ω_2, which simply inverts the transformation of Eq. (1.15) (thanks to the fact that no $|0\rangle_B \otimes |g\rangle$ term is present), correcting in this way the terms of Eq. (1.17) in which the second cavity contains two photons. As a consequence, the state of the system becomes

$$
\Big[\alpha_1|000\rangle + \alpha_2|001\rangle + \alpha_3|010\rangle + \alpha_4|011\rangle\Big] \otimes |e\rangle
$$

$$+\left[\alpha_5|100\rangle + \alpha_6|101\rangle + \alpha_7|111\rangle + \alpha_8|110\rangle\right] \otimes |g\rangle. \qquad (1.18)$$

Finally the atom enters A, where it is subjected to a π pulse resonant with the transition $|\mathcal{V}_+^{(0)}(0)\rangle \rightarrow |g,0\rangle$, exchanging $|0\rangle_A \otimes |g\rangle$ with $|0\rangle_A \otimes |e\rangle$, so that the second atom is disentangled from the cavities and one gets the desired generic output of a Toffoli gate, i.e.,

$$[\alpha_1|000\rangle + \alpha_2|001\rangle + \alpha_3|010\rangle + \alpha_4|011\rangle \qquad (1.19)$$
$$+\alpha_5|100\rangle + \alpha_6|101\rangle + \alpha_7|111\rangle + \alpha_8|110\rangle] \otimes |g\rangle.$$

Notice that in this way we have implemented the Toffoli gate with two atoms only, as in the C-NOT gate of Sec. II. Moreover this scheme can be easily extended to the case of $n \geq 4$ cavity qubits, for the implementation of the n-qubit generalization of the Toffoli gate. We need only two atoms crossing the aligned cavities in opposite directions also in this more general case. The pulse sequence is similar to that discussed above: both atoms undergo a π pulse resonant with the transition $|\mathcal{V}_+^{(0)}(0)\rangle \rightarrow |g,0\rangle$ in the first cavity, while in the following $n-2$ cavities they are submitted to a π pulse at the frequency ω_2. In the last cavity, the target qubit, the first atom experiences a C-NOT(atom $\rightarrow$ cavity) while the second atom crosses it undisturbed.

References

[1] C. H. Bennett and D. P. DiVincenzo, Nature (London) **404**, 247 (2000).

[2] C. Monroe *et al.*, Phys. Rev. Lett. **75**, 4714 (1995).

[3] I. L. Chuang *et al.* Nature (London) **393**, 143 (1998); J.A. Jones *et al.* Nature (London) **393**, 344 (1998).

[4] Q.A. Turchette *et al.*, Phys. Rev. Lett. **75**, 4710 (1995).

[5] A. Rauschenbeutel *et al.*, Phys. Rev. Lett. **83**, 5166 (1999).

[6] A. Barenco *et al.*, Phys. Rev. Lett. **74**, 4083 (1995).

[7] T. Sleator and H. Weinfurter, Phys. Rev. Lett. **74**, 4087 (1995).

[8] P. Domokos *et al.*, Phys. Rev. A **52**, 3554 (1995).

[9] G. Nogues, *et al.*, Nature (London) **400**, 239 (1999).

[10] S. Haroche and J.M. Raimond, in *Cavity Quantum Electrodynamics*, edited by P. Berman (Academic Press, New York, 1994), p. 123.

[11] A. Messiah, *Quantum Mechanics* (North Holland, Amsterdam, 1962).

[12] X. Maitre *et al.*, Phys. Rev. Lett. **79**, 769 (1997).

DUAL JOSEPHSON PHENOMENA: INTERACTION OF VORTICES WITH NON-CLASSICAL MICROWAVES

A. Vourdas

Department of Electrical Engineering and Electronics,
The University of Liverpool, Brownlow Hill,
Liverpool L69 3BX, United Kingdom
ee21@liv.ac.uk

A. Konstadopoulou

Department of Electrical Engineering and Electronics,
The University of Liverpool, Brownlow Hill,
Liverpool L69 3BX, United Kingdom
A.Konstadopoulou@liv.ac.uk

J. M. Hollingworth

Department of Electrical Engineering and Electronics,
The University of Liverpool, Brownlow Hill,
Liverpool L69 3BX, United Kingdom
J.M.Hollingworth@liv.ac.uk

Keywords: vortices, Josephson arrays, sinusoidal non-linearity, non-classical microwaves

Abstract   A ring made from a Josephson array in the insulating phase and a weak link (eg a constriction) is considered. The center of the ring is fed by a current $I(t)$ and contains a time-dependent charge $Q(t)$. Vortices circulate the ring and tunnel through the weak link. At frequencies where $\hbar\omega >> K_B T$ the quantum nature of the electromagnetic field is taken into account. A two-mode Hamiltonian is studied which describes quantum mechanically both the ring and the electromagnetic field. The sinusoidal non-linearity in this Hamiltonian leads to interesting quantum phenomena like entanglement between the ring and the electromagnetic field, squeezing of the elctromagnetic field, etc. The use of such nonlinear devices as amplifiers and frequency converters of non-classical microwaves is discussed.

1.   INTRODUCTION

Josephson junction arrays in the insulating phase, conduct vortices and insulate electric charges. In this phase, vortices behave as quantum objects with low mass and move with a high mobility; whilst electric charges are confined. This is dual (Girvin, 1996 and Widom et al., 1982) to the superconducting phase where electric charge pairs move with a high mobility and vortices are confined. Interference of vortices in a ring made from a Josephson array in the insulating phase, has been studied in (van Wees, 1990, Orlando and Delin, 1991, Elion et al., 1993). The center of the ring contains a static charge Q, and the interference demonstrates the Aharonov-Casher effect (Aharonov and Casher, 1984, Reznik and Aharonov, 1989, Goldhaber, 1989).

Vourdas, 1999a has studied a similar ring with a weak link (e.g., a constriction) which obstructs the movement of vortices and plays the role of a dual Josephson junction for vortices. The link couples weakly the two insulators where vortices move with a high mobility and provides an obstruction to vortices which go through it. A current feeds the center of the ring with a time-dependent charge. At low frequencies this charge can be generated by an ac source. For higher frequencies which is our main interest here, the ac source is a 'suitable' waveguide carrying microwaves. For example, we can put the device in a cylindrical waveguide in the TM_{01} mode, with the plane of the device perpendicular to the axis z of the waveguide. Since the wavelength of the microwaves is much greater than the length of the vortices, we can ignore the z-dependence of the various quantities and treat the problem as two-dimensional.

In this paper we study the interaction of this device with non-classical electromagnetic fields. Electromagnetic fields from a 'good' source (e.g. a maser) and at low temperatures($\hbar\Omega_2 \gg k_B T$) can be prepared in a particular quantum state. In section 2 we present the Hamiltonian and its quantization. In section 3 we use this Hamiltonian to calculate the time evolution of the system. The emphasis is on quantum phenomena such as entanglement. We conclude in section 4 with a discussion of our results.

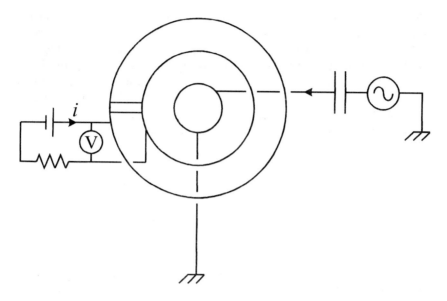

Figure 1 A ring from a Josephson array in the insulating phase with a weak link. A weak magnetic field perpendicular to the plane of the diagram creates magnetic vortices, which are forced by the current i to circulate around the ring.

2. THE HAMILTONIAN AND ITS QUANTIZATION

The Hamiltonian describing this system is:

$$H = \frac{1}{2}L(I - I_{mw})^2 + E_{dJ}(1 - \cos\delta) + \frac{(Q - Q_{mw})^2}{2C} . \qquad (1.1)$$

L and C are the inductance and capacitance of the system, correspondingly. I and I_{mw} are the total and external (microwave) current correspondingly, flowing in the radial direction of the device. Q and Q_{mw} are the total and external (microwave) charge correspondingly, in the inner boundary of the ring. E_{dJ} is the dual Josephson coupling constant and $\delta = \Phi_0 Q$ the dual phase. We use units where $\hbar = k_B = c = 1$. The Hamiltonian leads to the equations of motion:

$$L\partial_t I = -\Phi_0 E_{dJ} \sin\delta - \frac{Q - Q_{mw}}{C} \qquad (1.2)$$

$$\frac{1}{\Phi_0}\partial_t\delta = I - I_{mw} \qquad (1.3)$$

which are the 'dual Josephson equations'. Equ. (2) relates the various voltages in the radial direction of the device. $L\partial_t I = \partial_t \Phi$ is the rate with which the vortices cross the weak link and is proportional to the Faraday law voltage V between the inner and outer boundaries. $\Phi_0 E_{dJ} \sin \delta$ is the dual Josephson voltage and $\frac{Q - Q_{mw}}{C}$ is the voltage due to the capacitance between the inner and outer boundaries. Equ. (3) expresses the fact that the time derivative of the charge is equal to $I - I_{mw}$.

Quantization of the device is done with the creation and annihilation operators

$$a_1 = \left(\frac{1}{2\Omega_1 C'}\right)^{1/2} [Q + i\Omega_1^{-1} I], \tag{1.4}$$

$$a_1^\dagger = \left(\frac{1}{2\Omega_1 C'}\right)^{1/2} [Q - i\Omega_1^{-1} I], \tag{1.5}$$

$$[a_1, a_1^\dagger] = 1 \tag{1.6}$$

where $C' = (C^{-1} + \Phi_0^2 E_{dJ})^{-1}$ and $\Omega_1 = (LC')^{-1/2}$ is the frequency of the device. The electromagnetic field is quantized with the operators:

$$a_2 = \left(\frac{1}{2\Omega_2 C}\right)^{1/2} [Q_{mw} + i\Omega_2^{-1} I_{mw}], \tag{1.7}$$

$$a_2^\dagger = \left(\frac{1}{2\Omega_2 C}\right)^{1/2} [Q_{mw} - i\Omega_2^{-1} I_{mw}], \tag{1.8}$$

$$[a_2, a_2^\dagger] = 1 \tag{1.9}$$

where $\Omega_2 = (LC)^{-1/2}$ is the frequency of the microwaves. We note that $\Omega_1^2 = \Omega_2^2 + \frac{\Phi_0 E_{dJ}}{L}$.

The Hamiltonian can now be written as

$$
\begin{aligned}
H = {} & \Omega_1 a_1^\dagger a_1 + \Omega_2 a_2^\dagger a_2 - \left\{ \lambda \cos[\mu(a_1^\dagger + a_1)] + \lambda \mu^2 a_1^\dagger a_1 \right\} \\
& + \nu a_1^2 + \nu^* a_1^{\dagger 2} + \epsilon_{ampl}(a_1^\dagger a_2^\dagger + a_1 a_2) \\
& + \epsilon_{conv}(a_1^\dagger a_2 + a_1 a_2^\dagger)
\end{aligned} \tag{1.10}
$$

where $\lambda = E_{dJ}$, $\mu = \Phi_0(\Omega_1 C'/2)^{1/2}$, $\nu = -(E_{dJ}\Phi_0^2\Omega_1 C')/4$, $\epsilon_{ampl} = (1/2)(\Omega_1\Omega_2)^{1/2}(1 - \Omega_2/\Omega_1)$, $\epsilon_{conv} = (-1/2)(\Omega_1\Omega_2)^{1/2}(1 + \Omega_2/\Omega_1)$. Note that there is a $-\lambda\mu^2 a_1^\dagger a_1$ term inside the cosine term which is cancelled by the $+\lambda\mu^2 a_1^\dagger a_1$ term. Additional constant terms have been ignored.

3.   TIME EVOLUTION AND QUANTUM ENTANGLEMENT

Assuming that at time $t = 0$ the system is described by a density matrix $\rho(0)$, and that there is no dissipation, we have calculated numerically the density matrix $\rho(t)$

$$\rho(t) = \exp[iHt]\rho(0)\exp[-iHt] \tag{1.11}$$

The infinite dimensional matrix $\langle M_1, M_1 | H | N_1, N_2 \rangle$ has been truncated in the numerical calculations, with M_1, N_1 taking values from 0 up to K_{1max} and M_2, N_2 taking values from 0 up to K_{2max}. K_{1max} and K_{2max} were taken to be much greater than $\langle N_1 \rangle$ and $\langle N_2 \rangle$, correspondingly. As a measure of the accuracy of the approximation we calculated the traces of the truncated matrices. In the limit $K_{1max} \to \infty$ and $K_{2max} \to \infty$ they equal to 1; and in the truncated case they should be very close to 1. In all our results the above sum was greater than 0.98.

We have also calculated the reduced density matrices

$$\rho_1(t) = Tr_2\rho(t); \quad \rho_2(t) = Tr_1\rho(t) \tag{1.12}$$

and used them to calculate the

$$\langle N_i \rangle = Tr[a^\dagger a \rho_i] \tag{1.13}$$

as functions of time. We have also calculated the second order correlations

$$g_{ii}^{(2)} = \frac{\langle N_i^2 \rangle - \langle N_i \rangle}{\langle N_i \rangle^2}; \qquad i = 1, 2 \tag{1.14}$$

$$g_{12}^{(2)} = \frac{\langle N_1 N_2 \rangle}{\langle N_1 \rangle \langle N_2 \rangle} \tag{1.15}$$

and the ratio

$$r = \frac{[g_{12}^{(2)}]^2}{g_{11}^{(2)} g_{22}^{(2)}} \tag{1.16}$$

As the system evolves in time we get entanglement between the device and the electromagnetic field. In order to quantify this we calculate the entanglement entropy (Lindbland, 1973, Wehrl, 1978, Barnett and Phoenix, 1991, Vourdas, 1992):

$$I = S(\rho_1) + S(\rho_2) - S(\rho) \tag{1.17}$$

where $S(\rho) = -Tr(\rho \ln \rho)$ is the von Neumann entropy. We use natural logarithms and express the results in natural units (nats).

We consider two examples, both with $L = 8.3 \times 10^4$, $C = 1.2 \times 10^3$ and $E_{dJ} = 10^{-4}$ (in units where $\hbar = c = k_B = 1$). In this case $\Omega_1 = 1.5 \times 10^{-4}$, $\Omega_2 = 10^{-4}$, $\lambda = 10^{-4}$, $\nu = -2.161 \times 10^{-4}$, $\epsilon_{ampl} = 2.0412 \times 10^{-5}$, $\epsilon_{conv} = -1.0206 \times 10^{-4}$, $\mu = 2.079$. In the first example we consider the case that at $t = 0$ the device is in the number state $|N = 1\rangle$ $(a_1^\dagger a_1 |N\rangle = N|N\rangle)$ and the microwaves are in the coherent state $|A = 1.5\rangle$ $(a_2 |A\rangle = A|A\rangle)$:

$$\rho(t = 0) = |N = 1\rangle\langle N = 1| \otimes |A = 1.5\rangle\langle A = 1.5| \qquad (1.18)$$

In Fig. 2 we plot in the first graph $\langle N_1\rangle$ (solid line) and $\langle N_2\rangle$ (broken line) as functions of time, in the second graph $g_{11}^{(2)}$ (solid line), $g_{22}^{(2)}$ (broken line) and r (line of dots), and in the third graph the entanglement entropy I in natural units (nats) as a function of time.

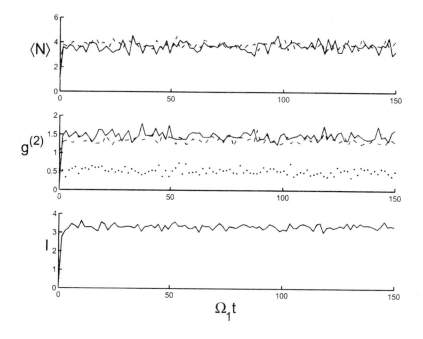

Figure 2 The system at $t = 0$ is in the state of Eq. (1.18). The first graph shows $\langle N_1\rangle$ (solid line) and $\langle N_2\rangle$ (broken line) as functions of time. The second graph shows $g_{11}^{(2)}$ (solid line), $g_{22}^{(2)}$ (broken line) and r (line of dots) as functions of time. The third graph shows the entanglement entropy I (in nats) as a function of time.

In the second example we consider the case that at $t = 0$ the device is in the thermal state

$$\rho_{th}(\beta) = (1 - e^{-\beta\Omega_1}) \sum_{N=0}^{\infty} e^{-\beta N\Omega_1} |N\rangle\langle N| \qquad (1.19)$$

with $\beta\Omega_1 = 1$, and the microwaves are in the coherent state $|A = 1.5\rangle$:

$$\rho(t = 0) = \rho_{th}(\beta) \otimes |A = 1.5\rangle\langle A = 1.5| \qquad (1.20)$$

In Fig. 3 we plot $\langle N_1 \rangle$ (solid line) and $\langle N_2 \rangle$ (broken line) as functions of time; $g_{11}^{(2)}$ (solid line), $g_{22}^{(2)}$ (broken line) and r (line of dots) as functions of time; and the entanglement entropy I in natural units (nats) as a function of time.

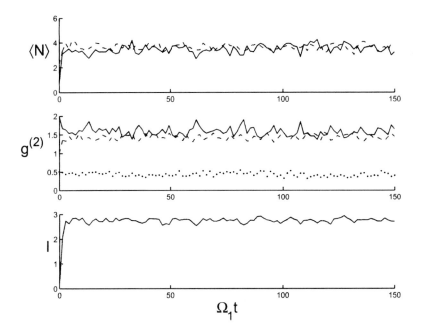

Figure 3 The system at time $t = 0$ is in the state of Eq. (1.20) with $\beta\Omega_1 = 1$. The first graph shows $\langle N_1 \rangle$ (solid line) and $\langle N_2 \rangle$ (broken line) as functions of time. The second graph shows $g_{11}^{(2)}$ (solid line), $g_{22}^{(2)}$ (broken line) and r (line of dots) as functions of time. The third graph shows the entanglement entropy I (in nats) as a function of time.

4.   DISCUSSION

There have been many advances in the last few years in the area of mesoscopic Josephson junctions (for reviews see Schon and Zaikin, 1987, Kastner, 1992 and Spiller et al., 1992). In this paper we have studied the interaction of vortices in Josephson arrays in the insulating phase, with non-classical microwaves (see also Vourdas, 1999b). More specifically, we have considered a ring made from a Josephson array in the insulating phase with a weak link. The interaction of the ring with non-classical microwaves has been modeled with the Hamiltonian of Eq. (1.1). Using this Hamiltonian we have calculated the evolution of the system for the case of no dissipation. In figs. 2, 3 we have presented various quantities that characterize the system as a function of time. The next step is to include dissipation in the calculations (Leggett et al., 1987) using various techniques. Work is currently in progress in this direction. These non-linear devices can be used as amplifiers and frequency converters of non-classical electromagnetic fields at microwave and THz frequencies (Vourdas, 1996).

Acknowledgments

We would like to thank Professor T.D. Clark (Sussex) and Dr. J. Ralph (Liverpool) for many helpful discussions. AV gratefully acknowledges financial support from the Royal Academy of Engineering. JMH and AK gratefully acknowledge financial support from the EPSRC.

References

Aharonov, Y. and Casher, A. (1984). *Phys. Rev. Lett.*, 53:319.

Barnett, S. M. and Phoenix, S. J. D. (1991). *Phys. Rev. A*, 44:535.

Elion, W. J., Wachters, J. J., Sohn, L. L., and Mooij, J. E. (1993). *Phys. Rev. Lett.*, 71:2311.

Girvin, S. M. (1996). *Science*, 274:524.

Goldhaber, A. S. (1989). *Phys. Rev. Lett.*, 62:482.

Kastner, M. (1992). *Rev. Mod. Phys.*, 64:849.

Leggett, A. J. et al. (1987). *Rev. Mod. Phys.*, 59:1.

Lindbland, G. (1973). *Commun. Math. Phys.*, 33:305.

Orlando, T. P. and Delin, K. A. (1991). *Phys. Rev. B*, 43:8717.

Reznik, B. and Aharonov, Y. (1989). *Phys. Rev. D*, 40:4178.

Schon, G. and Zaikin, A. D. (1987). *Phys. Rep.*, 148:1.

Spiller, T. P. et al. (1992). *Prog. Low Temp. Phys.*, 13:219.

van Wees, B. J. (1990). *Phys. Rev. Lett.*, 65:255.

Vourdas, A. (1992). *Phys. Rev. A*, 46:442.

Vourdas, A. (1996). *J. Mod. Optics*, 43:2105.

Vourdas, A. (1999a). *Europhys. Lett.*, 48:201.

Vourdas, A. (1999b). *Phys. Rev. B*, 60:620.

Wehrl, A. (1978). *Rev. Mod. Phys.*, 50:221.

Widom, A. et al. (1982). *J. Phys. A*, 15:3877.

QUANTUM COMPUTATION: THEORY, PRACTICE, AND FUTURE PROSPECTS

Isaac L. Chuang
IBM Almaden Research Center
650 Harry Road
San Jose, CA 95120

Abstract   We introduce quantum computation, summarize its theoretical promise and current experimental progress, and describe future prospects for the field.

1.   INTRODUCTION

Quantum computation and quantum information are new fields which offer new ways to understand and control physical systems at the level of quantum phenomena. These fields are fundamentally relevant to to the study of macroscopic quantum coherence (MQC), and *vice versa*: one of the most important obstacles in the construction of quantum information processing systems is decoherence, which has long been a topic of interest in the MQC community.

In this article, we review the basic constructions of quantum computation, including the ideas of quantum logic gates and quantum algorithms. We then briefly summarize the present experimental art, and future prospects for the field.

2.   FUNDAMENTAL CONCEPTS

We begin by describing the states which we shall consider, then a set of operations which act on these states is presented.

Quantum states

The basic unit of classical information is the *bit*, which is physically represented by a two-state classical system. Corresponding to this, the basic unit of quantum information is the *qubit*[1], which is physically represented by a two-state quantum system, such as two energy levels of an atom, or the spin of a nucleus.

A qubit $|\psi\rangle$ may be written as $|\psi\rangle = a|0\rangle + b|1\rangle$, where a and b are complex numbers satisfying the normalization condition $|a|^2 + |b|^2 = 1$. Because this state has only two degrees of freedom (the overall phase is unobservable), we can also write a qubit state as

$$|\psi\rangle = \cos\frac{\theta}{2}|0\rangle + e^{i\phi}\sin\frac{\theta}{2}|1\rangle\,, \qquad (1.1)$$

where θ and ϕ represent a point on a sphere. This *Bloch sphere* representation is quite useful in providing intuition about single qubit states and operations. Similarly, a two qubit state may be written as an arbitrary superposition of the four basis states $|00\rangle$, $|01\rangle$, $|10\rangle$, and $|11\rangle$,

$$|\psi\rangle = c_{00}|00\rangle + c_{01}|01\rangle + c_{10}|10\rangle + c_{11}|11\rangle\,. \qquad (1.2)$$

An n qubit state has $O(4^n)$ degrees of freedom; this exponential descriptive complexity is at the origin of Feynman's suggestion that quantum physics might be able to solve certain problems faster than is possible with classical systems[2].

Quantum operations

One of the most important operations which can be performed to a quantum state is *measurement*, and this operation immediately shows the difficulty of utilizing the complexity of quantum states. When a qubit is measured, we obtain a classical bit; if the qubit is initially $|\psi\rangle = a|0\rangle + b|1\rangle$, then the measurement result is 0 with probability $|a|^2$, and 1 with probability $|b|^2$. Furthermore, after the measurement, the qubit is either $|0\rangle$ or $|1\rangle$, in correspondence with the measurement result. Thus, determining k bits of $|a|^2$ requires $O(2^k)$ measurements of identical copies of $|\psi\rangle$. Similarly, accessing the $O(2^n)$ parameters in an n qubit state also requires $O(2^n)$ measurements.

The collapse performed by measurements apparently renders the exponential complexity of quantum states useless, because that complexity is not efficiently accessible. However, there is much more to the story, because there are other operations which can be performed to quantum states, to manipulate the information stored in the probability amplitudes *before* a measurement is done. These operations are the key to quantum computation.

The set of classical operations which can be performed to bits include logic gates such as NOT, AND, and XOR. These operations are described using truth tables, which specify the output for a given input state. Similarly, operations on qubits are described by matrices, which describe how basis states transform. These matrices are *unitary*, corresponding to solutions of the time-dependent Schrödinger equation. Four important single qubit operations are

$$I = \begin{bmatrix} 1 & 0 \\ 0 & 1 \end{bmatrix} \tag{1.3}$$

$$X = \begin{bmatrix} 0 & 1 \\ 1 & 0 \end{bmatrix} \tag{1.4}$$

$$Y = \begin{bmatrix} 0 & -i \\ i & 0 \end{bmatrix} \tag{1.5}$$

$$Z = \begin{bmatrix} 1 & 0 \\ 0 & -1 \end{bmatrix}. \tag{1.6}$$

These act by matrix multiplication on two-component vectors, where

$$|0\rangle = \begin{bmatrix} 1 \\ 0 \end{bmatrix} \quad \text{and} \quad |1\rangle = \begin{bmatrix} 0 \\ 1 \end{bmatrix}. \tag{1.7}$$

I is the identity operation; it leaves the qubit unchanged. X is the equivalent of a NOT operation, and is also called a bit-flip. Z is a phase flip, because it interchanges the two states

$$|+\rangle \equiv \frac{|0\rangle + |1\rangle}{\sqrt{2}} \quad \text{and} \quad |-\rangle \equiv \frac{|0\rangle - |1\rangle}{\sqrt{2}}. \tag{1.8}$$

and Y is a bit-phase flip, $ZX = iY$. Additional important single qubit operations include

$$H = \frac{1}{\sqrt{2}} \begin{bmatrix} 1 & 1 \\ 1 & -1 \end{bmatrix} \tag{1.9}$$

$$S = \begin{bmatrix} 1 & 0 \\ 0 & i \end{bmatrix} \tag{1.10}$$

$$T = \begin{bmatrix} 1 & 0 \\ 0 & e^{i\pi/8} \end{bmatrix}, \tag{1.11}$$

the Hadamard, Phase, and $\pi/8$ gates. Note $H|0\rangle = |+\rangle$ and $H|1\rangle = |-\rangle$.

Perhaps the single most useful two qubit gate is the controlled-NOT gate, which has the matrix representation

$$\text{CNOT} = \begin{bmatrix} 1 & 0 & 0 & 0 \\ 0 & 1 & 0 & 0 \\ 0 & 0 & 0 & 1 \\ 0 & 0 & 1 & 0 \end{bmatrix}, \tag{1.12}$$

where the basis states are

$$|00\rangle = \begin{bmatrix} 1 \\ 0 \\ 0 \\ 0 \end{bmatrix} \quad |01\rangle = \begin{bmatrix} 0 \\ 1 \\ 0 \\ 0 \end{bmatrix} \quad |10\rangle = \begin{bmatrix} 0 \\ 0 \\ 1 \\ 0 \end{bmatrix} \quad |11\rangle = \begin{bmatrix} 0 \\ 0 \\ 0 \\ 1 \end{bmatrix}. \tag{1.13}$$

These discrete sets of quantum operations are important because of the following theorem[3, 4]:

Theorem 1 (Solovay-Kitaev) *For any single qubit operation U, and any $\epsilon > 0$, a unitary operator V such that $\|U - V\| < \epsilon$ can be constructed using at most $O(\log^c(1/\epsilon))$ single qubit gates from the set $\{X, Y, Z, H, S, T\}$.*

The precise set that is used isn't important: the point is that the set is finite, and any U can be *efficiently* approximated. Furthermore, adding the controlled-NOT gate to the set allows operations on any number of qubits to be approximated efficiently. This includes the set of permutations, thus capturing the well-known result that any Boolean function can be implemented with a finite set of (classical) logic gates.

The Quantum Fourier Transform

Consider the classical discrete Fourier transform $\mathcal{F}$. This function reversible maps a vector of 2^N complex numbers x_n to $\tilde{x}_n$ as

$$x_n \overset{\mathcal{F}}{\to} \tilde{x}_n = \frac{1}{\sqrt{2^N}} \sum_{k=0}^{2^N-1} x_k e^{2\pi i kn/2^N}. \tag{1.14}$$

For example, $\mathcal{F}$ transforms the impulse $x_n = \delta_{n0}$ into an equally weighted superposition $\tilde{x}_n = 1/\sqrt{2^N}$. Translating the impulse causes the output to be modulated by a complex phase. Note that the input is specified by 2^{N+1} continuous parameters.

The Fourier transform can be performed using a simple "quantum circuit" composed of gates including some of those introduced above. Consider a register of N qubits, and let us denote the state using the basis vectors $|n\rangle$, where the bits of the decimal number n specify the state of each qubit. Recall that quantum registers may exist in a superposition of states, which we may write in general as $\sum_n x_n |n\rangle$. This input may be transformed by the quantum Fourier transform U_{FT} to give us

$$\sum_{n=0}^{2^N-1} x_n |n\rangle \overset{U_{FT}}{\to} \frac{1}{\sqrt{2^N}} \sum_{n=0}^{2^N-1} \left[\sum_{k=0}^{2^N-1} x_k e^{2\pi i kn/2^N} \right] |n\rangle. \tag{1.15}$$

Note that the bracketed expression on the right is simply $\tilde{x}_n$. Thus, we see that the state of the qubit register labels the index of the vector, and the complex numbers x_n are "stored" as the probability amplitudes of each state! Importantly, U_{FT} acting on N qubits may be constructed from $O(\text{poly}(N))$ local operations[5]; the quantum circuit is given in Figure 1.

This result seems to indicate that N qubits can be used to perform a Fourier transform of a vector of 2^N complex numbers. Unfortunately,

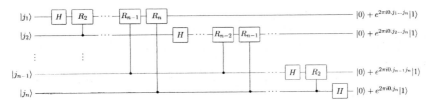

Figure 1 Efficient circuit for the quantum Fourier transform.

that is not exactly the case. The catch here, of course, is that if the output state is measured (assume we measure in the computational basis), it will collapse into any one of the basis states. Only by repeating the procedure an exponential number of times with the same input preparation, to accumulate statistics which reveal the probability distribution, would we obtain the transform result $\tilde{x}_n$.

This example speaks to the heart of the conundrum of devising a quantum algorithm. On one hand, we have a system which can explore a phase space of size 2^N using only N qubits, and within this space perform complex calculations But on the other hand, the results of such a calculation are not available to us if we go about it in such a straightforward manner. More cleverness is required in order to harness the power of quantum computation.

3. QUANTUM ALGORITHMS

Quantum computation is best known for two singular results, which show that certain mathematical problems (factoring and search) can be solved more efficiently with quantum resources than with classical resources. Here, we describe a simple prototypical quantum algorithm known as Deutsch's algorithm, and summarize the steps involved in the quantum factoring algorithm.

Deutsch's Algorithm

Consider a classical function $f(x)$ which acts on one bit and returns one bit, and suppose we have a black box U_f which performs the unitary transform $U_f|x,y\rangle = |x, y \oplus f(x)\rangle$, where $\oplus$ is addition modulo two. This box accepts x as the argument to the function, and stores the result in the second bit, originally with value y. Classically, there is no way to obtain the sum of the function evaluated on all possible inputs, $f(0) \oplus f(1)$, without evaluating the function twice. On the other hand, as has been shown[6, 7], this is possible quantum-mechanically with only one function evaluation.

The algorithm works shown in Figure 2. We prepare the initial state $|\psi_0\rangle = |01\rangle$. We then perform two Hadamard transforms to obtain $|\psi_1\rangle = |+-\rangle$. This state is then fed into U_f, which returns $|\psi_2\rangle = \pm|+-\rangle$ if $f(0) \oplus f(1) = 0$, and $|\psi_2\rangle = \pm|--\rangle$ otherwise. Finally, we perform one Hadamard to the first qubit, obtaining a final state in which the first qubit is $|0\rangle$ if and only if $f(0) \oplus f(1) = 0$.

Quantum Factoring

The above problem is artificially constructed to favor quantum computation. Upon what principle does it work, and are there more realistic problems which may be addressed? The current understanding is that quantum algorithms are good for efficiently solving only certain problems, which involve computing *global properties* of functions, such as the period. A key role is played by the quantum Fourier transform U_{FT} in such computations. In particular, Shor's factoring algorithm[8] may be summarized as follows:

Problem: Let N be the product of two unknown prime numbers, and choose $y \perp N$. What is the smallest r such that $y^r \bmod N = 1$?

Solution: Choose $q \approx N^2$, and do:

$$|0, 0\rangle \qquad\qquad \text{initialize state}$$

$$\rightarrow \sum_{0 \leq x \leq q} |x, 0\rangle \qquad\qquad \text{create superposition}$$

$$\rightarrow \sum_{0 \leq x \leq q} |x, y^x \bmod N\rangle \qquad\qquad \text{calculate modular exponentiation}$$

$$\rightarrow \sum_{0 \leq k < q/r} |l + kr, y^l \bmod N\rangle \qquad\qquad \text{measure second label}$$

$$\rightarrow \sum_{0 \leq c < r} e^{2\pi i l c / r} |cq/r, y^l \bmod N\rangle \qquad\qquad \text{perform } U_{FT}^\dagger; \text{ measure first label}$$

Since, q is known and c is an integer, then a continued fraction expansion can be used to determine r from the ratio cq/r with high probability. Standard results from number theory may then be employed to determine from r the factors of N with high probability.

---

4.   IMPLEMENTATION

An enormous range of concepts for experimental realization of quantum computers has been explored; here, we shall only touch on a few of the fundamental ideas and summarize the current state of affairs. There are four basic requirements, the abilities to: (1) robustly represent quantum information, (2) perform a universal family of unitary transforms, (3) prepare a fiducial initial state, and (4) measure the output result.

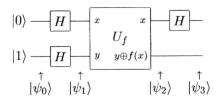

Figure 2 Quantum circuit implementing Deutsch's algorithm.

Nuclear spins and electronic spins make excellent choices as physical representations of qubits[9], because they typically have long coherence times compared with timescales for coupling and manipulation. For example, nuclear spins in molecules[10] can have $T_1 \approx T_2 \approx 1$ s, while single spin operations can be performed in about 10^{-6} s, and coupled operations in 10^{-3} s. Such a wide hierarchy of timescales is quite important in selecting a good qubit representation.

A universal set of quantum gates can be implemented by controlling many known Hamiltonians. For example, the Hamiltonian

$$\mathcal{H} = \sum_k \omega_k Z_k + \sum_{jk} J_{jk}(t) Z_j Z_k + \sum_k P_k^x(t) X_k + P_k^y(t) Y_k \qquad (1.16)$$

allows all of the gates described in the Introduction to be implemented by controlling the classical parameters $J(t)$ and $P(t)$ appropriately. It has been proposed that photons[11], trapped ions[12], spins in quantum dots[13], charged cooper pairs in superconducting boxes[14], and even impurities doped into silicon[15] controlled with surface gates can have such a Hamiltonian.

Of course, a significant challenge is to properly initialize the system into a well known (i.e. "fiducial") state, such as with all spins in their ground state, $|0\rangle$. This is particularly difficult for systems such as spins in molecules, where $\hbar\omega \ll k_B T$. Similarly, small quantum systems are often difficult to measure, or produce such small signals that averaging over large ensembles is required. The development of procedures for circumventing these difficulties has enabled small quantum computers to be realized with NMR[16, 17].

5. CONCLUSION

The ability to perform quantum computation implies the ability to create macroscopic quantum superpositions. Experimentally, quantum computation techniques have been used to entangle four trapped ions [18], and seven nuclear spins[19], and quantum algorithms such as Shor's seem to require entanglement in order to function. On the other hand,

if quantum computation fails to be experimentally feasible at some large
length scale (for reasons we do not understand today), we may thereby
uncover new principles of physics which prohibit the existence of macroscopic quantum coherence. Either way, these are two fields whose futures
are intimately related, and I hope the introduction provided here will be
useful in opening the doors of this field to you.

References

[1] B. W. Schumacher, Phys. Rev. A **54**, 2614 (1996).

[2] R. P. Feynman, Int. J. Theor. Phys. **21**, 467 (1982).

[3] A. Y. Kitaev, Russ. Math. Surv. **52**, 1191 (1997).

[4] M. Nielsen and I. Chuang, *Quantum Computation and Quantum Information* (Cambridge University Press, Cambridge, 2000).

[5] D. Coppersmith, IBM Research Report RC 19642 (1994).

[6] D. Deutsch and R. Jozsa, Proc. R. Soc. London A **439**, 553 (1992).

[7] R. Cleve, A. Ekert, C. Macchiavello, and M. Mosca, Proc. R. Soc. London A **454**, 339 (1998).

[8] P. W. Shor, SIAM J. Comp. **26**, 1484 (1997).

[9] D. P. DiVincenzo, Science **270**, 255 (1995), arXive e-print quant-ph/9503016.

[10] N. Gershenfeld and I. L. Chuang, Science **275**, 350 (1997).

[11] I. L. Chuang and Y. Yamamoto, Phys. Rev. Lett. **76**, 4281 (1996).

[12] J. I. Cirac and P. Zoller, Phys. Rev. Lett. **74**, 4091 (1995).

[13] D. P. DiVincenzo and D. Loss, Superlattices and Microstructures **23**, 419 (1998).

[14] Y. Nakamura, Y. A. Pashkin, and J. S. Tsai, Nature **398**, 786 (1999).

[15] B. Kane, Nature **393**, 133 (1998).

[16] D. G. Cory, A. F. Fahmy, and T. F. Havel, Proc. Nat. Acad. Sci. USA **94**, 1634 (1997).

[17] I. L. Chuang *et al.*, Nature **393**, 143 (1998).

[18] C. A. Sackett *et al.*, Nature **404**, 256 (2000).

[19] E. Knill, R. Laflamme, R. Martinez, and C. Tseng, Nature **404**, 368 (2000).

READING-OUT A QUANTUM STATE: AN ANALYSIS OF THE QUANTUM MEASUREMENT PROCESS

Yu. Makhlin

Institut für Theoretische Festkörperphysik, Universität Karlsruhe, D-76128 Karlsruhe

Landau Institute for Theoretical Physics, Kosygin st. 2, 117940 Moscow

makhlin@tfp.physik.uni-karlsruhe.de

G. Schön

Institut für Theoretische Festkörperphysik, Universität Karlsruhe, D-76128 Karlsruhe

Forschungszentrum Karlsruhe, Institut für Nanotechnologie, D-76021 Karlsruhe

schoen@tfp.physik.uni-karlsruhe.de

A. Shnirman

Institut für Theoretische Festkörperphysik, Universität Karlsruhe, D-76128 Karlsruhe

sasha@tfp.physik.uni-karlsruhe.de

Keywords: quantum measurement, quantum computation, dephasing

Abstract   We analyze the quantum measurement process in mesoscopic systems, using the example of a Cooper-pair box (an effective two-state quantum system) observed by a single-electron transistor. To study this process we investigate the time evolution of the density matrix of the coupled system of qubit and meter. This evolution is characterized by three time scales. On a fast dephasing time scale the meter destroys the phase coherence of the qubit. After a longer time the resolution becomes sufficient to deduce the information about the initial quantum state from the output signal, the current in the SET. On a third, mixing time scale the measurement-induced transitions between qubit's states destroy the information about their initial occupations. We study the statistics of current and demonstrate that these time scales appear in its noise spectrum.

Introduction

The quantum measurement is an essential ingredient of investigations of quantum coherent effects. In particular, it is needed to probe macroscopic quantum coherence or to read out the result of a quantum computation. In this paper we discuss the measurement of a quantum state of a mesoscopic two-state system (qubit). To study the measurement process we analyze the dynamics of the density matrix of the coupled system of a Cooper-pair box (Josephson charge qubit [1, 2]) and a SET [3]. Although the details of the derivation may differ, our analysis of the measurement process and the long-time dynamics of the qubit and meter can be also applied to other systems, for instance, to a Cooper-pair box measured by a superconducting SET [4], a double dot observed by a quantum point contact [5, 6, 7, 8], or a Josephson flux qubit coupled to a dc-SQUID-magnetometer [4, 9, 10].

One possibility to observe a quantum system is to couple to the system weakly and perform a *continuous* measurement [4]. Such weak measurement reveals typical time scales of the system's dynamics but not the information about its initial quantum state. To acquire this information a *strong* measurement is needed. The analysis demonstrates the mutual influence of detector and qubit in the course of the measurement.

A certain 'pointer' basis of the qubit is always associated with a quantum measurement. This basis, in which the measurement is performed, is the eigenbasis of the measured observable. Our analysis demonstrates how the pointer basis, $|0\rangle , |1\rangle$, emerges as a result of the interaction between the qubit and detector.

The analysis reveals three characteristic time scales. On the shortest, the dephasing time τ_φ, the detector destroys phase coherence between the states $|0\rangle$ and $|1\rangle$. At the same time the information about the qubit's state is transferred to the SET. After the second time scale, τ_{meas}, it can be read out by monitoring the tunneling current. In accordance with the laws of quantum mechanics the read-out gives one of two results, 0 or 1, with probabilities $|a|^2$ and $|b|^2$, determined by the initial quantum state $a |0\rangle + b |1\rangle$. Finally, the back-action of the detector onto the qubit destroys information about the initial quantum state. The detector-induced transitions between the pointer states mix these states and change their occupation probabilities on a time scale τ_{mix}.

We study the statistics of the output signal (the current). The characteristic time scales appear in the current noise spectrum. In particular, in the limit of strong measurement we find the telegraph-noise long-time behavior, with the jumps between the pointer states at a typical rate τ_{mix}^{-1}.

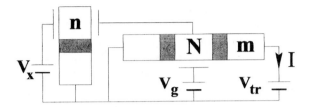

Figure 1 The circuit of a qubit and a SET used as a meter.

1. MEASUREMENT BY A SET

The system under consideration is shown in Fig. 1. The qubit is a superconducting single-charge box with Josephson junction in the Coulomb blockade regime. Its dynamics is limited to a two-dimensional Hilbert space spanned by two charge states, with $n = 0$ or 1 extra Cooper pair on a superconducting island. The island is coupled capacitively to the SET, influencing the tunneling current. During manipulations of the qubit [3] the SET is kept in the off-state ($V_{tr} = 0$), i.e. no dissipative currents causing decoherence are flowing. To perform the measurement, the transport voltage V_{tr} is switched to a sufficiently high value, so that the current starts to flow in the SET. As we will show, monitoring the current provides information about the qubit's state [11].

The Hamiltonian of the system is given by

$$\mathcal{H} = \mathcal{H}_{\text{SET}} + \mathcal{H}_{\psi} + \mathcal{H}_{\text{T}} + \mathcal{H}_{\text{qb}} + \mathcal{H}_{\text{int}} . \qquad (1.1)$$

The first three terms describe the single-electron transistor. Here $\mathcal{H}_{\text{SET}} = E_{\text{SET}}(N - N_{\text{g}})^2$ is its charging energy, quadratic in the charge eN on the middle island. The gate charge eN_{g} is defined by the gate voltage V_{g} and other voltages in the circuit. The term $\mathcal{H}_{\psi}$ describes the Fermions in the island and electrodes, while $\mathcal{H}_{\text{T}}$ governs the tunneling in the SET. The Hamiltonian of the qubit is given, in the eigenbasis of the charge $\hat{n}$, by $\mathcal{H}_{\text{qb}} = E_{\text{ch}}\hat{n} - \frac{1}{2}E_{\text{J}}\hat{t}$. Here $\hat{n} = \frac{1}{2}(1 - \hat{\sigma}_z) = \begin{pmatrix} 0 & 0 \\ 0 & 1 \end{pmatrix}$, while $\hat{t} = \hat{\sigma}_x$ is the tunneling term restricted to two lowest charge states. Finally, $\mathcal{H}_{\text{int}} = 2E_{\text{int}}N\hat{n}$ is the Coulomb coupling between the SET and the qubit. In the figure me denotes the charge which has tunneled through the SET. The charging energy scales E_{SET}, E_{ch}, E_{int} are determined by capacitances in the circuit, and E_{J} is the Josephson coupling.

The full density matrix can be reduced by tracing over microscopic degrees of freedom while keeping track only of the qubit's state, N and m. Moreover, a closed set of equations can be derived for $\rho_N^{ij}(m)$, the entries of the density matrix, which are diagonal in N and m [12] ($i, j = 0, 1$ refer to a qubit's basis). From this density matrix we obtain by further reduction the 2×2 density matrix of the qubit, $\hat{\varrho}(t) \equiv \sum_{N,m} \hat{\rho}_N(m,t)$, the charge distribution $P(m,t) \equiv \sum_N \text{tr}\,\hat{\rho}_N(m,t)$, as well as other statistical characteristics of the current in the SET.

At low temperatures and transport voltages only two charge states of the middle island of the SET, with $N = 0$ and $N + 1 = 1$ electrons, contribute to the dynamics. Expanding in the tunneling term to lowest order, we obtain after the Fourier transformation $\hat{\rho}_N(k) \equiv \sum_m e^{-ikm}\hat{\rho}_N(m)$ the following master equation [3, 11]:

$$\frac{d}{dt}\begin{pmatrix} \hat{\rho}_N \\ \hat{\rho}_{N+1} \end{pmatrix} + \frac{i}{\hbar}\begin{pmatrix} [\mathcal{H}_{\text{qb}}, \hat{\rho}_N] \\ [\mathcal{H}_{\text{qb}} + 2E_{\text{int}}\hat{n}, \hat{\rho}_{N+1}] \end{pmatrix} = \begin{pmatrix} -\check{\Gamma}_L & e^{ik}\check{\Gamma}_R \\ \check{\Gamma}_L & -\check{\Gamma}_R \end{pmatrix}\begin{pmatrix} \hat{\rho}_N \\ \hat{\rho}_{N+1} \end{pmatrix}$$

(1.2)

The operators $\check{\Gamma}_{L/R}$ are the tunneling rates in the left and right junctions. They are defined by

$$\check{\Gamma}_L\hat{\rho} \equiv \Gamma_L\hat{\rho} - \tfrac{1}{\hbar}\pi\alpha_L\left\{2E_{\text{int}}\hat{n}, \hat{\rho}\right\},$$ (1.3)

$$\check{\Gamma}_R\hat{\rho} \equiv \Gamma_R\hat{\rho} + \tfrac{1}{\hbar}\pi\alpha_R\left\{2E_{\text{int}}\hat{n}, \hat{\rho}\right\}.$$ (1.4)

Here $\alpha_{\text{L/R}} \equiv R_{\text{K}}/(8\pi^3 R_{\text{L/R}}^{\text{T}})$ is the tunnel conductance of the junctions in units of the resistance quantum $R_{\text{K}} = h/e^2$. The rates are fixed by the potentials μ_L and $\mu_R = \mu_L + V_{\text{tr}}$ of the leads: $\hbar\Gamma_L = 2\pi\alpha_L[\mu_L - (1 - 2N_{\text{g}})E_{\text{SET}}]$ and $\hbar\Gamma_R = 2\pi\alpha_R[(1 - 2N_{\text{g}})E_{\text{SET}} - \mu_R]$. They define the tunneling rate $\Gamma = \Gamma_L\Gamma_R/(\Gamma_L+\Gamma_R)$ through the SET. The anticommutators in Eqs. (1.3,1.4) make these rates (and hence the current) sensitive to the qubit's state, and thus allow the measurement.

2.   EVOLUTION IN THE POINTER BASIS

We find several regimes where the analysis simplifies because there exists a (pointer) qubit's basis in which one can treat off-diagonal elements perturbatively. In particular, under suitable conditions *dephasing* (decay of the off-diagonal entries of the qubit's density matrix in this basis) is much faster than *mixing* (relaxation of the diagonal to their stationary values), which is the prerequisite for a measurement process.

When the transport voltage is turned on, the charge N on the middle island of the SET fluctuates, randomly switching between N and $N+1$ at high rates Γ_L and Γ_R. The Hamiltonian of the qubit $\mathcal{H}_{\text{qb}} + 2E_{\text{int}}N\hat{n}$ follows this random dynamics. In the weak-coupling regime, $E_{\text{int}} \ll \hbar(\Gamma_L + \Gamma_R)$, the qubit's dynamics is described by the mean value of the Hamiltonian $\bar{\mathcal{H}}_{\text{qb}} \equiv \mathcal{H}_{\text{qb}} + 2\bar{N}E_{\text{int}}\hat{n}$ and the fluctuating part $2(N - \bar{N})E_{\text{int}}\hat{n}$, which destroys coherence. [The average charge $\bar{N} \equiv \Gamma_L/(\Gamma_L + \Gamma_R)$ fixes also the average energy $\bar{E}_{\text{ch}} \equiv E_{\text{ch}} + 2\bar{N}E_{\text{int}}$.] Comparing the bare dephasing rate due to these fluctuations, $\Gamma_\varphi^0 = 4\Gamma E_{\text{int}}^2/\hbar^2(\Gamma_L + \Gamma_R)^2$, with the level spacing $\Delta E \equiv (E_{\mathrm{J}}^2 + \bar{E}_{\mathrm{ch}}^2)^{1/2}$ of $\bar{\mathcal{H}}_{\mathrm{qb}}$, we find two different physical limits: In the Hamiltonian-dominated limit, $\Delta E \gg \hbar \Gamma_\varphi^0$, the measurement is performed in the eigenbasis of $\bar{\mathcal{H}}_{\mathrm{qb}}$, while in the fluctuation-dominated regime, $\hbar \Gamma_\varphi^0 \gg \Delta E$, it is performed in the charge basis. In both limits one can treat non-diagonal entries of $\bar{\mathcal{H}}_{\mathrm{qb}}$, $\mathcal{H}_{\mathrm{int}}$ perturbatively.

2.1. HAMILTONIAN-DOMINATED REGIME

In this regime $\Delta E \gg \hbar \Gamma_\varphi^0$, and the pointer basis coincides with the eigenbasis of $\bar{\mathcal{H}}_{\mathrm{qb}}$. In this basis $2 E_{\mathrm{int}} \hat{n} = E_{\mathrm{int}}^{\parallel}(1 - \hat{\sigma}_z) - E_{\mathrm{int}}^{\perp} \hat{\sigma}_x$, where $E_{\mathrm{int}}^{\parallel} \equiv E_{\mathrm{int}} \bar{E}_{\mathrm{ch}}/\Delta E$ and $E_{\mathrm{int}}^{\perp} \equiv E_{\mathrm{int}} E_{\mathrm{J}}/\Delta E$. In zeroth order, we analyze the dynamics without off-diagonal mixing terms, $E_{\mathrm{int}}^{\perp} = 0$. In this case the entries ρ_N^{ij} with different pairs of indices ij are decoupled.

For the diagonal modes the absence of mixing implies the conservation of occupations of the eigenstates $\varrho^{ii} = \rho^{ii}(k=0)$ [here $i = 0, 1$ and $\hat{\rho}(k) \equiv \sum_N \hat{\rho}_N(k)$]. The eigenvalues of two corresponding Goldstone modes,

$$\lambda^{ii}(k) \approx \mathrm{i}\,\Gamma^i k - \frac{1}{2} f^i \Gamma^i k^2 , \qquad k \ll 1 , \tag{1.5}$$

give the tunneling rates through the SET for two pointer states, $\Gamma^i \equiv \Gamma_L^i \Gamma_R^i/(\Gamma_L^i + \Gamma_R^i)$. Here the tunneling rates in the junctions are $\Gamma_L^{0/1} = \Gamma_L \pm 2\pi \alpha_L E_{\mathrm{int}}^{\parallel}/\hbar$ and $\Gamma_R^{0/1} = \Gamma_R \mp 2\pi \alpha_R E_{\mathrm{int}}^{\parallel}/\hbar$. The Fano factors $f^0 \approx f^1 \approx f \equiv 1 - 2\Gamma/(\Gamma_L + \Gamma_R)$ describe the reduction of the shot noise.

The analysis of the dynamics of ρ_N^{01} reveals the dephasing of the qubit by the measurement, with rate $\tau_\varphi^{-1} = 4\Gamma E_{\mathrm{int}}^{\parallel\,2}/\hbar^2(\Gamma_L + \Gamma_R)^2$.

Taking finite $E_{\mathrm{int}}^{\perp}$ into account modifies the picture and introduces mixing: In second order the long-time evolution of the occupations $\rho^{ii}(k)$ is given by a reduced master equation,

$$\frac{d}{dt} \begin{pmatrix} \rho^{00}(k) \\ \rho^{11}(k) \end{pmatrix} = M(k) \begin{pmatrix} \rho^{00}(k) \\ \rho^{11}(k) \end{pmatrix} , \tag{1.6}$$

$$M(k) = \begin{pmatrix} \lambda^{00}(k) & 0 \\ 0 & \lambda^{11}(k) \end{pmatrix} + \frac{1}{2\tau_{\mathrm{mix}}} \begin{pmatrix} -1 & 1 \\ 1 & -1 \end{pmatrix} . \tag{1.7}$$

For the mixing time, τ_{mix}, we obtain:

$$\tau_{\mathrm{mix}} = \frac{\Delta E^2 + \hbar^2(\Gamma_L + \Gamma_R)^2}{4\Gamma E_{\mathrm{int}}^{\perp\,2}} . \tag{1.8}$$

Besides, the second order correction to the dephasing rate is $(2\tau_{\mathrm{mix}})^{-1}$.

To describe the read-out we consider first short times $t \ll \tau_{\text{mix}}$ and neglect the last term in Eq. (1.7). Then, for the qubit initially in a superposition $a|0\rangle + b|1\rangle$ of eigenstates of $\tilde{\mathcal{H}}_{\text{qb}}$, the distribution $P(m, t)$ develops two peaks at $m = \Gamma^0 t$ and $m = \Gamma^1 t$. The peaks' weights $|a|^2$ and $|b|^2$ are determined by the initial qubit's state. Their widths are growing as $\sqrt{2f^i\Gamma^i t}$, and the peaks separate after the time

$$\tau_{\text{meas}} = \left(\frac{\sqrt{2f^0\Gamma^0} + \sqrt{2f^1\Gamma^1}}{\Gamma^0 - \Gamma^1} \right)^2 . \tag{1.9}$$

At longer times $t > \tau_{\text{mix}}$ the mixing modifies this picture: the occupations relax to the equal-weight mixture: $\varrho^{00}(t) - \varrho^{11}(t) \propto \exp(-t/\tau_{\text{mix}})$, and the double-peak structure is smeared, as we discuss in the next section. Thus the two peaks appear only in the time interval between τ_{meas} and τ_{mix}. Therefore, a strong measurement requires $\tau_{\text{meas}} \ll \tau_{\text{mix}}$.

2.2.   FLUCTUATION-DOMINATED REGIME

In this regime $\hbar\Gamma_\varphi^0 \gg \Delta E$. The analysis is similar to that in the previous subsection. In this regime the pointer basis coincides with the basis of charge states. One can expand in E_J which is the only off-diagonal term in the charge basis. The dephasing rate is Γ_φ^0, while for the mixing we get: $\tau_{\text{mix}}^{-1} = E_J^2/\hbar^2\Gamma_\varphi^0$. A phenomenon, termed the Zeno or watchdog effect, can be seen [5, 6]: the stronger the dephasing, the weaker is the rate τ_{mix}^{-1} of jumps between the charge states.

The measurement time is given by the same expression (1.9) where now the tunneling rates through the SET, Γ^0, Γ^1, are defined by the tunneling rates in the junctions $\Gamma_L^{0/1} = \Gamma_L \pm 2\pi\alpha_L E_{\text{int}}/\hbar$ and $\Gamma_R^{0/1} = \Gamma_R \mp 2\pi\alpha_R E_{\text{int}}/\hbar$ (note the replacement of $E_{\text{int}}^{\|}$ by E_{int}).

3.   STATISTICS OF CURRENT

The statistical quantities studied depend on the initial density matrix, $P(m, t \,|\, \rho_0)$. In the two-mode approximation (1.6,1.7) this reduces to a dependence on $|a|^2 - |b|^2$. We solve Eq. (1.6) to obtain the distribution $P(m, t \,|\, \rho_0) = \text{tr}_{\text{qb}}[U(m, t)\rho_0]$. Here $U(m, t)$ is the inverse Fourier transform of the evolution operator $U(k, t) \equiv \exp\left[M(k)\, t\right]$ and tr_{qb} denotes tracing over qubit's states. If the tunneling rates Γ^0, Γ^1 in two pointer states are close, the resulting distribution is

$$P(m, t \,|\, \rho_0) = \sum_{\delta m} \tilde{P}(m - \delta m, t \,|\, \rho_0) \frac{e^{-\delta m^2/2f\bar{\Gamma}t}}{\sqrt{2\pi f\bar{\Gamma}t}} . \tag{1.10}$$

Here $\bar{\Gamma} = (\Gamma^0 + \Gamma^1)/2$ and $\delta\Gamma = \Gamma^0 - \Gamma^1$.

The first term in the convolution (1.10) contains two delta-peaks, corresponding to two qubit's pointer states:

$$\tilde{P}(m, t \,|\, \rho_0) = P_{\mathrm{pl}}\left(\frac{m - \bar{\Gamma}t}{\delta\Gamma t/2}, \frac{t}{2\tau_{\mathrm{mix}}} \,\Big|\, \rho_0\right)$$
$$+ e^{-t/2\tau_{\mathrm{mix}}}\left[|a|^2\delta\left(m - \Gamma^0 t\right) + |b|^2\delta\left(m - \Gamma^1 t\right)\right]. \quad (1.11)$$

On the time scale τ_{mix} the peaks' weights vanish; instead a plateau arises between the peaks. It is described by

$$P_{\mathrm{pl}}(x, \tau \,|\, \rho_0) = e^{-\tau}\frac{1}{2\,\delta\Gamma\,\tau_{\mathrm{mix}}}\left\{I_0\left(\tau\sqrt{1 - x^2}\right)\right.$$
$$\left. + \left[1 + x(|a|^2 - |b|^2)\right] I_1\left(\tau\sqrt{1 - x^2}\right)/\sqrt{1 - x^2}\right\}, \quad (1.12)$$

at $|x| < 1$ and $P_{\mathrm{pl}} = 0$ for $|x| > 1$. Here I_0, I_1 are the modified Bessel functions. At longer times the plateau transforms into a narrow peak centered around $m = \bar{\Gamma}t$. This peak does not contain any information about the initial state of the qubit. The Gaussian in Eq. (1.10) arises due to shot noise. Its effect is to smear out the distribution (see Fig. 2).

Similarly, one can analyze the distribution of possible values of the tunneling current in the SET. Since instantaneous values of the current fluctuate strongly, we study the current averaged over a finite time interval Δt, i.e. $\bar{I} \equiv \int_t^{t+\Delta t} I(t')dt'$. The analysis shows that the probability $p(\bar{I}, \Delta t, t)$ to measure the current $\bar{I}$ at the time t can be expressed in terms of the charge distribution (1.10) for different initial conditions:

$$p\left(\bar{I}, \Delta t, t \,|\, |a|^2 - |b|^2\right) = P\left(m = \bar{I}\Delta t, \Delta t \,|\, e^{-t/\tau_{\mathrm{mix}}}\left[|a|^2 - |b|^2\right]\right). \quad (1.13)$$

A strong quantum measurement is achieved if $\tau_{\mathrm{meas}} < \Delta t < \tau_{\mathrm{mix}}$ (Fig. 2). In this case the current, measured at $t < \tau_{\mathrm{mix}}$, is close to Γ^0 or Γ^1, with probabilities $|a|^2$ and $|b|^2$, respectively. At longer t a typical current pattern is a telegraph signal jumping between Γ^0 and Γ^1 on a time scale τ_{mix}. If $\Delta t \ll \tau_{\mathrm{meas}}$ the meter does not have enough time to extract the signal from the shot-noise background. Averaging over longer intervals $\Delta t > \tau_{\mathrm{mix}}$ erases the information due to the meter-induced mixing.

The investigation of the stationary current noise also reveals the telegraph noise behavior. At low frequencies $\omega\tau_\varphi \ll 1$ one can use the two-mode approximation (1.6,1.7), and the noise spectrum is the sum of the shot- and telegraph-noise contributions:

$$S_I(\omega) = 2e^2 f\bar{\Gamma} + \frac{e^2\delta\Gamma^2\tau_{\mathrm{mix}}}{\omega^2\tau_{\mathrm{mix}}^2 + 1}. \quad (1.14)$$

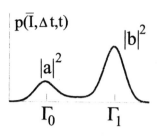

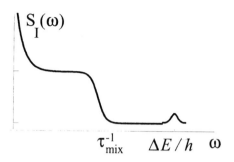

Figure 2 Distribution of possible values of the current averaged over a finite time interval Δt, at times $\tau_{\text{meas}} < t < \tau_{\text{mix}}$.

Figure 3 Current noise spectrum has two Lorentzian peaks, at $\omega = 0$ and $\omega = \Delta E/\hbar$. We also show the $1/f$-noise at low frequencies (note the log-scale of the ω-axis).

At low frequencies $\omega \tau_{\text{mix}} \ll 1$ the latter becomes visible on top of the shot noise (Fig. 3) as we approach the regime of the strong measurement: $S_{\text{telegraph}}/S_{\text{shot}} \approx 4\tau_{\text{mix}}/\tau_{\text{meas}}$. To study the noise at higher frequencies $\omega > \tau_\varphi^{-1}$ one needs to incorporate off-diagonal modes into the calculation. In the Hamiltonian-dominated regime coherent oscillations of the qubit induce an additional peak at its eigenfrequency, $\omega = \Delta E/\hbar$ (cf. Ref. [13]), with the width given by the dephasing rate. The height of the peak with respect to the shot noise is suppressed by a factor $\tau_\varphi/\tau_{\text{meas}}$, the detector's efficiency. In addition, the weights of both peaks depend on the ratio of qubit energy scales, $E_{\text{J}}/\bar{E}_{\text{ch}}$, with $\bar{E}_{\text{ch}}$ favoring the telegraph peak and E_{J} favoring the peak at ΔE.

4.   DISCUSSION

Several parameters can be used to characterize the efficiency of a quantum detector. As expected from the basic principles of quantum mechanics the measurement process above all disturbs the quantum state; hence $\tau_{\text{meas}} \geq \tau_\varphi$. In the sense that the measurement takes longer than the dephasing, it can be called non-efficient. The parameter $\tau_{\text{meas}}/\tau_\varphi$ which quantifies the efficiency is of order $\alpha_{L/R}^2$ if the bias is close to symmetric, $\Gamma_L \sim \Gamma_R$. However, the efficiency can reach values of the order of 100% close to the Coulomb threshold or in the cotunneling regime. Note that $\tau_{\text{meas}} = \tau_\varphi$ for a symmetric QPC coupled to a double dot, symmetric SSET or a dc-SQUID [4]. When the read-out is performed in the time domain, the efficiency lower than 100% implies that a longer time is needed to obtain the result than to destroy quantum coherence. On the other hand, if the stationary current noise is studied, the efficiency determines the height of the peak at $\omega = \Delta E/\hbar$.

Furthermore, the mixing renders the measurement non-ideal. The measurement is only useful if the mixing is slow, $\tau_{\text{mix}} \gg \tau_{\text{meas}}$. In the absence of mixing, if $\tau_{\text{mix}}/\tau_{\text{meas}} \to \infty$, an ideal projective quantum measurement is realized which leaves the qubit in one of two pointer states, $|0\rangle$ or $|1\rangle$, corresponding to the outcome of the measurement. The ratio $\tau_{\text{meas}}/\tau_{\text{mix}} \ll 1$ determines inaccuracy of the read-out procedure. In the opposite limit $\tau_{\text{meas}}/\tau_{\text{mix}} \gg 1$ the mixing quickly erases the information about the qubit's state and prevents a successful read-out at τ_{meas}.

Another important requirement to the detector is that its dephasing effect in the off-state should be negligible. Let $\tau_{\varphi}^{\text{off}}$ be the dephasing time of the qubit's state by the detector. A dimensionless figure of merit is its value relative to the measurement time, $\tau_{\text{meas}}/\tau_{\varphi}^{\text{off}}$. This ratio should be much smaller than unity. For the SET coupled to a Cooper-pair box the dephasing by the switched-off detector is associated with cotunneling processes in the transistor. Straightforward estimates show that $\tau_{\text{meas}}/\tau_{\varphi}^{\text{off}} \sim \alpha(T/E_{\text{SET}})^3$.

So far in our considerations we neglected the effect of the environment (other degrees of freedom apart from the qubit and meter) on the qubit's dynamics during the measurement process. Under conditions which are suitable for investigation of quantum coherence, the coupling to the environment is weak and does not affect the choice of the pointer basis. The environment only contributes to the dephasing and mixing of the qubit's states but does not change the measurement time τ_{meas}. If the environment-induced relaxation is faster than the detector-induced mixing, it can change the long-time dynamics. First, it can spoil the read-out if $\tau_{\text{rel}} < \tau_{\text{meas}}$. Second, it can change the stationary state of the qubit coupled to the detector: instead of an equal-weight mixture the environment relaxes the qubit to the ground state (at $T = 0$). This was demonstrated in the experiments of the Saclay group [14] where the ground state charge was measured.

References

[1] A. Shnirman, G. Schön, and Z. Hermon, Phys. Rev. Lett. **79**, 2371 (1997).

[2] Yu. Makhlin, G. Schön, and A. Shnirman, Nature **386**, 305 (1999).

[3] A. Shnirman, G. Schön, Phys. Rev. B **57**, 15400 (1998).

[4] D.V. Averin, to be published in "Exploring the Quantum-Classical Frontier." Eds. J.R. Friedman and S. Han, cond-mat/0004364.

[5] S.A. Gurvitz, Phys. Rev. B **56**, 15215 (1997).

[6] S.A. Gurvitz, preprint, quant-ph/9808058.

[7] L. Stodolsky, Phys. Lett. B **459**, 193 (1999).

[8] A.N. Korotkov, Phys. Rev. B **60**, 5737 (1999).

[9] J. Mooij, T. Orlando, L. Levitov, L. Tian, C. van der Wal, and S. Lloyd, Science **285**, 1036 (1999).

[10] J.R. Friedman, V. Patel, W. Chen, S.K. Tolpygo, and J.E. Lukens, Nature, **406**, 43 (2000).

[11] Yu. Makhlin, G. Schön, and A. Shnirman, to be published in "Exploring the Quantum-Classical Frontier." Eds. J.R. Friedman and S. Han, cond-mat/9811029.

[12] H. Schoeller, G. Schön, Phys. Rev. B **50**, 18436 (1994).

[13] A.N. Korotkov, D.V. Averin, preprint cond-mat/0002203

[14] V. Bouchiat, D. Vion, P. Joyez, D. Esteve, and M.H. Devoret, Physica Scripta **T76**, 165 (1998).

ADIABATIC INVERSION IN THE SQUID, MACROSCOPIC COHERENCE AND DECOHERENCE

P. Silvestrini[1] and L. Stodolsky[2]

[1]Instituto di Cibernetica del CNR, via Toiano 6, I-80072, Arco Felice, Italy and MQC group, INFN/Naples, Italy

[2]Max-Planck-Institut für Physik (Werner-Heisenberg-Institut), Föhringer Ring 6, 80805 München, Germany

Presented by LS at the Conference on Macroscopic Coherence and Quantum Computing, Naples, June 2000

Abstract   A procedure for demonstrating quantum coherence and measuring decoherence times between differentfluxoid states of a SQUID by using "adiabatic inversion" is discussed. One fluxoid state is smoothly transferred into the other, like a spin reversing direction by following a slowly moving magnetic field. This is accomplished by sweeping an external applied flux, and depends on a well-defined quantum phase between the two macroscopic states. Varying the speed of the sweep relative to the decoherence time permits one to move from the quantum regime, where such a well-defined phase exists, to the classical regime where it is lost andthe inversion is inhibited. Thus observing whether inversion has taken place or not as a function of sweep speed offers the possibility of measuring the decoherence time. Estimates with some typical SQUID parameters are presented and it appears that such a procedure should be experimentally possible. The main requirement for the feasibility of the scheme appears to be thatthe low temperature relaxation time among the quantum levels of the SQUID be long compared to other time scales of the problem, including the readout time. Applications to the "quantum computer," with the level system of the SQUID playing the role of the qbit, are briefly examined.

1. INTRODUCTION

One of the first and certainly probably the most discussed systems in which one tries to demonstrate coherence between apparently macroscopically different states is theSQUID [1].

In the last decade a number of beautiful experiments atlow temperature [2] have seen effects connected with the quantized energy levels [3] expected in the SQUID, showing that it does in fact in many ways resemble a "macroscopic atom." A fast sweeping method [4] has also seen the effects of these quantized levels even at relatively high temperature.

271

If we could further show in some way that the quantum phase between two fluxoid states of the SQUID, where a great number of electrons go around a ring in one direction or the other, is physically meaningful and observable it would certainly provide even more impressive evidence for the applicability of quantum mechanics on large scales and help put to sleep any ideas about the existence of some mysterious scale where quantum mechanics stops working.

At this meeting, results using microwave spectroscopy have indeed been reported showing the repulsion of energy levels expected from the quantum mechanical mixing of different fluxoid states [5]. These results indicate by implication that two opposite fluxoid states can exist in meaningful linear combinations and so that their relative phase is significant.

Here we would like to propose another method for demonstrating the coherence between opposite fluxoid states and the meaningfulness of the phase between them. Furthermore the method to be described allows a measurement of the "decoherence time," the time in which the definite phase or the coherence between the two states is lost. This is interesting of itself since it never has been done and is also relevant to the "quantum computer" where decoherence is the main obstacle to be overcome.

Our basic idea is to use the process of adiabatic inversion or level crossing where a slowly varying field is used to reverse the states of a quantum system. In its most straightforward realization, our proposal consists of starting with the SQUID in its lowest state, making a fast but adiabatic sweep, and reading out to see if the final state is the same or the opposite fluxoid state. If the state has switched to the other fluxoid, the system has behaved quantum mechanically, with phase coherence between the two states. If it stays in the original fluxoid state, the phase coherence between the states was lost and the system behaved classically (see Figs. 3 and 4).

In order for these effects to be unambiguously observable, relaxation towards thermal equilibrium must be small on the time scales involved in the experiment. As is discussed in section IV, this seems to be obtainable at low temperature with suitably chosen sweep times.

A very interesting aspect of the present proposal, as we shall discuss below, is the possiblity of passing between these two regimes, quantum and classical, by simply varying the sweep speed. This allows us to obtain, for adiabatic conditions, the decoherence time as the longest sweep time for which the inversion is successful.

2. ADIABATIC INVERSION IN THE rf SQUID

Most workers in this field are probably accustomed to discussing level crossing problems such as ours in a Landau-Zener picture where one exhibits the spacing of the two crossing energy levels as some external parameter or field is varied. This is of course a perfectly good and useful picture for many applications. However I would like to urge consideration of another perhaps more intuitive visualization, one which is particularly well suited to time dependent problems as we have here, and which also has the advantage that it gives a complete representation of the state of the system at a given time and not just the level splitting.

This picture utilizes the fact that any two-state system may viewed as constituting the two components of a "spin." Looking at this "spin" and its motions then provides an easy visualization, one which has been used in many contexts [6]. That is, if we have two states $|L\rangle$ and $|R\rangle$, which for us will be the two lowest opposing fluxoid states, then the general state

$$\alpha|L\rangle + \beta|R\rangle \tag{1}$$

has the interpretation of a spin pointing in some direction. Even the relative phase is represented: the numbers α and β are complex numbers and the spin points in different directions when we change their relative phase. Thus the spin visualizastion gives a complete picture of the state (up to an irrelevant overall phase).

To identify $|L\rangle$ and $|R\rangle$ here, consider (Fig. 1) the familiar double well potential [7] for the rfSQUID biased with an external flux Φ_x

$$U = U_0[1/2(\Phi - \Phi_x)^2 + \beta L \cos \Delta]. \tag{2}$$

The horizontal axis, the "position" coordinate of the present problem, is Φ the flux in the SQUID ring. Depending on whether the system is in the left or right well, the flux through the ring has a different sign and the current goes around in opposite directions. The potential can be varied by altering the external Φ_x, becoming symmetric for $\Phi_x = 0$. We identify the lowest level in each well with $|L\rangle$ and $|R\rangle$. Quantum mechanical linear combinations of them result from tunneling through the barrier. In the adiabatic inversion procedure to be discussed, the external Φ_x is swept from a maximum to an minimum value, passing through zero, such that the initially asymmetric left and right wells exchange roles, the originally higher well becoming the lower one and vice versa. The asymmetry of theconfigurations, however, is kept small so that we effectively have only a two-state system, composed of the lowest state in each well.

Various influences such as the tunneling energy or the external flux may affect

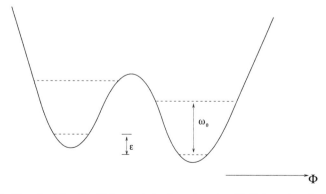

Figure 1. Double potential well with harmonic level spacing ω_0 and initial spacing ϵ between lowest levels.

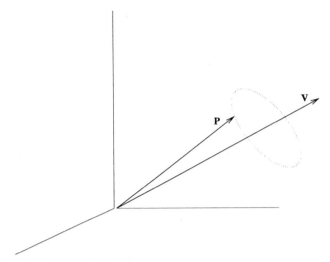

Figure 2. Precession of the "spin" vector **P** around the pseudo-magnetic field vector **V**(t). In adiabatic inversion **V** swings from "up" to "down" and carries **P** with it.

the orientation of the spin. Pursuing the analogy, these influences may be thought of as creating a kind of pseudo-magnetic field **V** which causes the spin to precess. Representing the spin by a "polarization" **P** we get the picture of Fig. 2. The equation governing the motion of **P**

$$\dot{\mathbf{P}} = \mathbf{V} \times \mathbf{P} - D\mathbf{P}_{tr}, \tag{3}$$

where **V** can be time dependent. The quantity D is the decoherence parameter, which we neglect for the moment and will deal with below. Note that in the absence of D, according to Eq. (3) P cannot change its length, so the density matrix, which **P** parameterizes, retains its degree of purity. In using Eq. (3) one assumes the temperature low enough so that relaxation processes like barrier hoping or jumps between prinicpal levels may be neglected, thus there is damping of only the transverse components, $\mathbf{P}_{tr}$ (see below).

Now it is a familiar fact under adiabatic conditions, where **V** varies slowly, that the "spin" **P** will tend to "follow" a moving magnetic field **V**(t). This is a completely familiar procedure when rotating the spin of say an atom or a neutron by a magnetic field. If we wish, we can completely invert the state by having **V** swing from "up" to "down."

In the present problem, we can create a moving **V** by sweeping Φ_x. This is because the vertical component of **V**, V_{vert} corresponds to the difference in the two lowest energy levels. Hence we can induce a level crossing and reverse the potential wells (see Figs. 3 and 4) by reversing Φ_x and so the direction of V_{vert}. For small asymmetry of the wells the splitting is linear in Φ_x. Thus if σ is the initial level splitting, obtained when $\Phi_x = \Phi_x^{max}$, we can write $V_{vert}(t) = \epsilon(\Phi_x(t)/\Phi_x^{max})$. As Φ_x sweeps from its positive maximum value to its negative minimum value, V_{vert} reverses direction.

In doing so, V_{vert} passes through zero where $\mathbf{V}$ is horizontal with only a transverse component $\mathbf{V_{tr}}$. $\mathbf{V_{tr}}$ corresponds to the tunneling energy between the two quasi-degenerate states, $V_{tr} = \omega_{tunnel}$. As V_{vert} passes through zero at $\Phi_x = 0$, the $|L\rangle$ and $|R\rangle$ states are strongly mixed and the splitting of the resulting energy eigenstates is determined by V_{tr} alone, as is also familiar in the Landau-Zener picture. The magnitude of $V(t)$ at a given time, $|V| = \sqrt{V_{vert}^2 + V_{tr}^2}$ gives the instantaneous splitting of the two levels. This varies from approximately ϵ in the vertical position of $\mathbf{V}$ to ω_{tunnel} in the horizontal position.

Having identified the components of $\mathbf{V}$, our next task is to ascertain the meaning of "adiabatic" or "slow" for the motion of $\mathbf{V}$. Adiabatic conditions obtain when the time variation in question does not contain significant frequencies or fourier components corresponding to the energy splitting between levels. Expressed in terms of time, the rate of variation of $\mathbf{V}$ should be slow on the time scale corresponding to the tunneling time between the two states. Thus we have the requirement on $\mathbf{V}$ that its relative rate of variation $\dot{V}/V$ always be small compared to V itself. Since the varying component of $\mathbf{V}$ is $V_{vertical}$ (neglecting the indirect effect of ϕ_x on ω_{tunnel}) we require $\dot{V}_{vertical}/V \ll V$. Thus taking the near degenerate configuration where $V \approx \omega_{tunnel}$ we find

$$\epsilon \frac{\dot{\Phi}_x(t)}{\Phi_x^{max}} \approx \epsilon \omega_{sweep} \ll \omega_{tunnel}^2 \qquad (4)$$

as the condition of adiabaticity.

If the adiabatic condition is violated then $\mathbf{P}$ cannot follow $\mathbf{V}$ (see Fig. 6 below). We stress that for adiabatic inversion there must be a well-defined quantum phase between the two states, and that when this phase is lost the inversion is suppressed.

In Figs. 3 and 4 we give a schematic representation of the whole procedure. In Fig. 3 a succesful inversion takes place. Starting (upper sketch) with the system in the lowest state sincewe are at low temperature, a sweep is performed. After the levelcrossing has been performed (lower sketch), the system hasreversed flux, remaining in the lowest energy state. In thevisualization of Fig. 2, P_{vert} has reversed direction. This is the behavior to be expected of a quantum mechanical system with well-defined phase relations.

In Fig. 4 we have the same starting situation but the inversion is unsuccessful. The system ends up in the same fluxoid state, P_{vert} does not reverse. This is the behavior to be expected classically, when decoherence is significant and there is no well-defined quantum phase between the two states.

More detailed information, as well as intermediate cases will be discussed in terms of the behavior of $\mathbf{P}$, as shown in Figs. 5 and 6 below. For this a discussion of the role of damping ordecoherence is necessary. This is the topic of the next section.

3. DAMPING

We now turn to a quantitative discusssion of dissipative or damping effects, those effects tending to destroy the quantum coherence of the system; that is we

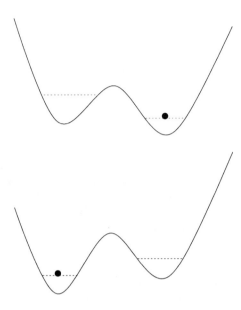

Figure 3. A succesful inversion, starting from the upper figure andending with the lower figure. The black dot indicates which stateis occupied. The system starts in the lowest energy level and staysthere, reversing fluxoid states. It behaves as a quantum systemwith definite phase relations between the two states.

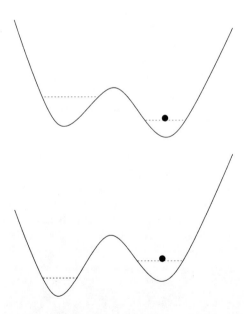

Figure 4. An inhibited inversion, starting from the upper figure and ending with the lower figure. Due to the lack of phase coherence the system behaves classically and stays in the same state, the flux is not reversed.

consider the role of the D term in Eq. (3). While the vertical component of **P** characterizes the relative amounts of the two states which are present in a probabilistic sense, the quantity $\mathbf{P}_{tr}$ measures the degree of phase coherence between the two states. D gives the rate of loss of this phase coherence.

One could also have included a damping parameter for the vertical component of **P**. Instead of the loss of phase coherence, this would represent direct transitions from one well to the other. Such relaxation processes, like jumping the potential barrier due to thermal effects, or "radiative transitions" from one well to another with emission or absorption of some energy can be minimized by low temperatures and fast sweep times, as discussed in the next section.

In the present problem we are in particular interested in the effect of D on the inversion process. With an increasing loss of phase coherence, we expect the situation to become more and more "classical" and finally when the D is large, for the inversion to be inhibited. Indeed, in solving Eq. (3) in thelimit of large D one finds the inversion is strongly blocked and that one arrives in the "Turing-Watched Pot-Zeno" regimewhere $\mathbf{P}_{vert}$ is essentially "frozen" [6]. (See the lower right panels of Figs. 5 and 6.)

These general expectations are confirmed by a numerical study [8] of Eq. (3) with a moving $\mathbf{V(t)}$. Figs. 5 and 6 show results of the numerical study. The horizontal axis represents the time, running from the beginning to the end of the sweep. The two curves in each picture represent $\mathbf{P}_{vert}$ (labeled P_3) and $\mathbf{P}_{trans}$. $\mathbf{P}_{trans}$ is represented in absolute value, while $\mathbf{P}_{vert}$ can change sign. In Fig. 5 an adiabatic sweep is performed. In the first panel, where $D = 0$, one sees the successful inversion of $\mathbf{P}_{vert}$ as it moves from $+1$ to -1. $\mathbf{P}_{trans}$ passes through 1 as **P** passes

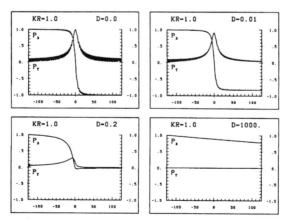

Figure 5. The vertical and horizontal components of the vector P as a function of time, for a sweep of the psuedofield V_{vert} (not shown). The four panels show increasing values of the damping paramater D. Adiabatic inversion is seen to occur for $D = 0$ and $D = .01$, while further increasing D blocks the inversion. For very large D the "Turing-Watched Pot-Zeno" behavior sets in, where P_{vert} evolves extremely slowly. The curve starting at 1 is P_{vert} and the curve starting at zero is P_{tr}. The time and $1/D$ are in units of $1/V_{tr} = 1/\omega_{tunnel}$. An adiabaticity parameter (KR) is kept at a moderate value corresponding, in these units, to a sweep time of about 20.

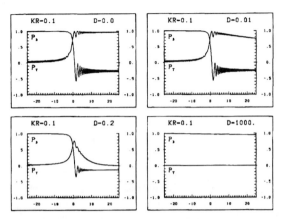

Figure 6. Same as Fig. 5, but with a faster sweep (smaller *KR*), showing the effects of a mild non-adiabaticity. **P** fails to follow **V**. When there is no damping **P** remains with wide oscillations. With moderate damping **P** gradually shrinks to zero. Finally with very strong damping the "Turing-Watched Pot-Zeno" behavior keeps **P** essentially constant.

through the horizontal position. In the next panel, where D is increased somewhat, to 0.01, the picture is essentially unchanged. In the third panel where $D = 0.2$, decoherence takes effect since the decoherence time, given by $1/D$, is on the order of the sweep time. **P** shrinks to zero during the sweep, showing the loss of phase coherence. Finally, with very large D, $\mathbf{P}_{vert}$ evolves hardly at all during the sweep and $\mathbf{P}_{trans}$ stays at zero. This final case is the "Turing-Watched Pot-Zeno" behavior.

We see three regimes for the result of the sweep: A) the system stays in the original state (classical case), B) goes to the opposite state (quantum case), or C) ends sometimes in one state and sometimes in the other (intermediate case). These correspond to P_{vert} starting as $+1$ and ending as $+1$, -1, or 0, respectively. Or in terms of Fig. 5, A) corresponds to the last panel (lower right), B) to the first two panels, and C) to the middle panel (lower left).

We stress that all this arises simply from solving Eq. (3) and that no further notions or assumptions are necessary.

In Fig. 6 we show the effects of non-adiabaticity, using a sweep afactor of ten faster than in Fig. 5. Even with $D = 0$ the inversionis incomplete, with **P** keeping unit length, but ending up rotating roughly in the horizontal plane. As D is turned on, **P** shrinks and finally for very large D the "Turing-Watched Pot-Zeno" behavior sets in.

Given adiabatic conditions, the important relation is that between the sweep speed and the decoherence time $1/D$. As one sees from Fig. 5, there is no effective decoherence until $1/D$ is on the order of the sweep time. This relation determines if the system has time to "decohere" during the sweep. Roughly speaking, one enters the classical regime when P_{tr}, characterizing the phase coherence, has time to shrink during the inversion. This is very interesting for us, since it means, if the experimental situation isfavorable, that we can pass from the quantum regime to the classical regime simply by varying the sweep speed.

4. PARAMETER VALUES

We now discuss some some typical parameters of our system. In particular since we envision working at time scales shorter than have been common in this field, it may be useful to give a qualitative discussion of the various time scales involved. The highest frequency or shortest time present is the ordinary harmonic frequency ω_0 (Fig. 1) giving the approximate level spacing in the well. For typical conditions this spacing may be on the order of several Kelvin (K), or some hundred *GHZ*. Since we suppose working in the Kelvin to milliKelvin range, this implies that for the equilibrium system at the start of the sweep only the lowest level is populated. The next parameter is the tunneling frequency through the barrier. This is a sensitive function of the SQUID parameters, with sample values [$\beta_L = 1.19$, $C = 0.1$ *pF*, $L = 400$ *pH*, $R = 5$ *MΩ*] we have $\omega_{tunnel} \approx 600$ *MHZ*. This corresponds to $\theta_{tunnel}/\omega_0 \sim 10^{-2}$, the tunneling introduces small energy shifts on the scale of the principal level splitting.

The next two parameters concern the extent and speed of the sweep. The initial asymmetry ϵ (Fig. 1) should be small compared to β_0 in order to retain the approximate two-state character of the system, but large compared to ω_{tunnel} to avoid initial mixing of the two states, say $\epsilon/\omega_0 \sim 10^{-1}–10^{-2}$. The speed of the sweep, ω_{sweep} will be the easiest experimental parameter to control, and the behavior of the results as ω_{sweep} is varied will be an important check on the theory. It must not be so fast as to lose adiabaticity, but not so slow as to allow relaxation processes to mask the results. We may suppose it to be in the range $10^{-3}\omega–10^{-4}\omega$ $= 10–100$ *MHZ*. With these numbers it seems possible to satisfy the adiabatic condition Eq. (4) on the one hand and to sweep fast relative to the relaxation time (see below) on the other.

We now come to the dissipative parameters: D, and ω_{relax}, the relaxation rate for transitions among the SQUID levels. The calculation of dissipative effects in the SQUID is usually approached in terms of the Caldeira-Leggett model [9], where a coupling to a pseudoboson field, related to the resistance of the device, represents the dissipative effects. The distinction between relaxation and decoherence in the SQUID is principally a question of energy scales. For the former, influences (e.g., the pseudobosons), involving jumps between levels, energies on the order of the level splitting or more are involved. For the latter, where there is only a "dephasing," low energies, those below the level splitting, are important. In general, of course, both processes are present, but at low enough temperature relaxation should become small. For this reason the damping term in Eq. (3) is taken to only affect the transverse components of **P**.

Weiss and Grabert [10], have given a calculation of the effects of dissipation on coherence, and we may identify their "decay rate" at weak dissipation with D. This gives the estimate $D = T/Re^2$ (their Γ). For $R = 5$ *MΩ*, we find $D = 0.08$ *mk* $= 9.6$ *MHZ* at $T = 100$ *mk*, and $D = 0.008$ *mk* $= 960$ *kHZ* for $T = 10$ *mK*. (Units: $1\,K = 8$ *meV* $= 120$ *GHZ*, $1/e^2 = 4$ *kΩ*). With these estimates, say 1–10 MHZ, we are in an interesting range since as mentioned, this offers the interesting possiblity of being able to choose sweep speeds either slow or fast with respect to the decoherence time, while still retaining the adiabatic condition. The resulting switches between classical and quantum behavior would provide persuasive evidence for the correctness of our general picture, and allow the measurement of D and its temperature dependence.

Our last parameter is the relaxation rate Ω_{relax}, characterizing, as said, the rate of conventional kinetic relaxation processes. We wish this to be small for two reasons. One is that relaxation should not take place during the sweep, which would obviously obscure our effects. Secondly, relaxation should also be small during the readout time, otherwise for example, the final configuration of Fig. 4 might turn into that of Fig. 3 before it could be detected.

Here we may use calculations which have been made in connection with the tunneling to the voltage state (11). The rate one finds depends on the separation of the states in question. If we call Ω_{relax} the value with the level seperation at the beginning or end of the sweep we obtain in our example ~20 *kHZ*. This characterizes the time in which the readout must occur. During the sweep itself the level spacing is changing. Taking this into account we obtain the curve of Fig. 7, which shows the probability of an inter-level jump as a function of Ω_{sweep}. It appears that for sweeps shorter than a microsecond relaxation is indeed negligible. To indicate the general relationship of the various time scales, the frequencies ω_{relax}, $\omega_{dec} = 1/D$ and ω_{tunnel} are marked for the horizontal axis. As may be seen, we may indeed be able to obtain the favorable situation of being able to sweep slow or fast relative to the decoherence time, while retaining small relaxation.

Concerning the readout, we have examined a scheme involving a switchable flux linkage to a DC SQUID, which would be sufficiently fast for the above

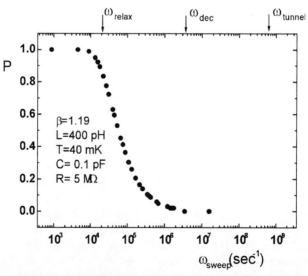

Figure 7. The curve shows the probability of relaxation during a sweep, as a function of the sweep frequency (inverse sweep time). Important time scales or frequencies are marked for the parameters indicated. It appears possible to have sweeps either slow or fast relative with respect to the decoherence time, while retaining small relaxation.

estimates [12]. Here one profits from the fact that observation of the system is only necessary after the procedure is completed.

V. NONADIABATIC PROCEDURES

In general, we need not necessarily limit ourselves to slow, adiabatic processes. There is, say, the opposite case of the"sudden approximation" where **V** is changed very quickly and **P** tends to stay put, as was beginning to happen in Fig. 6. This could be of interest, for example, if we find experimentally that we always get an inversion (Fig. 3), and never a failed inversion (Fig. 4) for the range of parameters available to us operationally. There could be two reasons for this behavior. It could be a true result in the sense that the decoherence is very small; we are always in the quantum situation and the adiabatic inversion always works.

But we might worry that instead we have a rapid relaxation from the upper state to the lower one after or during the sweep. That is, perhaps we really had no inversion and Fig. 4 applies. The relaxation rate is greater than we think and the system just falls back to the lowest energy state with the emission of some energy before we can read out.

A way to clarify the situation would be to start our adiabatic inversion procedure from the *upper* state. Then a succesful inversion means we end in the upper state. If relaxation was not the problem, we should then detect the upper state. On the other hand if relaxation is significant, we should always end in the lower state, regardless of where we start.

Since at low temperature everything is in the lowest state, in order to perform this check we need a method of getting the system into the upper state at the start of the sweep. We can accomplish this by a sudden, nonadiabatic, reversal of $V_{vertical}$, that is, of the applied flux. Since the wave function or **P** will (approximately) not change, we will have the system in the upper state to start with. We now proceed with the adiabatic sweep as before and we can check if the results were due to relaxation or true quantum coherence. Other interesting configurations and procedures can undoubtedly be arrived at by combining various operations in this way.

VI. QBITS AND THE QUANTUM COMPUTER

The two-state systemunder discussion here suggests itself as a physical embodiment of the "quantum computer." The "qbit" itself is naturally represented by L and R playing the role of 0 and 1. A linear combination of L and R may be created by adiabatically rotating **P** from some starting position. Adiabatic inversion is evidently an embodiment of NOT since it will turn one linear combination into another one with the weights of L and R interchanged. As for CNOT, the other basic operation, a NOT is performed or not performed on a"target bit" according to the state of a second, "control bit," which itself does not change its state. One straightforward realization of this would be to perform the NOT operation as just described in the presence of an additional linking flux supplied by a second SQUID nearby. The magnitude and direction of this linking flux would be so arranged that the inversion of the first SQUID is or is not successful depending on the state of the second SQUID. This and many other interesting combinations of junctions and flux linkages, not to mention other devices [13], may be contemplated and are under study [14].

SQUID systems like these would seem to be particularly well suited for the embodiment of the quantum computer, where we wish to generate a series of unitary transformations for the various steps of computation. This may be done by creating a "moving landscape" of potential maxima and minima, as in our simplest one-dimensional example of the adiabatic inversion. This imaginary landscape can be produced and manipulated by controlling various external parameters (as with our sweeping flux), performing the various operations in one physical device. Naturally, practicality will depend very much on the relation between the speed of these operations and the decoherence/relaxation times which we hope to determine by the present methods. Although our interest here has been the SQUID, it will be evident that the principle of determining decoherence times through the inhibition of adiabatic inversion could be applied to many other types of systems as well.

REFERENCES

1. See A. J. Leggett, Les Houches, session XLVI (1986) *Le Hasard et La Matiere*; North-Holland (1987), references cited therein, and introductory talk, this volume (MQC2: Conference on Macroscopic Quantum Coherence and Computing, Naples, June 2000, Proceedings to be published by Kluwer Academic / Plenum.)
2. R. Rouse, S. Y. Han, and J. E. Lukens, Phys. Rev. Let. **75**, 1615 (1995).
3. J. Clarke, A. N. Cleland, M. H. Devoret, D. Esteve, J. M. Martinis, Science **239**, 992 (1988).
4. P. Silvestrini, V. G. Palmieri, B. Ruggiero, and M. Russo, Phys. Rev. Let. **79**, 3046 (1997).
5. J. Friedman, V. Patel, W. Chen, S. K. Tolpygo and J. E. Lukens, Nature *406*, 43 (2000). C. S. van der Wal, A. C. ter Haar, F. K. Wilhelm, R. N. Schouten, C. J. P. M. Harmans, T. P. Orlando, Seth Lloyd, and J. E. Mooij, this volume, (MQC2: Conference on Macroscopic Quantum Coherence and Computing, Naples, June 2000, Proceedings to be published by Kluwer Academic / Plenum.)
6. R. A. Harris and L. Stodolsky, Phys. Lett. **116B**, 464 (1982). For a general introduction and review of these concepts see L. Stodolsky, "Quantum Damping and Its Paradoxes" in *Quantum Coherence*, J. S. Anandan ed. World Scientific, Singapore (1990). For recent work involving mesoscopic devices see L. Stodolsky, Phys. Lett. **B459**, 193 (1999) and for the "fluctuation-decoherence relation," Physics Reports **320**, 51 (1999).
7. A. Barone and G. Paternó, *Physics and Applications of the Josephson Effect*, Wiley, New York, 1982.
8. These plots are from the diploma thesis of J. Flaig, *Neutrino-Oszillationen in thermischer Umgebung*, University of Munich (1989), unpublished. Adiabatic inversion has been extensively studied in neutrino physics where it is called the "MSW effect." See L. Stodolsky, Phys. Rev. **D36** 2273 (1987), and chapter 9 of G. G. Raffelt, *Stars as Laboratories for Fundamental Physics*, Univ. Chicago Press, 1996.
9. A. O. Caldeira and A. J. Leggett, Phys. Rev. Lett. **46**, 211 (1981).
10. U. Weiss and H. Grabert, Europhys. Lett. 2 667 (1986) use a Brownian motion approach. See Eq. (16) of this paper.

11. We use the formulas in B. Ruggiero, P. Silvestrini, and Yu. Ovchinnikov, Phys. Rev. **B54**, 1246 (1996), based on Yu. Ovchinnikov and A. Schmid, Phys. Rev. **B50**, 6332 (1994).
12. B. Ruggiero, V. Corato, E. Esposito, C. Granata, M. Russo, P. Silvestrini, and L. Stodolsky, presented at SATT 10, Intl. Jnl. Mod. Phys. **B**, in press.
13. D. V. Averin, Solid State Communications **105** 659 (1998), has discussed related ideas using adiabatic operations on the charge states of small Josephson junctions.
14. P. Silvestrini and L. Stodolsky, CNOT Operation with SQUIDS, to be published.

QUANTUM COMPUTING WITH SEPARABLE STATES?

Rüdiger Schack

Department of Mathematics, Royal Holloway, University of London,
Egham, Surrey TW20 0EX, UK

r.schack@rhbnc.ac.uk

Abstract   It has been shown recently that all nuclear spin states occuring in current NMR (nuclear magnetic resonance) quantum computing experiments are separable, i.e., the spins are unentangled and can be described as a collection of classical tops. Attempts to find a classical description for the quantum gate operations in NMR quantum computation, however, have had only limited success, and it may be conjectured that such a classical description does not exist in general. This raises the intriguing question of whether quantum computing with separable states is possible.

NMR quantum computing [1, 2] has made a lot of headlines in the last few years. An impressive list of experiments have been performed; for example there have been reports of the implementation of the Deutsch-Jozsa algorithm [3], Grover's search algorithm [4], quantum teleportation [5], quantum error correction [6], and the realization of the Greenberger-Horne-Zeilinger (GHZ) state [7]. Nuclear magnetic resonance seems to be the most advanced quantum information processing technology.

On the other hand, there is a recent debate [8] questioning to which extent current NMR experiments demonstrate genuine quantum information processing. It has been shown that all states that occur in current NMR quantum computing experiments are separable [9]. Separable states are unentangled and equivalent to ensembles of classical spinning tops. Does this imply that NMR experiments are just classical simulations of quantum information processing?

The answer to this question is no. To establish that NMR experiments are equivalent to classical simulations, one would need to provide a classical description not only of the states, but also of the dynamics, i.e., the gate operations. Such a classical description of NMR dynamics has been provided in Ref. [10], but only in the case that both the number of spins and the number of gates are small. The failure to find a classical model for the general case is interesting in itself. In this paper, we give a simple account of these arguments, based on Refs. [9–11].

The states of a D-level quantum system are described by density operators, i.e., positive Hermitian operators of trace one on a D-dimensional Hilbert space. Thus a state ρ is given by a $D \times D$ Hermitian matrix with nonnegative eigenvalues that sum to 1. If there is maximal information about the system, the density operator has exactly one nonzero eigenvalue, $\lambda = 1$, and the density operator is a one-dimensional projection operator, $\rho = |\psi\rangle\langle\psi|$, onto the corresponding eigenvector $|\psi\rangle$. The state of the system is then called a pure state; otherwise, it is called a mixed state.

A qubit is two-level system, for which $D = 2$. There is a one-to-one correspondence between the pure states of a qubit and the points on the unit sphere, or *Bloch sphere*. Any pure state of a qubit can be written in terms of the Pauli matrices,

$$
\sigma_1 = \begin{pmatrix} 0 & 1 \\ 1 & 0 \end{pmatrix} ,
$$

$$
\sigma_2 = \begin{pmatrix} 0 & -i \\ i & 0 \end{pmatrix} ,
$$

$$
\sigma_3 = \begin{pmatrix} 1 & 0 \\ 0 & -1 \end{pmatrix} , \tag{1.1}
$$

as

$$
P_{\vec{n}} = |\vec{n}\rangle\langle\vec{n}| = \frac{1}{2}(1 + \vec{\sigma} \cdot \vec{n}) \tag{1.2}
$$

$$
= \frac{1}{2}(1 + \sigma_1 n_1 + \sigma_2 n_2 + \sigma_3 n_3) ,
$$

where $\vec{n} = (n_1, n_2, n_3)$ is a unit vector, and 1 denotes the unit matrix. An arbitrary state ρ, mixed or pure, of a qubit can be expressed as

$$
\rho = \frac{1}{2}(1 + \vec{S} \cdot \vec{\sigma}) , \tag{1.3}
$$

where $0 \leq |\vec{S}| \leq 1$. Furthermore, there is an infinite number of ways in which ρ can be expanded as a convex mixture of pure states,

$$
\rho = \int d\Omega \, w(\vec{n}) P_{\vec{n}} , \tag{1.4}
$$

where $w(\vec{n}) \geq 0$.

The Hilbert space of N qubits is the N-fold tensor product space

$$
\mathcal{H}^{\otimes N} \equiv \mathcal{H} \otimes \cdots \otimes \mathcal{H} , \tag{1.5}
$$

where $\mathcal{H}$ is the two-dimensional Hilbert space of a single qubit. A state of N qubits is given by an $2^N \times 2^N$ Hermitian matrix with nonnegative eigenvalues that sum to 1. We define the pure-product-state projector

$$
\begin{aligned}
P(\tilde{n}) &\equiv P_{\vec{n}_1} \otimes \cdots \otimes P_{\vec{n}_N} \\
&= \frac{1}{2^N} (1 + \vec{\sigma} \cdot \vec{n}_1) \otimes \cdots \otimes (1 + \vec{\sigma} \cdot \vec{n}_N) ,
\end{aligned} \tag{1.6}
$$

where $\tilde{n}$ stands for the collection of unit vectors $\vec{n}_1, \ldots, \vec{n}_N$. This state can be interpreted as N classical tops pointing in the directions given by $\tilde{n}$.

Definition: A *separable* state ρ of N qubits is one that can be written as a convex mixture of pure-product-state projectors, i.e., as

$$
\begin{aligned}
\rho &= \int d\Omega_{\tilde{n}}\, w(\tilde{n}) P(\tilde{n}) \\
&\equiv \int d\Omega_{n_1} \cdots d\Omega_{n_N}\, w(\tilde{n}) P(\tilde{n}) ,
\end{aligned} \tag{1.7}
$$

where $w(\tilde{n}) \geq 0$. A separable state of N qubits is therefore equivalent to a classical ensemble of spinning tops.

We now turn to the principles of current NMR quantum computing experiments (see, e.g., Refs. [1–7]). The state of the sample is described by a density operator ρ for the N spins (qubits) in each molecule. The molecules are prepared in an initial state of the form

$$
\rho = (1 - \epsilon)M + \epsilon \rho_1 , \tag{1.8}
$$

where $M = 1/2^N$ is the maximally mixed density operator for N qubits and ρ_1 is a density operator usually chosen to be the projector onto a pure state in which all spins are aligned with the external magnetic field. The parameter ϵ scales like

$$
\epsilon = \frac{\alpha N}{2^N} , \tag{1.9}
$$

where $\alpha = h\bar{\nu}/2kT$ determines the polarization of the sample, $\bar{\nu}$ being the average resonant frequency of the active spins in the strong magnetic field. If ρ_1 is a pure state, ρ is called a pseudo-pure state [1]. A typical value for α is 2×10^{-6}.

One interpretation of ϵ is that it specifies the fraction of molecules in the sample that occupy the desired initial state ρ_1. This interpretation is not unique, and it is not mandated or even preferred by quantum mechanics. It becomes a physical fact only if one actually prepares a fraction of the molecules in a particular state, a situation that does not apply to a high-temperature NMR experiment. The freedom to interpret ρ in terms of different ensembles underlies the results of this paper.

After synthesis of the desired initial state, a sequence of radio-frequency pulses, alternating with continuous evolution, is applied to the sample. We restrict attention to the case where the evolution is described by a unitary transformation U. Applying the unitary operator U to ρ results in the state

$$
\begin{aligned}
\rho^{\text{out}} &\equiv U\rho U^\dagger \\
&= (1-\epsilon)M + \epsilon U\rho_1 U^\dagger \\
&\equiv (1-\epsilon)M + \epsilon\rho_1^{\text{out}} .
\end{aligned}
\tag{1.10}
$$

The totally mixed part of the state is unaffected by the unitary transformation. The output state retains the form (1.8) with the same value of ϵ, but—and this is the essence of the bulk-ensemble paradigm for quantum computation—ρ_1 undergoes the desired unitary transformation.

The last step in any NMR experiment is the readout. By applying radio-frequency pulses and then measuring the transverse magnetization of the sample, one can determine all expectation values of the form

$$
\text{tr}\left(\rho^{\text{out}}\sigma_{\beta_1} \otimes \cdots \otimes \sigma_{\beta_N}\right) = \epsilon\,\text{tr}\left(\rho_1^{\text{out}}\sigma_{\beta_1} \otimes \cdots \otimes \sigma_{\beta_N}\right) .
\tag{1.11}
$$

The tensor product in this expression includes one operator for each spin; σ_{β_k} denotes the unit operator 1 for the kth spin if $\beta_k = 0$, and it denotes a Pauli matrix if $\beta_k = 1, 2$, or 3. In writing Eq. (1.11) and Eq. (1.20) below, it is assumed that there is at least one Pauli matrix in the tensor product (not all the β's are zero). The maximally mixed part of the density operator does not contribute to the measured expectation values, which are determined by the state ρ_1^{out} that undergoes the desired evolution. The parameter ϵ appears naturally as a measure of the strength of the magnetization signal (or of the signal-to-noise ratio).

For small ϵ, the states (1.8) are close to the maximally mixed state M. It seems intuitively clear that there exist separable states arbitrarily close to M. One might imagine, however, that even though M is close to other separable states, these separable states lie in a low-dimensional subspace within the space of all states. By leaving this subspace, even infinitesimally, one could reach entangled states. In [12], this problem is addressed by an existence proof; namely, it is shown that there exists a sufficiently small neighborhood of M inside which all states are separable. In [13], a lower bound on the size of the neighborhood is given.

We will now give a constructive proof that provides an explicit representation of any state sufficiently close to M as a mixture of product states, and we give a much improved lower bound on the size of the neighborhood. We show that states of the form (1.8) are separable provided that [9]

$$\epsilon \leq \eta \equiv \frac{1}{1 + 2^{2N-1}} . \tag{1.12}$$

Any density operator $\rho = \frac{1}{2}(1 + \vec{S} \cdot \vec{\sigma})$ for a single qubit can be represented in the form

$$\rho = \frac{1}{4\pi} \int d\Omega \, P_{\tilde{n}} (1 + 3\vec{S} \cdot \vec{n}) . \tag{1.13}$$

For N qubits, using the pure-product-state projectors Eq. (1.6), any density operator can be expanded as

$$\begin{aligned}
\rho &= \int d\Omega_{\tilde{n}} \, w^{\rho}(\tilde{n}) P(\tilde{n}) \\
&\equiv \int d\Omega_{n_1} \cdots d\Omega_{n_N} w^{\rho}(\tilde{n}) P(\tilde{n}) ,
\end{aligned} \tag{1.14}$$

where

$$w^{\rho}(\tilde{n}) \equiv \mathrm{tr}\left(\rho Q(\tilde{n})\right) \tag{1.15}$$

is a canonical expansion function, with

$$Q(\tilde{n}) = \frac{1}{(4\pi)^N}(1 + 3\vec{\sigma} \cdot \vec{n}_1) \otimes \cdots \otimes (1 + 3\vec{\sigma} \cdot \vec{n}_N) . \tag{1.16}$$

If the coefficients $w^{\rho}(\tilde{n})$ are nonnegative, the density operator ρ is separable, and the representation (1.14) provides an explicit product-state ensemble for the density operator.

For any N-qubit density operator ρ, the expansion coefficients (1.15) obey the bound

$$w^{\rho}(\tilde{n}) \geq \left(\begin{array}{c} \text{smallest eigenvalue} \\ \text{of } Q(\tilde{n}) \end{array}\right) = -\frac{2^{2N-1}}{(4\pi)^N} . \tag{1.17}$$

This follows from the fact that $1 + 3\vec{\sigma} \cdot \vec{n}$ has eigenvalues -2 and 4. The most negative eigenvalue of the product operator (1.16) is therefore $(4\pi)^{-N}(-2)4^{N-1}$, from which Eq. (1.17) follows.

Consider now an N-qubit density operator ρ of the form (1.8), and let $w^{\rho_1}(\tilde{n})$ be the coefficients (1.15) for the density operator ρ_1. The coefficients (1.15) for ρ are given by

$$w^{\rho}(\tilde{n}) = \frac{1 - \epsilon}{(4\pi)^N} + \epsilon w^{\rho_1}(\tilde{n}) \geq \frac{1 - \epsilon(1 + 2^{2N-1})}{(4\pi)^N} , \tag{1.18}$$

which is nonnegative for

$$\epsilon \leq \frac{1}{1 + 2^{2N-1}} . \tag{1.19}$$

This bound, derived for the first time in [9], implies that for $\epsilon \leq 1/(1 + 2^{2N-1})$, all density operators of the form (1.8) are separable. It sets a lower bound on the size of the separable neighborhood surrounding the maximally mixed state.

According to this result, states of the form (1.8) are separable, i.e., they can be written in terms of probability distributions (1.7) over spin orientations for N classical tops, provided that the inequality (1.12) holds. If we assume that $\alpha = 2 \times 10^{-6}$ in Eq. (1.9), this inequality holds for $N < 16$ qubits.

The importance of separability for NMR is that a separable state of N qubits can be interpreted in terms of an ensemble of classical tops, because the expectation values (1.11) have the standard form for an ensemble with probability distribution $w(\tilde{n})$:

$$\mathrm{tr}\Big(\rho\, \sigma_{\beta_1} \otimes \cdots \otimes \sigma_{\beta_N}\Big) \tag{1.20}$$

$$= \int d\Omega_{\tilde{n}}\, w(\tilde{n})\, \underbrace{\mathrm{tr}\Big(P(\tilde{n})\sigma_{\beta_1} \otimes \cdots \otimes \sigma_{\beta_N}\Big)}_{= (n_1)_{\beta_1} \cdots (n_N)_{\beta_N}} .$$

In this expression, $(n_j)_{\beta_j} = 1$ if $\beta_j = 0$, and $(n_j)_{\beta_j}$ is a Cartesian component of the vector $\vec{n}_j$ if $\beta_j = 1, 2$, or 3.

Moreover, even completely general measurements, described by a positive-operator-valued measure [14] or POVM, $\{E_r\}$, can be interpreted classically, since

$$\mathrm{tr}(\rho\, E_r) = \int d\Omega_{\tilde{n}}\, w(\tilde{n})\, \underbrace{\mathrm{tr}(P(\tilde{n})E_r)}_{\equiv\, w(r|\tilde{n})} . \tag{1.21}$$

The expression $w(r|\tilde{n})$ is nonnegative and hence can be viewed as a classical probability of observing the outcome r given that the spin directions are $\tilde{n}$. Any measurement on the system, including measurements in entangled bases, can be described by a POVM.

The results given above suggest that current NMR experiments are not true quantum computations, since no entanglement appears in the physical states at any stage. Yet the fact that all states—initial, intermediate, and final—that occur in a given NMR experiment are equivalent to ensembles of classical tops does not mean, by itself, that there is a classical model for the entire experiment. To see this, consider the following naïve attempt to describe a unitary transformation U by classical transition probabilities:

$$w^{U\rho U^\dagger}(\tilde{n}) = \text{tr}\Big(U\rho U^\dagger Q(\tilde{n})\Big) = \int d\Omega_{\tilde{m}}\, w^\rho \underbrace{\text{tr}\Big(UP(\tilde{m})U^\dagger Q(\tilde{n})\Big)}_{\equiv\, t_U(\tilde{n}|\tilde{m})} . \quad (1.22)$$

The transition function $t_U(\tilde{n}|\tilde{m})$ is not a transition probability because it assumes negative values; for example, for the trivial case of the identity transformation, $U = I$, we have

$$t_I(\tilde{n}|\tilde{m}) = \text{tr}\Big(P(\tilde{m})Q(\tilde{n})\Big) = \frac{1}{(4\pi)^N}\prod_{j=1}^{N}(1 + 3\vec{m}_j\cdot\vec{n}_j) . \quad (1.23)$$

The expansion coefficients $w^{P(\tilde{m})}(\tilde{n}) = \text{tr}\Big(P(\tilde{m})Q(\tilde{n})\Big)$ in Eq. (1.23) are the coefficients for the pure product state $P(\tilde{m})$. That these coefficients can be negative might seem to be a drawback of our representation, but in fact it is an essential feature of a program to construct a classical model. To see this, suppose that we could use a representation w^ρ, linear in ρ, to construct a classical model for all unitary operators acting on density operators sufficiently close to the maximally mixed state. This means that there is a nonnegative transition probability $w_U(\tilde{n}|\tilde{m}) \geq 0$ such that

$$w^{U\rho U^\dagger}(\tilde{n}) = \int d\Omega_{\tilde{m}}\, w_U(\tilde{n}|\tilde{m})w^\rho(\tilde{m}) \quad (1.24)$$

for all unitary operators U and for all density operators ρ of the form (1.8), provided ϵ is sufficiently small. Applying Eq. (1.24) to the case where ρ_1 is a pure state P and using the linearity of w^ρ, we get immediately that

$$w^{UPU^\dagger}(\tilde{n}) = \int d\Omega_{\tilde{m}}\, w_U(\tilde{n}|\tilde{m})w^P(\tilde{m}) . \quad (1.25)$$

If $w^P(\tilde{m})$ is everywhere nonnegative, we get a nonnegative representation for $UPU^\dagger$ and, hence, conclude that $UPU^\dagger$ is separable. This can't be true, however, because we can always choose U to map P to an entangled state. The conclusion is that w^P must go negative for all pure states P.

This discussion of the difficulty of finding classical transition probabilities for a general unitary transformation U is not conclusive. It shows clearly, however, why the existence of a classical model for the dynamics does not follow from the absence of entanglement in the physical states. For a small number of spins, such a classical model has been constructed using additional spin degrees of freedom [10], but this method does not generalize easily. A full analysis of the power of general unitary operations in their action on separable states remains to be carried out.

References

[1] D. G. Cory, A. F. Fahmy, and T. F. Havel, Proc. Nat. Acad. Sci. USA **94**, 1634 (1997).

[2] N. A. Gershenfeld and I. L. Chuang, Science **275**, 350 (1997).

[3] N. Linden, H. Barjat, and R. Freeman, Chem. Phys. Lett. **296**, 61 (1998).

[4] J. A. Jones, M. Mosca, and R. H. Hansen, Nature **393**, 344 (1998).

[5] M. A. Nielsen, E. Knill, and R. Laflamme, Nature **396**, 52 (1998).

[6] D. G. Cory, M. D. Price, W. Maas, E. Knill, R. Laflamme, W. H. Zurek, T. F. Havel, and S. S. Somaroo, Phys. Rev. Lett. **81**, 2152 (1998).

[7] R. Laflamme, E. Knill, W. H. Zurek, P. Catasti, and S. V. S. Mariappan, Phil. Trans. Roy. Soc. London A **356**, 1743 (1998).

[8] R. Fitzgerald, Physics Today **53**, 20 (2000).

[9] S. L. Braunstein, C. M. Caves, R. Jozsa, N. Linden, S. Popescu, and R. Schack, Phys. Rev. Lett. **83**, 1054 (1999).

[10] R. Schack and C. M. Caves, Phys. Rev. A **60**, 4354 (1999).

[11] R. Schack and C. M. Caves, in Proceedings of the X. International Symposium on Theoretical Electrical Engineering, edited by W. Mathis and T. Schindler (Otto-von-Guericke University of Magdeburg, Germany, 1999), p. 73.

[12] K. Życzkowski, P. Horodecki, A. Sanpera, and M. Lewenstein, Phys. Rev. A **58**, 883 (1998).

[13] G. Vidal and R. Tarrach, Phys. Rev. A **59**, 141 (1999).

[14] K. Kraus, *States, Effects, and Operations. Fundamental Notions of Quantum Theory* (Springer, Berlin, 1983), lecture Notes in Physics Vol. 190.

Spintronics and Quantum Computing with Quantum Dots

Patrik Recher and Daniel Loss
*Department of Physics and Astronomy, University of Basel, Klingelbergstrasse 82,
CH-4056 Basel, Switzerland*

Jeremy Levy
*Department of Physics University of Pittsburgh 3941 O'Hara St., Pittsburgh,
PA 15260, USA*

Abstract. The creation, coherent manipulation, and measurement of spins in nanostructures open up completely new possibilities for electronics and information processing, among them quantum computing and quantum communication. We review our theoretical proposal for using electron spins in quantum dots as quantum bits. We present single- and two qubit gate mechanisms in laterally as well as vertically coupled quantum dots and discuss the possibility to couple spins in quantum dots via superexchange. We further present the recently proposed schemes for using a single quantum dot as spin-filter and spin read-out/memory device.

1. Introduction

Theoretical research on electronic properties in mesoscopic condensed matter systems has focussed primarily on the charge degrees of freedom of the electron, while its spin degrees of freedom have not yet received the same attention. But an increasing number of spin-related experiments[1, 2, 3, 4, 5, 6] show that the spin of the electron offers unique possibilities for finding novel mechanisms for information processing and information transmission most notably in quantum-confined nanostructures with unusually long spin dephasing times[2, 3, 4] approaching microseconds, as well as long distances of up to 100 μm [2] over which spins can be transported phase-coherently. Besides the intrinsic interest in spin-related phenomena, there are two main areas which hold promises for future applications: Spin-based devices in conventional[1] as well as in quantum computer hardware[7]. In conventional computers, the electron spin can be expected to enhance the performance of quantum electronic devices, examples being spin-transistors (based on spin-currents and spin injection), non-volatile memories, single spin as the ultimate limit of information storage etc.[1]. On the one hand, none of these devices exist yet, and experimental progress as well as theoretical investigations are needed to provide guidance and support in the search for realizable implementations. On the other hand, the emerging field of quantum computing[8, 9] and quantum communication[9, 10] requires a radically new approach to the design of the necessary hardware. As first pointed out in Ref.[7], the spin of the electron is a most natural candidate for the qubit—the fundamental unit of quantum information. We have shown[7] that these spin qubits, when located in quantum-confined structures such as semiconductor quantum dots, atoms or molecules, satisfy all requirements needed for a scalable quantum computer. Moreover, such spin-qubits—being attached to an electron with orbital degrees of freedom—can be transported along conducting wires between different subunits in a quantum network[9]. In particular, spin-entangled electrons can be created in coupled quantum dots and—as mobile Einstein-Podolsky-Rosen (EPR) pairs[9]—provide then the necessary resources for quantum communication.

It follows a short introduction of quantum computing and quantum communication and we will then present our current theoretical efforts towards a realization of quantum computing. We thereby focus on the implementation of the necessary gate and read-out operations schemes with quantum dots.

1.1. Quantum Computing and Quantum Communication

We give a brief description of the emerging research field of quantum computation. It has attracted much interest recently as it opens up the possibility of outperforming classical computation through new and more powerful quantum algorithms such as the ones discovered by Shor[11] and by Grover[12]. There is now a growing list of quantum tasks[9, 10] such as cryptography, error correcting schemes, quantum teleportation, etc. that have indicated even more the desirability of experimental implementations of quantum computing. In a quantum computer each quantum bit (qubit) is allowed to be in any state of a quantum two-level system. All quantum algorithms can be implemented by concatenating one- and two-qubit gates. There is a growing number of proposed physical implementations of qubits and quantum gates. A few examples are: Trapped ions[13], cavity QED[14], nuclear spins[15, 16], superconducting devices[17, 18, 19, 20], and our qubit proposal[7] based on the spin of the electron in quantum-confined nanostructures.

1.2. Quantum Dots

Since quantum dots are the central objects of this work we shall make some general remarks about these systems here. Semiconductor quantum dots are structures where charge carriers are confined in all three spatial dimensions, the dot size being of the order of the Fermi wavelength in the host material, typically between 10 nm and 1 μm [21]. The confinement is usually achieved by electrical gating of a two-dimensional electron gas (2DEG), possibly combined with etching techniques. Precise control of the number of electrons in the conduction band of a quantum dot (starting from zero) has been achieved in GaAs heterostructures[22]. The electronic spectrum of typical quantum dots can vary strongly when an external magnetic field is applied[21, 22], since the magnetic length corresponding to typical laboratory fields $B \approx 1$ T is comparable to typical dot sizes. In coupled quantum dots Coulomb blockade effects[23], tunneling between neighboring dots[21, 23], and magnetization[24] have been observed as well as the formation of a delocalized single-particle state[25].

2. Quantum Gate Operations with Coupled Quantum Dots

One and two qubit gates are known to be sufficient to carry out any quantum algorithm. For electron spins in nearby coupled quantum dots the desired two qubit coupling is provided by a combination of Coulomb interaction and the Pauli exclusion principle.

At zero magnetic field, the ground state of two coupled electrons is a spin singlet, whereas the first excited state in the presence of strong Coulomb repulsion is usually a triplet. The remaining spectrum is separated from these two states by a gap which is either given by the Coulomb repulsion or the single particle confinement. The low-energy physics of such a system can then be described by the Heisenberg spin Hamiltonian

$$H_s(t) = J(t) \, \mathbf{S}_1 \cdot \mathbf{S}_2, \qquad (1)$$

where $J(t)$ is the exchange coupling between the two spins $\mathbf{S}_1$ and $\mathbf{S}_2$, and is given by the energy difference between the singlet and triplet states. If we pulse the exchange coupling such that $\int dt J(t)/\hbar = J_0 \tau_s/\hbar = \pi \pmod{2\pi}$, the associated unitary time evolution $U(t) = T \exp(i \int_0^t H_s(\tau) d\tau/\hbar)$ corresponds to the "swap" operator U_{sw} which exchanges the quantum states of qubit 1 and 2 [7]. Having an array of dots it is therefore possible to couple any two qubits. Furthermore, the quantum XOR gate can be constructed by applying the sequence[7]

$$U_{\text{XOR}} = e^{i(\pi/2)S_1^z} e^{-i(\pi/2)S_2^z} U_{sw}^{1/2} e^{i\pi S_1^z} U_{sw}^{1/2}, \qquad (2)$$

i.e. a combination of "square-root of swap" $U_{sw}^{1/2}$ and single-qubit rotations $\exp(i\pi S_i^z)$. Since U_{XOR} (combined with single-qubit rotations) is proven to be a universal quantum gate[26], it can be used to assemble any quantum algorithm. The study of universal quantum computation in coupled quantum dots is thus essentially reduced to the study of single qubit rotations and the *exchange mechanism*, in particular how the exchange coupling $J(t)$ can be controlled experimentally. Note that the switchable coupling mechanism described below need not be restricted to quantum dots: the same principle can be used in other systems, e.g. coupled atoms in a Bravais lattice, supramolecular structures, or overlapping shallow donors in semiconductors.

2.1. LATERALLY COUPLED QUANTUM DOTS

We first discuss a system of two laterally coupled quantum dots defined by depleted regions in a 2DEG containing one (excess) electron each[27]. The electrons are allowed to tunnel between the dots (if the tunnel barrier is low) leading to spin correlations via their charge (orbital) degrees of freedom. We model the coupled system with the Hamiltonian $H = H_{\text{orb}} + H_Z$, where $H_{\text{orb}} = \sum_{i=1,2} h_i + C$ with

$$h_i = \frac{1}{2m} \left(\mathbf{p}_i - \frac{e}{c} \mathbf{A}(\mathbf{r}_i) \right)^2 + V(\mathbf{r}_i), \quad C = \frac{e^2}{\kappa \left| \mathbf{r}_1 - \mathbf{r}_2 \right|} \ . \tag{3}$$

Here, h_i describes the single-electron dynamics in the 2DEG confined to the xy-plane, with m being the effective electron mass. We allow for a magnetic field $\mathbf{B} = (0,0,B)$ applied along the z-axis that couples to the electron charge via the vector potential $\mathbf{A}(\mathbf{r}) = \frac{B}{2}(-y,x,0)$, and to the spin via a Zeeman coupling term H_Z. The single dot confinement as well as the tunnel-coupling is modeled by a quartic potential, $V(x,y) = \frac{m\omega_0^2}{2} \left(\frac{1}{4a^2} \left(x^2 - a^2 \right)^2 + y^2 \right)$, which, in the limit $a \gg a_B$, separates into two harmonic wells of frequency ω_0 where $2a$ is the interdot distance and $a_B = \sqrt{\hbar/m\omega_0}$ is the effective Bohr radius of a dot. This choice for the potential is motivated by the experimental observation[22] that the low-energy spectrum of single dots is well described by a parabolic confinement potential. The (bare) Coulomb interaction between the two electrons is described by C where κ denotes the dielectric constant of the semiconductor. The screening length λ in almost depleted regions like few-electron quantum dots can be expected to be much larger than the bulk 2DEG screening length (about 40 nm for GaAs). Therefore, λ is large compared to the size of the coupled system, $\lambda \gg 2a \approx 40$ nm for small dots, and we will consider the limit of unscreened Coulomb interaction ($\lambda/a \gg 1$). At low temperatures $kT_B \ll \hbar\omega_0$ we are allowed to restrict our analysis to the two lowest orbital eigenstates of H_{orb}, leaving us with a symmetric (spin-singlet) and an antisymmetric (three triplets T_0, $T_\pm$) orbital state. In

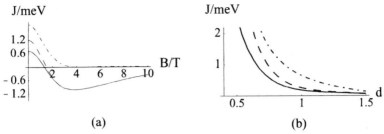

Figure 1. The exchange coupling J (full line) for GaAs quantum dots with confinement energy $\hbar\omega = 3\,\text{meV}$ and $c = 2.42$. For comparison we plot the usual short-range Hubbard result $J = 4t^2/U$ (dashed-dotted line) and the extended Hubbard result[27] $J = 4t^2/U + V$. In (a), J is plotted as a function of magnetic field B at fixed inter-dot distance ($d = a/a_B = 0.7$), and in (b) as a function of the inter-dot distance $d = a/a_B$ at $B = 0$.

this reduced (four-dimensional) Hilbert space, H_{orb} can be replaced by the effective Heisenberg spin Hamiltonian Eq. (1) where the exchange coupling $J = \epsilon_t - \epsilon_s$ is given by the difference between the triplet and singlet energy. We make use of the analogy between atoms and quantum dots (artificial atoms) and caculate ϵ_t and ϵ_s with variational methods similiar to the ones used in molecular physics. With the Heitler-London approximation using single-dot groundstate orbitals we find[27],

$$
J = \frac{\hbar\omega_0}{\sinh\left(2d^2 \frac{2b-1}{b}\right)} \left\{ \frac{3}{4b}\left(1 + bd^2\right) \right.
$$
$$
\left. + c\sqrt{b}\left[e^{-bd^2} I_0\left(bd^2\right) - e^{d^2(b-1)/b} I_0\left(d^2 \frac{b-1}{b}\right)\right] \right\},
\tag{4}
$$

where we introduce the dimensionless distance $d = a/a_B$ and the magnetic compression factor $b = B/B_0 = \sqrt{1 + \omega_L^2/\omega_0^2}$, where $\omega_L = eB/2mc$ is the Larmor frequency. I_0 denotes the zeroth Bessel function. The first term in Eq. (4) comes from the confinement potential. The terms proportional to $c = \sqrt{\pi/2}(e^2/\kappa a_B)/\hbar\omega_0$ are due to the Coulomb interaction C, where the exchange term appears with a minus sign. Note that typically $|J/\hbar\omega_0| \ll 1$ which makes the exclusive use of ground-state single-dot orbitals in the Heitler-London ansatz a self-consistent approach. The exchange J is given as a function of B and d in Fig. 1. We observe that $J > 0$ for $B = 0$, which is generally true for a two-particle system with time reversal invariance. The most remarkable feature of $J(B)$, however, is the change of sign from positive (ferromagnetic) to negative (antiferromagnetic), which occurs at some finite B over a wide range of parameters c and a. This singlet-triplet crossing is caused by the long-range Coulomb interaction and is therefore absent in the standard Hubbard model that takes only into account short range interaction and, in the limit $t/U \ll 1$, is given by $J = 4t^2/U > 0$ (see Fig. 1). Large magnetic fields ($b \gg 1$) and/or large interdot distances ($d \gg 1$) reduce the overlap between the dot orbitals leading to an exponential decay of J contained in the $1/\sinh$ prefactor in Eq. (4). This exponential suppression is partly compensated by the exponentially growing exchange term $\propto \exp(2d^2(b - 1/b))$. As a consequence, J decays exponentially as $\exp(-2d^2b)$ for large b or d. Thus, J can be tuned through zero and then exponentially suppressed to zero by a magnetic field in a very efficient way (exponential switching is highly desirable to minimize gate errors). Further, working around the singlet-triplet crossing provides a smooth exchange switching, requiring only small local magnetic fields. Qualitatively similar results are obtained[27] when we extend the Heitler-London result by taking into account higher levels and double occupancy of the dots (using a Hund-Mullikan approach). In the absence of tunneling ($J = 0$) direct Coulomb interaction between the electron charges can still be present. However the spins (qubit) remain unaffected provided the spin-orbit coupling is sufficiently small, which is the case for s-wave electrons in GaAs structures with unbroken inversion symmetry. Finally, we note that a spin coupling can also be achieved on a long distance scale by using a cavity-QED scheme[28] or superconducting leads to which the quantum dots are attached[29].

2.2. Vertically coupled quantum dots

We also investigated vertically coupled Quantum dots[30]. This kind of coupling can be implemented with multilayer self-assembled quantum dots[31] as well as with etched mesa heterostructures[32].

We model the vertical coupled dot system by a potential $V = V_l + V_v$ where V_l describes the parabolic lateral confinement and V_v models the vertical dot coupling assumed to be a quartic potential similar to the one introduced above for the lateral coupling. We allow for different dot sizes $a_{B\pm} = \sqrt{\hbar/m\alpha_{0\pm}\omega_z}$ with ω_z being the vertical confinement (see Fig. 2), implying an effective Bohr radius $a_B = \sqrt{\hbar/m\omega_z}$ and a dimensionless interdot distance $2d = 2a/a_B$. By applying an in-plane electric field $E_\parallel$ (see Fig. 2) an interesting new switching mechanism arises. The dots are shifted parallel to the field by $\Delta x_\pm = E_\parallel/E_0\alpha_{0\pm}^2$, where $E_0 = \hbar\omega_z/ea_B$. Thus, the larger dot is shifted a greater distance $\Delta x_- > \Delta x_+$ and so the mean distance between the electrons grows as $d' = \sqrt{d^2 + A^2(E_\parallel/E_0)^2} > d$, taking $A = (\alpha_{0+}^2 - \alpha_{0-}^2)/2\alpha_{0+}^2\alpha_{0-}^2$.

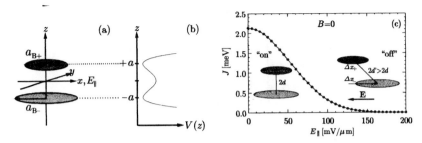

Figure 2. (a) Sketch of the vertically coupled double quantum-dot system. The two dots may have different lateral diameters, a_{B+} and a_{B-}. We consider an in-plane electric field $E_{\parallel}$. (b) The model potential for the vertical confinement is a double well, which is obtained by combining two harmonic wells with frequency ω_z at $z = \pm a$. (c) Switching of the spin-spin coupling between dots of different size by means of an in-plane electric field $E_{\parallel}$ ($B = 0$). The exchange coupling is switched "on" at $E = 0$ (see text). We have chosen $\hbar\omega_z = 7\,\text{meV}$, $d = 1$, $\alpha_{0+} = 1/2$ and $\alpha_{0-} = 1/4$. For these parameters, $E_0 = \hbar\omega_z/ea_B = 0.56\,\text{mV/nm}$ and $A = (\alpha_{0+}^2 - \alpha_{0-}^2)/2\alpha_{0+}^2\alpha_{0-}^2 = 6$. The exchange coupling J decreases exponentially on the scale $E_0/2A = 47\,\text{mV/}\mu\text{m}$ for the electric field.

Since the exchange coupling J is exponentially sensitive to the inter-dot distance d' (see Eq. (4)) we have another exponential switching mechanism for quantum gate operations at hand.

2.3. COUPLING TWO SPINS BY SUPEREXCHANGE

There is a principal problem if one wants to couple two "extended" dots whose energy levels are closely spaced (i.e. smaller than k_BT), as would be the case if there is a sizable distance between the two confined qubits before the barrier is lowered. In this case, the singlet-triplet splitting becomes vanishingly small, and it would not take much excitation energy to get states which are not entangled at all. In other words, the adiabatic switching time[27] which is proportional to the inverse level spacing becomes arbitrarily large. A better scenario for coupling the two spin-qubits is to make use of a superexchange mechanism to obtain a Heisenberg interaction[7]. Consider three aligned quantum dots where the middle dot is empty and so small that only its lowest levels will be occupied by 1 or 2 electrons in a virtual hopping process (see Fig. 3). The left and right dots can be much larger but still small enough such that the Coulomb charging energies $U_L \approx U_R$ are high enough (compared to k_BT) to suppress any double occupancy. Let us assume now that the middle dot has energy levels higher than the ground states of right and left dots, assumed to be approximately the same. These levels include single particle energy (set to zero) and Coulomb charging energy $N^2e^2/2C$, with N the number of electrons and C the

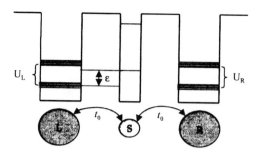

Figure 3. Geometry for superexchange method of coupling two quantum dots.

capacitance of the middle dot, and thus the ground state energy of the middle dot is 0 when empty, $\epsilon = e^2/2C$ for one electron, and 4ϵ for 2 electrons. The tunnel coupling between the dots is denoted by t_0. Now, there are two types of virtual processes possible which couple the spins but only one is dominant. First, the electron of the left (right) dot hops on the middle dot, and then the electron from the right (left) dot hops on the *same* level on the middle dot, and thus, due to the Pauli principle, the two electrons on the middle dot form a singlet, giving the desired entanglement. And then they hop off again into the left and right dots, respectively. (Note that U must be larger than $k_B T$, otherwise real processes involving 2 electrons in the left or right dot will be allowed). It is not difficult to see that this virtual process leads to an effective Heisenberg exchange interaction with exchange constant $J = 4t_0^4/4\epsilon^3$, where the virtual energy denominators follow the sequence $1/\epsilon \rightarrow 1/4\epsilon \rightarrow 1/\epsilon$.

In the second type of virtual process the left (right) electron hops via the middle dot into the right (left) dot and forms there a singlet, giving $J = 4t_0^4/U_R\epsilon^2$. However, this process has vanishing weight because there are also many nearby states available in the outer dots for which there is no spin correlation required by the Pauli principle. Thus, most of the virtual processes, for which we have 2 electrons in the left (right) dot, do not produce spin correlations, and thus we can neglect these virtual processes of the second type altogether. It should be possible to create ferroelectrically defined nanostructures for which superexchange is the dominant mechanism for coupling neighboring electrons. The geometry will resemble closely that of Fig. 3, except that the central barrier becomes a narrow well.

3. Single-Spin Rotations

In order to perform one qubit gates single-spin rotations are required. This is done by exposing a single spin to a time-varying Zeeman coupling $(g\mu_B \mathbf{S} \cdot \mathbf{B})(t)$ [27], which can be controlled through both the magnetic field $\mathbf{B}$ and/or the g-factor g. We have proposed a number of possible implementations[7, 27, 9, 33] for spin-rotations: Since only relative phases between qubits are relevant we can apply a homogeneous $\mathbf{B}$-field rotating all spins at once. A local change of the Zeeman coupling is then possible by changing the Larmor frequency $\omega_L = g\mu_B B/\hbar$. The equilibrium position of an electron can be changed through electrical gating, therefore if the electron wavefunction is pushed into a region with different magnetic field strength or different (effective) g-factor, the relative phase of such an electron then becomes $\phi = (g'B' - gB)\mu_B\tau/2\hbar$. Regions with an increased magnetic field can be provided by a magnetic (dot) material while an effective magnetic field can be produced e.g. with dynamically polarized nuclear spins (Overhauser effect)[27]. We shall now explain a concept for using g-factor-modulated materials[9, 33]. In bulk semiconductors the free-electron value of the Landé g-factor $g_0 = 2.0023$ is modified by spin-orbit coupling. Similarly, the g-factor can be drastically enhanced by doping the semiconductor with magnetic impurities[4, 3]. In confined structures such as quantum wells, wires, and dots, the g-factor is further modified and becomes sensitive to an external bias voltage[34]. We have numerically analyzed a system with a layered structure (AlGaAs-GaAs-InAlGaAs-AlGaAs), in which the effective g-factor of electrons is varied by shifting their equilibrium position from one layer to another by electrical gating[35]. We have found that in this structure the effective g-factor can be changed by about $\Delta g_{\text{eff}} \approx 1$ [33].

Alternatively one can use electron-spin-resonance (ESR) techniques [27] to perform single-spin rotations, e.g. if we want to flip a certain qubit (say from $|\uparrow\rangle$ to $|\downarrow\rangle$) we apply an ac-magnetic field perpendicular to the $\uparrow$- axis that matches the Larmor frequency of that particular electron. Due to paramagnetic resonance[36] the spin can flip.

Furthermore, localized magnetic fields can be generated with the magnetic tip of a scanning force microscope, a magnetic disk writing head, by placing the dots above a grid of current-carrying wires, or by placing a small wire coil above the dot etc.

4. Read-Out of a Single Spin

The final step of each (quantum) computation, consists in reading out the state of each qubit, i.e. if the electron spin is in the $| \uparrow \rangle$ or $| \downarrow \rangle$ state. It is very hard to detect an electron spin over its tiny (of the order of μ_B) magnetic moment directly. We proposed several devices for read out like tunneling of the electron into a supercooled paramagnetic dot[7, 9], thereby inducing a magnetization nucleation from the metastable phase into a ferromagnetic domain. The domain's magnetization direction is along the measured spin and can be detected by conventional methods and provides a 75%-reliable result for the read out of the electron spin. Another possibility is to use a spin selective tunnelbarrier (conventional spin filter), that let pass only one spin direction. If an electron passes the barrier to enter another dot an electrometer can detect the charge[7].

4.1. QUANTUM DOT AS SPIN FILTER AND READ-OUT/MEMORY DEVICE

We recently proposed[37] another setup using a quantum dot attached to in and outgoing current leads $l = 1, 2$ that can work either as a spin filter or as a read-out device, or as a spin memory where the spin stores the information. A new feature of this proposal is that we lift the spin-degeneracy with *different* Zeeman splittings for the dot and the leads, e.g. by using materials with different effective g-factors for leads and dot[37]. This results in Coulomb blockade oscillation peaks and spin-polarized currents which are uniquely associated with the spin state on the dot.

The setup is described by a standard tunneling Hamiltonian $H_0 + H_T$ [38], where $H_0 = H_L + H_D$ describes the leads and the dot. H_D includes the charging and interaction energies of the electrons in the dot as well as their Zeeman energy $\pm g\mu_B B/2$ in an external magnetic field $\mathbf{B}$. Tunneling between leads and the dot is described by $H_T = \sum_{l,k,p,\sigma} t_{lp} c_{lk\sigma}^{\dagger} d_{p\sigma} + \text{h.c.}$, where $c_{lk\sigma}$ annihilates electrons with spin σ and momentum k in lead l and $d_{p\sigma}$ annihilates electrons in the dot. We work in the Coulomb blockade regime[21] where the charge on the dot is quantized. We use a stationary master equation approach[21, 37] for the reduced density matrix of the dot and calculate the transition rates in a "golden-rule" approach up to 2nd order in H_T. The first-order contribution to the current is the sequential tunneling (ST) current I_s[21], where the number of electrons on the dot fluctuates and thus the processes of an electron tunneling from the lead onto the dot and vice versa are allowed by energy conservation. The second-order contribution is the cotunneling (CT) current I_c[39], where charge is transported over intermediate virtual states of the dot.

We now consider a system, where the Zeeman splitting in the leads is negligible (i.e. much smaller than the Fermi energy) while on the dot it is given as $\Delta_z = \mu_B|gB|$. We assume a small bias $\Delta\mu = \mu_1 - \mu_2 > 0$ between the leads at chemical potential $\mu_{1,2}$ and low temperatures so that $\Delta\mu, k_B T < \delta$, where δ is the characteristic energy-level distance on the dot. First we tune the system to the ST resonance $\mu_1 > \Delta E > \mu_2$ where the number of electrons can fluctuate between N and $N+1$. ΔE is the energy difference between the $N+1$ and N-particle groundstates (GS) of the dot. We first consider a quantum dot with N odd and total spin $s = 1/2$ with the N-particle GS to be $|\uparrow\rangle$ and to have energy $E_\uparrow = 0$. In this state the dot can receive an electron from the leads and, depending on the spin of the incoming electron form a singlet $|S\rangle$ with energy E_S (for spin down) or a triplet $|T_+\rangle$ with energy E_{T_+} (for spin up). The singlet is (usually) the GS for N even, whereas the three triplets $|T_\pm\rangle$ and $|T_0\rangle$ are excited states. In the regime $E_{T_+} - E_S, \Delta_z > \Delta\mu, k_B T$, energy conservation only allows ground state transitions. Thus, spin-up electrons are not allowed to tunnel from lead 1 via the dot into lead 2, since this would involve virtual states $|T_+\rangle$ and $|\downarrow\rangle$, and so we have $I_s(\uparrow) = 0$ for ST. However, spin down electrons may pass through the dot in the process $\downarrow\textcircled{\uparrow}_i \to \textcircled{\uparrow\downarrow}_f$, followed by $\textcircled{\uparrow\downarrow}_i \to \textcircled{\uparrow}\downarrow_f$. Here the state of the quantum dot is drawn inside the circle, while the states in the leads are drawn to the left and right, *resp.*, of the circle. This leads to a *spin-polarized* ST current $I_s = I_s(\downarrow)$, which we have calculated as[37]

$$I_s(\downarrow)/I_0 = \theta(\mu_1 - E_S) - \theta(\mu_2 - E_S), \quad k_B T < \Delta\mu, \quad (5)$$

$$I_s(\downarrow)/I_0 = \frac{\Delta\mu}{4k_B T} \cosh^{-2}\left[\frac{E_S - \mu}{2k_B T}\right], \quad k_B T > \Delta\mu, \quad (6)$$

where $\mu = (\mu_1 + \mu_2)/2$ and $I_0 = e\gamma_1\gamma_2/(\gamma_1 + \gamma_2)$. Here $\gamma_l = 2\pi\nu|A_{lnn'}|^2$ is the tunneling rate between lead l and the dot. n and n' denote the N and $N+1$ particle eigenstates of H_D involved in the tunnel process. The dependence of $A_{ln'n} = \sum_{p\sigma} t_{lp}\langle n'|d_{p\sigma}|n\rangle$ on n and n' is weak compared to the resonant character of the tunneling current considered here[37]. Similarly, for N even we find $I_s(\downarrow) = 0$ while for $I_s(\uparrow)$ a similar result holds[37] as in Eqs. (5), (6).

Even though I_s is completely spin-polarized, a leakage of current with opposite polarization arises through cotunneling processes[37]; still the leakage is small, and the efficiency for $\Delta_z < |E_{T_+} - E_S|$ for spin filtering in the sequential regime becomes[37]

$$I_s(\downarrow)/I_c(\uparrow) \sim \frac{\Delta_z^2}{(\gamma_1 + \gamma_2)\max\{k_B T, \Delta\mu\}}, \quad (7)$$

and equivalently for $I_s(\uparrow)/I_c(\downarrow)$ at the even-to-odd transition. In the ST regime we have $\gamma_i < k_BT, \Delta\mu$, thus, for $k_BT, \Delta\mu < \Delta_z$, we see that the spin-filtering is very efficient. Above or below a ST-resonance the system is in the CT regime where the current is solely due to CT-processes. Again, in the regime $E_{T_+} - E_S, \Delta_z > \Delta\mu, k_BT$ the current is *spin-polarized* and the spin filter also works in the CT regime[37].

We discuss now the opposite case where the leads are fully spin polarized with a much smaller Zeeman splitting on the dot[37]. Such a situation can be realized with magnetic semiconductors (with effective g-factors reaching 100 [3]) where spin-injection into GaAs has recently been demonstrated for the first time[3, 4]. Another possibility would be to work in the quantum Hall regime where spin-polarized edge states are coupled to a quantum dot[40]. In this setup the device can be used as read-out for the spin state on the dot. Assume now that the spin polarization in both leads is up, and the ground state of the dot contains an odd number of electrons with total spin 1/2. Now the leads can *provide* and *take up* only spin-up electrons. As a consequence, a ST current will only be possible if the dot state is $|\downarrow\rangle$ (to form a singlet with the incoming electron, whereas the triplet is excluded by energy conservation). Hence, the current is much larger for the spin on the dot being in $|\downarrow\rangle$ than it is for $|\uparrow\rangle$. Again, there is a small CT leakage current for the dot-state $|\uparrow\rangle$, with a ratio of the two currents given by Eq. (7) (assuming $E_S > \Delta_z$). Thus, we can probe (read out) the spin-state on the quantum dot by measuring the current which passes through the dot. Given that the ST current is typically on the order of $0.1 - 1$ nA [21], we can estimate the read-out frequency $I/2\pi e$ to be on the order of $0.1 - 1$ GHz. Combining this with the initialization and read-in techniques, i.e. ESR pulses to switch the spin state, we have a *spin memory* at the ultimate single-spin limit, whose relaxation time is just the spin relaxation time. This relaxation time can be expected to be on the order of 100's of nanoseconds[2], and can be directly measured via the currents when they switch from high to low due to a spin flip on the dot[37].

5. Conclusions

We have described a scalable scenario for the implementation of a solid state quantum computer based on the electron spin in quantum dots as the qubit. We have shown how electron spins can be manipulated through their charge (orbital) degrees of freedom to implement single and two-qubit gates as well as the possibility of read in/out a single qubit (spin).

References

1. G. Prinz, Phys. Today **45**(4), 58 (1995); G. A. Prinz, Science **282**, 1660 (1998).
2. J.M. Kikkawa, I.P. Smorchkova, N. Samarth, and D.D. Awschalom, Science **277**, 1284 (1997); J.M. Kikkawa and D.D. Awschalom, Phys. Rev. Lett. **80**, 4313 (1998); D.D. Awschalom and J.M. Kikkawa, Phys. Today **52**(6), 33 (1999).
3. R. Fiederling *et al.*, Nature **402**, 787 (1999).
4. Y. Ohno *et al.*, Nature **402**, 790 (1999).
5. F.G. Monzon and M.L. Roukes, J. Magn. Magn. Mater. **198**, 632 (1999).
6. S. Lüscher *et al.*, cond-mat/0002226.
7. D. Loss and D.P. DiVincenzo, Phys. Rev. A **57**, 120 (1998); cond-mat/9701055.
8. A. Steane, Rep. Prog. Phys. **61**, 117 (1998).
9. D.P. DiVincenzo and D. Loss, J. Magn. Magn. Mater. **200**, 202 (1999); cond-mat/9901137.
10. C. H. Bennett and D. P. DiVincenzo, Nature **404**, 247 (2000).
11. P.W. Shor, in *Proc. 35th Symposium on the Foundations of Computer Science*, (IEEE Computer Society Press), 124 (1994).
12. L.K. Grover, Phys. Rev. Lett. **79**, 325 (1997).
13. J.I. Cirac and P. Zoller, Phys. Rev. Lett. **74**, 4091 (1995); C. Monroe *et al.*, *ibid.* **75**, 4714 (1995).
14. Q.A. Turchette *et al.*, Phys. Rev. Lett. **75**, 4710 (1995).
15. D. Cory, A. Fahmy, and T. Havel, Proc. Nat. Acad. Sci. U.S.A. **94**, 1634 (1997); N. A. Gershenfeld and I. L. Chuang, Science **275**, 350 (1997).
16. B. Kane, Nature **393**, 133 (1998).
17. A. Shnirman, G. Schön, and Z. Hermon, Phys. Rev. Lett. **79**, 2371 (1997).
18. D.V. Averin, Solid State Commun. **105**, 659 (1998).
19. L.B. Ioffe *et al.*, Nature **398**, 679 (1999).
20. T.P. Orlando *et al.*, Phys. Rev. B **60**, 15398 (1999).
21. L. P. Kouwenhoven *et al.*, Wingreen, Proceedings of the ASI on *Mesoscopic Electron Transport*, eds. L.L. Sohn, L.P. Kouwenhoven, and G. Schön (Kluwer, 1997).
22. S. Tarucha *et al.*, Phys. Rev. Lett. **77**, 3613 (1996).
23. F.R. Waugh *et al.*, Phys. Rev. Lett. **75**, 705 (1995); C. Livermore *et al.*, Science **274**, 1332 (1996).
24. T. H. Oosterkamp *et al.*, Phys. Rev. Lett. **80**, 4951 (1998).
25. R.H. Blick *et al.*, Phys. Rev. Lett. **80**, 4032 (1998); *ibid.* **81**, 689 (1998). T.H. Oosterkamp *et al.*, Nature **395**, 873 (1998); I.J. Maasilta and V.J. Goldman, Phys. Rev. Lett. **84**, 1776 (2000).
26. A. Barenco *et al.*, Phys. Rev. A **52**, 3457 (1995).
27. G. Burkard, D. Loss, and D. P. DiVincenzo, Phys. Rev. B **59**, 2070 (1999).
28. A. Imamoglu, D.D. Awschalom, G. Burkard, D. P. DiVincenzo, D. Loss, M. Sherwin, and A. Small, Phys. Rev. Lett. **83**, 4204 (1999).
29. M.-S. Choi, C. Bruder, and D. Loss; cond-mat/0001011.
30. G. Burkard, G. Seelig, and D. Loss; Phys. Rev. B **62**, 2581 (2000)
31. R. J. Luyken *et al.*, preprint.
32. D. G. Austing *et al.*, Physica B **249-251**, 206 (1998).
33. G. Burkard, H.-A. Engel, and D. Loss, to appear in Fortschritte der Physik, special issue on *Experimental Proposals for Quantum Computation*, eds. S. Braunstein and K.L. Ho; cond-mat/0004182.
34. E.L. Ivchenko, A.A. Kiselev, M. Willander, Solid State Comm. **102**, 375 (1997).

35. K. Ensslin, private communication.
36. R. Shankar, *Principles of Quantum Mechanics*, Ch. 14, Plenum Press, New York, 1994.
37. P. Recher, E.V. Sukhorukov, and D. Loss, cond-mat/0003089, to appear in Phys. Rev. Lett.
38. G. D. Mahan, *Many Particle Physics*, 2nd Ed. (Plenum, New York, 1993).
39. D. V. Averin and Yu. V. Nazarov, in *Single Charge Tunneling*, eds. H. Grabert, M. H. Devoret, NATO ASI Series B: Physics Vol. 294, Plenum Press, New York, 1992.
40. M. Ciorga *et al.*, cond-mat/9912446.

DOUBLE QUANTUM DOT COUPLED TO TWO SUPERCONDUCTORS: TRANSPORT AND SPIN ENTANGLEMENT

Mahn-Soo Choi, Christoph Bruder, and Daniel Loss
Department of Physics and Astronomy, University of Basel, Klingelbergstrasse 82, CH-4056 Basel, Switzerland

Keywords: Quantum dot, superconductivity, entanglement, Josephson effect

Abstract   A double quantum dot coupled to two superconducting leads is studied in the Coulomb blockade regime. The double dot mediates an effective Josephson coupling between the two superconducting leads, which in turn induce an exchange coupling between the spin states of the dots. The Josephson current depends on the spins on the dots, while the spin exchange coupling can be tuned by the superconducting phase difference. This spin-dependent Josephson current can be used to probe directly the correlated spin states (singlet or triplets).

Superconductors and quantum dots have been subjects of intensive studies on their own. In view of transport phenomena on mesoscopic scales, the most interesting properties of them might be the existence of quantum coherence on the macroscopic level and superconducting energy gap in superconductors, and the Coulomb blockade effects and resonant tunneling in quantum dots (usually coupled to non-interacting leads) [1]. The idea of coupling quantum dots to superconducting leads dates back to the studies of an Anderson impurity [2] or a quantum dot coupled to superconductors [3, 4, 5]. Besides a number of experimental[3] and theoretical [4] papers on the spectroscopic properties of a quantum dot coupled to two superconductors, an effective dc Josephson effect through strongly interacting regions between superconducting leads has been analyzed [6, 7, 8, 9]. Furthermore, recent researches have shown that not only the charges but also the spins on quantum dots can be a valuable resource. In particular, in semiconducting nanostructures, it was found that the direct coupling of two quantum dots by a tunnel junction can be used to create entanglement between spins [10], and that

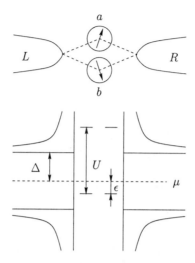

Figure 1 Sketch of the superconductor-double quantum dot-superconductor (S-DD-S) nanostructure (top), and schematic representation of the quasiparticle energy spectrum in the superconductors and the single-electron levels of the two quantum dots (bottom).

such spin correlations can be observed in charge transport experiments [11].

In this work, we consider a double quantum dot (DD) coupled *in parallel* to two superconducting leads (but not directly with each other). We first investigate two interesting aspects of this particular structure: Josephson current through the strongly interacting region (quantum dots), and effective exchange coupling between the spins on DD due to the superconductors. The Josephson current depends on the spin states whereas the spin exchange coupling can be tuned by the superconducting phase difference between the two superconducting leads. Exploiting such an interplay between the two effects, we also propose an experimental setup to probe spin entanglement on DD.

1.   MODEL

The double-dot (DD) system we propose is sketched in Fig. 1: Two quantum dots (a,b), each of which contains one (excess) electron and is connected to two superconducting leads (L, R) via tunnel junctions (indicated by dashed lines) [13]. There is no direct coupling between the two dots. The Hamiltonian describing this system consists of three parts, $H_S + H_D + H_T \equiv H_0 + H_T$. The leads are assumed to be conventional singlet superconductors that are described by the BCS Hamiltonian

$$
H_S = \sum_{j=L,R} \int_{\Omega_j} \frac{d\mathbf{r}}{\Omega_j} \left[\sum_{\sigma=\uparrow,\downarrow} \psi_\sigma^\dagger(\mathbf{r}) h(\mathbf{r}) \psi_\sigma(\mathbf{r}) + \Delta_j(\mathbf{r}) \psi_\uparrow^\dagger(\mathbf{r}) \psi_\downarrow^\dagger(\mathbf{r}) + h.c. \right],
$$

$$(1.1)$$

where Ω_j is the volume of lead j, $h(\mathbf{r}) = (-i\hbar\nabla)^2/2m - \mu$, and $\Delta_j(\mathbf{r}) = \Delta_j e^{-i\phi_j(\mathbf{r})}$ is the pair potential. For simplicity, we assume identical leads with same chemical potential μ, and $\Delta_L = \Delta_R = \Delta$. The two quantum dots are modeled as two localized levels ϵ_a and ϵ_b with strong on-site Coulomb repulsion U, described by the Hamiltonian

$$
H_D = \sum_{n=a,b} \left[-\epsilon \sum_\sigma d_{n\sigma}^\dagger d_{n\sigma} + U d_{n\uparrow}^\dagger d_{n\uparrow} d_{n\downarrow}^\dagger d_{n\downarrow} \right],
$$

$$(1.2)$$

where we put $\epsilon_a = \epsilon_b = -\epsilon$ ($\epsilon > 0$) for simplicity. U is typically given by the charging energy of the dots, and we have assumed that the level spacing of the dots is $\sim U$ (which is the case for small GaAs dots[1]), so that we need to retain only one energy level in H_D. Finally, the DD is coupled *in parallel* (see Fig. 1) to the superconducting leads, described by the tunneling Hamiltonian

$$
H_T = \sum_{j,n,\sigma} \left[t \psi_\sigma^\dagger(\mathbf{r}_{j,n}) d_{n\sigma} + h.c. \right],
$$

$$(1.3)$$

where $\mathbf{r}_{j,n}$ is the point on the lead j closest to the dot n. Unless mentioned otherwise, it will be assumed that $\mathbf{r}_{L,a} = \mathbf{r}_{L,b} = \mathbf{r}_L$ and $\mathbf{r}_{R,a} = \mathbf{r}_{R,b} = \mathbf{r}_R$.

Since the low-energy states of the whole system are well separated by the superconducting gap Δ as well as the strong Coulomb repulsion U ($\Delta, \epsilon \ll U - \epsilon$), it is sufficient to consider an effective Hamiltonian on the reduced Hilbert space consisting of singly occupied levels of the dots and the BCS ground states on the leads. To lowest order in H_T, the effective Hamiltonian is

$$
H_{eff} = P H_T \left[(E_0 - H_0)^{-1}(1 - P) H_T \right]^3 P,
$$

$$(1.4)$$

where P is the projection operator onto the subspace and E_0 is the ground-state energy of the unperturbed Hamiltonian H_0. (The 2nd order contribution leads to an irrelevant constant.) The lowest-order expansion (1.4) is valid in the limit $\Gamma \ll \Delta, \epsilon$ where $\Gamma = \pi t^2 N(0)$ and $N(0)$ is the normal-state density of states per spin of the leads at the Fermi energy. Thus, we assume that $\Gamma \ll \Delta, \epsilon \ll U - \epsilon$, and temperatures which are less than ϵ (but larger than the Kondo temperature).

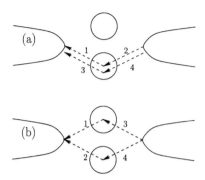

Figure 2 Two examples of virtual tunneling processes contributing to H_{eff} (1.4). The numbered arrows indicate the direction and the order of occurrence of the charge transfers. Processes of type (a) give a contribution proportional to J_0, whereas those of type (b) give contributions proportional to J. The two types give significantly different behavior in transport as discussed in the text.

2.   EFFECTIVE HAMILTONIAN

There are a number of virtual hopping processes that contribute to the effective Hamiltonian (1.4), see Ref. [12] for discussion in more detail. Out of all the possible virtual processes, we show in Fig. 2 the two most physically interesting and significant processes. In a process of type Fig. 2 (a), the two charges of a Cooper pair traverse through the same dot. Such a process, therefore, is not sensitive to spin configuration on DD. In Fig. 2 (b), on the other hand, the two charges take different paths. It turns out that those processes depend strongly on the spin correlation on DD.

Instead of the whole parameter region (see Ref. [12]), here we will focus on two regimes defined by $\zeta \equiv \epsilon/\Delta \gg 1$ and $\zeta \ll 1$. In the limit $\zeta \ll 1$, the most contributions come from the processes of Fig. 2 (b) or the similar and the effective Hamiltonian (1.4) is dominated by a single term

$$H_{eff} \approx J(1 + \cos\phi) \left\lceil \mathbf{S}_a \cdot \mathbf{S}_b - \frac{1}{4} \right\rceil \qquad (1.5)$$

up to terms of order ζ, where $\phi = \phi_L - \phi_R$ is the superconducting phase difference between the two leads and

$$J \approx 2\Gamma^2/\epsilon . \qquad (1.6)$$

Equation (1.5) is one of our main results. A remarkable feature of it is that a Heisenberg exchange coupling between the spin on dot a and on dot b is induced by the superconductor. This coupling is antiferromagnetic ($J > 0$) and thus favors a singlet ground state of spin a and

b. This in turn is a direct consequence of the assumed singlet nature of the Cooper pairs in the superconductor [14]. As discussed below, an immediate observable consequence of H_{eff} is a *spin-dependent* Josephson current from the left to right superconducting lead (see Fig. 1) which probes the correlated spin state on the DD.

Since the exchange coupling in Eq. (1.5) results from two charges of the same Cooper pair traversing to different dots, it is clear that the dot cannot be separated too far. When the distance between the contact points is given by $\delta r = |\mathbf{r}_{L,a} - \mathbf{r}_{L,b}| = |\mathbf{r}_{R,a} - \mathbf{r}_{R,b}|$, one can show that $J(\delta r) \approx J(0)e^{-2\delta r/\xi} \sin^2(k_F \delta r)/(k_F \delta r)^2$ up to order $1/k_F \xi$. Here ξ is the superconducting coherence length and k_F the Fermi wave vector in the leads. Thus, to have $J(\delta r)$ non-zero, δr should not exceed the superconducting coherence length ξ.

Another interesting but experimentally less feasible region is the limit $\zeta \gg 1$. In this region, the effective Hamiltonian (1.4) can be reduced to

$$H_{eff} \approx J_0 \cos\phi + 2J_0(1 + \cos\phi)\left\lceil \mathbf{S}_a \cdot \mathbf{S}_b - \frac{1}{4}\right\rceil , \qquad (1.7)$$

up to order $(\log\zeta)/\zeta$, where

$$J_0 \approx 0.1\frac{\Gamma^2}{\epsilon}\frac{\log\zeta}{\zeta} . \qquad (1.8)$$

As can be seen in Fig. 2 (a), the first term in (1.7) has the same origin as that in the single-dot case [2]: Each dot separately constitutes an effective Josephson junction with coupling energy $-J_0/2$ (i.e. π-junction) between the two superconductors. The two resulting junctions form a dc SQUID, leading to the total Josephson coupling in the first term of (1.7). The Josephson coupling in the second term in (1.7) has a spin-dependence similar to Eq. (1.5) but its origin is quite different (coming from processes other than either of Fig. 2). For the singlet state, it gives an ordinary Josephson junction with coupling $2J_0$ and competes with the first term, whereas it vanishes for the triplet states.

3. TO PROBE SPIN ENTANGLEMENT

The result (1.5) allows us to probe directly the correlated spin states on the double dot by measuring the Josephson current. But it would be useful if one could prepare the system in states other than the singlet ground state and compare their different behaviors in a single setup. We propose a dc-SQUID-like structure Fig. 3. We now propose a possible experimental setup to probe the correlations (entanglement) of the spins on the dots, based on the effective model (1.5). According to (1.5) the S-DD-S structure can be regarded as a *spin-dependent* Josephson junction.

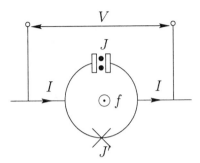

Figure 3   dc-SQUID-like geometry consisting of the S-DD-S structure (filled dots at the top) connected in parallel with another ordinary Josephson junction (cross at the bottom).

Moreover, this structure can be connected with an ordinary Josephson junction to form a dc-SQUID-like geometry, see Fig. 3. The Hamiltonian of the entire system is then given by

$$H = J[1 + \cos(\theta - 2\pi f)] \left(\mathbf{S}_a \cdot \mathbf{S}_b - \frac{1}{4} \right) \tag{1.9}$$
$$+ \alpha J(1 - \cos\theta) \,,$$

where $f = \Phi/\Phi_0$, Φ is the flux threading the SQUID loop, $\Phi_0 = hc/2e$ is the superconducting flux quantum, θ is the gauge-invariant phase difference across the auxiliary junction (J'), and $\alpha = J'/J$ with J' being the Josephson coupling energy of the auxiliary junction[16]. One immediate consequence of (1.9) is that at zero temperature, we can *effectively* turn on and off the spin exchange interaction: For half-integer flux ($f = 1/2$), singlet and triplet states are degenerate at $\theta = 0$. Even at finite temperatures, where θ is subject to thermal fluctuations, singlet and triplet states are almost degenerate around $\theta = 0$. On the other hand, for integer flux ($f = 0$), the energy of the singlet is lower by J than that of the triplets.

This observation allows us to probe directly the spin state on the double dot via a Josephson current across the dc-SQUID-like structure in Fig. 3. The supercurrent through the SQUID-ring is defined as $I_S = (2\pi c/\Phi_0)\partial\langle H\rangle/\partial\theta$, where the brackets refer to a spin expectation value on the DD. Thus, depending on the spin state on the DD we find

$$I_S/I_J = \begin{cases} \sin(\theta - 2\pi f) + \alpha \sin\theta & \text{(singlet)} \\ \alpha \sin\theta & \text{(triplets)} \end{cases}, \tag{1.10}$$

where $I_J = 2eJ/\hbar$. When the system is biased by a dc current I larger than the spin- and flux-dependent critical current, given by $\max_\theta\{|I_S|\}$, a finite voltage V appears. Then one possible experimental procedure might be as follows: Apply a dc bias current such that $\alpha I_J < I < (\alpha + 1)I_J$. Here, αI_J is the critical current of the triplet states, and $(\alpha + 1)I_J$ the critical current of the singlet state at $f = 0$, see (1.10). Initially prepare the system in an equal mixture of singlet and triplet states by tuning the flux around $f = 1/2$. (With electron g-factors $g \sim 0.5\text{–}20$ the Zeeman splitting on the dots is usually small compared with $k_B T$ and can thus be ignored.) Adopting the discussion of Ref. [17] to our case, we find that in this mixture, one will measure a dc voltage given by $(V_0 + 3V_1)/4$, where $V_0 = \pi\Delta[2e(\alpha + 1)]^{-1}\sqrt{(I/I_J)^2 - (\alpha - 1)^2}$ is associated with the singlet and $V_1 = \pi\Delta[2e\alpha]^{-1}\sqrt{(I/I_J)^2 - \alpha^2}$ with the triplet states. At a later time $t = 0$, the flux is switched off (i.e. $f = 0$), with I being kept fixed. The ensuing time evolution of the system is characterized by three time scales: the time $\tau_{coh} \sim \max\{1/\Delta, 1/\Gamma\} \sim 1/\Gamma$ it takes to establish coherence in the S-DD-S junction, the spin relaxation time τ_{spin} on the dot, and the switching time τ_{sw} to reach $f = 0$. We will assume $\tau_{coh} \ll \tau_{spin}, \tau_{sw}$, which is not unrealistic in view of measured spin decoherence times in GaAs exceeding 100 ns [18]. If $\tau_{sw} < \tau_{spin}$, the voltage is given by $3V_1/4$ for times less than τ_{spin}, i.e. the singlet no longer contributes to the voltage. For $t > \tau_{spin}$ the spins have relaxed to their ground (singlet) state, and the voltage vanishes. One therefore expects steps in the voltage versus time. If $\tau_{spin} < \tau_{sw}$, a broad transition region of the voltage from $(V_0 + 3V_1)/4$ to 0 will occur [19].

To our knowledge, there are no experimental reports on quantum dots coupled to superconductors. However, hybrid systems consisting of superconductors (e.g., Al or Nb) and 2DES (InAs and GaAs) have been investigated by a number of groups [20].

In conclusion, we have investigated a double quantum dot each dot of which is coupled to two superconductor leads. The superconductors induce an effective exchange coupling between spins on the double dot whereas the Josephson current from one lead to the other depends on the spin state. This leads to the possibility to probe the spin states of the dot electrons by measuring a Josephson current.

Acknowledgment

We would like to thank G. Burkard, C. Strunk, and E. Sukhorukov for discussions and the Swiss NSF for support.

References

[1] See, e.g., *Mesoscopic Electron Transport*, eds. L. L. Sohn, L. P. Kouwenhoven, G. Schön (Kluwer, Dordrecht, 1997).

[2] L. I. Glazman, K. A. Matveev, Pis'ma Zh. Eksp. Teor. Fiz. **49**, 570 (1989) [JETP Lett. **49**, 659 (1989)]; B. I. Spivak, S. A. Kivelson, Phys. Rev. B **43**, 3740 (1991).

[3] D. C. Ralph, C. T. Black, M. Tinkham, Phys. Rev. Lett. **74**, 3241 (1995).

[4] C. B. Whan, T. P. Orlando, Phys. Rev. B **54**, R5255 (1996); A. Levy Yeyati, J. C. Cuevas, A. Lopez-Davalos, A. Martin-Rodero, Phys. Rev. B **55**, R6137 (1997).

[5] S. Ishizaka, J. Sone, T. Ando, Phys. Rev. B **52**, 8358 (1995); A. V. Rozhkov, D. P. Arovas, Phys. Rev. Lett. **82**, 2788 (1999); A. A. Clerk and V. Ambegaokar, Phys. Rev. B **61**, 9109 (2000).

[6] K. A. Matveev *et al.*, Phys. Rev. Lett. **70**, 2940 (1993).

[7] R. Bauernschmitt, J. Siewert, A. Odintsov, Yu. V. Nazarov, Phys. Rev. B **49**, 4076 (1994).

[8] R. Fazio, F. W. J. Hekking, A. A. Odintsov, Phys. Rev. B **53**, 6653 (1996).

[9] J. Siewert, G. Schön, Phys. Rev. B **54**, 7424 (1996).

[10] D. Loss, D. P. DiVincenzo, Phys. Rev. A **57**, 120 (1998).

[11] D. Loss, E. V. Sukhorukov, Phys. Rev. Lett. **84**, 1035 (2000).

[12] Mahn-Soo Choi, C. Bruder, Daniel Loss, cond-mat/0001011.

[13] One could also consider atomic impurities embedded between the grains of a granular superconductor.

[14] If, instead we had assumed leads consisting of unconventional superconductors with triplet pairing, we would find a *ferromagnetic* exchange coupling favoring a triplet ground state of spin a and b on the DD. Thus, by probing the spin ground state of the dots (e.g. via its magnetic moment) we would have a means to distinguish singlet from triplet pairing. The magnetization could be made sufficiently large by extending the scheme from two to N dots coupled to the superconductor.

[15] See, e.g., M. Tinkham, *Introduction to Superconductivity*, 2nd ed. (McGraw-Hill, New York, 1996).

[16] Without restriction we can assume $\alpha > 1$, since J' could be adjusted accordingly by replacing the J'-junction by another dc SQUID and flux through it.

[17] V. Ambegaokar, B. I. Halperin, Phys. Rev. Lett. **22**, 1364 (1969).

[18] J. M. Kikkawa, D. D. Awschalom, Phys. Rev. Lett. **80**, 4313 (1998).

[19] Another experimental setup would be to use an rf-SQUID geometry, i.e., to embed the S-DD-S structure into a superconducting ring [15].

However, to operate such a device, ac fields are necessary, and the sensitivity is not as good as for the dc-SQUID geometry.

[20] B. J. van Wees, H. Takayanagi, in [1]; H. Takayanagi, E. Toyoda, T. Akazaki, Superlattices and Microstructures **25**, 993 (1999); A. Chrestin *et al.*, ibid., 711 (1999).

TRANSPORT THROUGH ARTIFICIAL KONDO IMPURITIES

S. De Franceschi,[†] S. Sasaki,[*] J. M. Elzerman,[†] W. G. van der Wiel,[†]
S. Tarucha,[*§] and L. P. Kouwenhoven[†]

[†]*Department of Applied Physics, DIMES, and ERATO Mesoscopic Correlation Project,*
Delft University of Technology
PO Box 5046
2600 GA Delft, The Netherlands

[*]*NTT Basic Research Laboratories*
Atsugi-shi, Kanagawa 243-0129, Japan

[§]*ERATO Mesoscopic Correlation Project*
University of Tokyo
Bunkyo-ku, Tokyo 113-0033, Japan

Key words: Quantum dot, Kondo effect

Abstract: We report an experiment performed on a few-electron quantum dot in which
the quantum numbers of the occupied electron states can be precisely
identified. Besides the usual Kondo behavior for spin=1/2 and odd electron
number, an unexpected Kondo effect is observed for an even electron number.
This effect, which is actually very strong, occurs at a spin-singlet/spin-triplet
transition, tuned by a magnetic field. An essentially new Kondo phenomenon
is proposed to explain our finding.

Some metals exhibit a minimum in the temperature dependence of their resistivity. This anomalous behaviour was discovered over sixty years ago and remained for a long time an open question. In 1964 J. Kondo explained the resistivity minimum as a consequence of the exchange interaction between a localised spin and conduction electrons[1]. The Kondo effect is now recognized as a key mechanism in a wide class of correlated electron systems[2,3]. Control over single, localised spins has become relevant also in fabricated structures due to the rapid developments in nano-electronics[4,5]. Experiments have already demonstrated artificial realisations of isolated magnetic impurities at metallic surfaces[6,7], nanometer-scale magnets[8], controlled transitions between two-electron singlet and triplet states[9], and a tunable Kondo effect in semiconductor quantum dots[10-13]. Here, we report an unexpected Kondo effect realised in a few-electron quantum dot containing singlet and triplet spin states whose energy difference can be tuned with a magnetic field. This effect occurs for an even number of electrons at the degeneracy between singlet and triplet states. The characteristic energy scale is found to be much larger than for the ordinary spin-1/2 case.

Quantum dots are small electronic devices[14], which confine a well-defined number of electrons, N. The total spin is zero or an integer for $N =$ even and half-integer for $N =$ odd. The latter case constitutes the canonical example for the Kondo effect[15,16] when all electrons can be ignored, except for the one with the highest energy; i.e. the case of a single, isolated spin, $S =$ 1/2 (see Fig. 1a). Although the energy level ε_o is well below the Fermi energies of the two leads, Heisenberg uncertainty allows the electron on the dot to tunnel to one of the leads when it is replaced quickly by another electron. The time scale for such a co-tunneling process[18] is $\sim\hbar/U$, where $h = 2\pi\hbar$ is Planck's constant and U is the on-site Coulomb energy. Figure 1a illustrates that particle exchange by co-tunneling can effectively flip the spin on the dot. At low temperature, the coherent superposition of all possible co-tunneling processes involving spin flip can result in a time-averaged spin equal to zero. The whole system, i.e. quantum dot plus electrodes, forms a spin singlet. The energy scale for this singlet state is the Kondo temperature, T_K. In terms of density of states, a narrow peak with a width $\sim k_B T_K$ develops at the Fermi energy (k_B is Boltzmann's constant). Note that for $N =$ even and $S = 0$, co-tunneling gives rise to a lifetime broadening of the confined state, without producing any Kondo resonance. Such even/odd behaviour corresponding to no-Kondo/Kondo has been observed in recent experiments[10,11].

It is also possible that a quantum dot with $N =$ even has a total spin $S = 1$; e.g. when the last two electrons have parallel spins. If the remaining N–2 electrons can be ignored, this corresponds to a triplet state. Parallel spin filling is a consequence of Hund's rule occurring when the gain in exchange energy exceeds the spacing between single-particle states[9]. The spin of the triplet state can also be screened by co-tunneling events. These are illustrated in the center-left side of Fig. 1b. In contrast to single-particle states that are considered in the spin-1/2 Kondo problem, the spin triplet consists of three degenerate two-particle states. Co-tunneling exchanges only one of the two

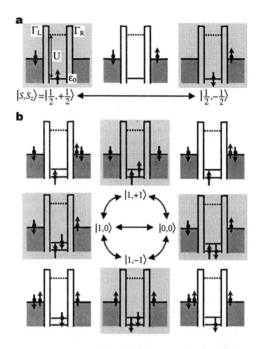

Figure 1. Spin-flip processes leading to ordinary and singlet-triplet Kondo effect in a quantum dot. (a) Co-tunneling event in a spin-1/2 quantum dot for N = odd. Only the highest-energy electron is shown occupying a single spin-degenerate level, ε_o. (The case of two, or more, closely spaced levels has also been considered theoretically within the context of the spin-1/2 Kondo effect[16].) The gray panels refer to $S_z = 1/2$ and $-1/2$ ground states, which are coupled by a co-tunneling event. The two tunnel barriers have tunneling rates Γ_R and Γ_L. In the Coulomb blockade regime ($|\varepsilon_o| \sim U$) adding or subtracting an electron from the dot implies an energy cost $\sim U$. Hence the intermediate step (diagram in the middle) is a high-energy, virtual state. The spin-flip event depicted here is representative of a large number of higher-order processes which add up coherently such that the local spin is screened. This Kondo effect leads to an enhanced linear-response conductance at temperatures $T \lesssim T_K$. (b) Co-tunneling in an integer-spin quantum dot for N = even at a singlet-triplet degeneracy. Two electrons can share the same orbital with opposite spins (singlet state in the gray panel on the right) or occupy two distinct orbitals in one of the three spin-triplet configurations (top, left, and bottom gray panels). The different spin states are coupled by virtual states (intermediate diagrams). Similar to the spin-1/2 case, spin-flip events can screen the local magnetic moment. Note that an $S = 1$ Kondo effect only involves $|1,+1\rangle$, $|1,0\rangle$, and $|1,-1\rangle$.

electrons with an electron from the leads. The total spin of the many-body Kondo state depends on how many modes in the leads couple effectively to the dot[19,20]. If there is only one mode, the screening is not complete and the whole system does not reach a singlet state. In this case the Kondo effect is called "underscreened". Calculations show that also for $S = 1$ a narrow Kondo resonance arises at the Fermi energy, however, the corresponding T_K is typically lower than in the case of $S = 1/2$ [21,22]. Some experiments have reported the absence of even/odd behaviour[23,24], which may be related to the formation of higher spin states.

Here, we investigate a quantum dot with $N = $ even where the last two electrons occupy a degenerate state of a spin singlet and a spin triplet. Figure 1b illustrates the different co-tunneling processes occurring in this special circumstance. Starting from $|S = 1, S_z = 1\rangle$, where S_z is the z-component of the total spin on the dot, co-tunneling via a virtual state $|1/2,1/2\rangle$, can lead either to the triplet state $|1,0\rangle$, or to the singlet state $|0,0\rangle$. Via a second co-tunneling event the state $|1, -1\rangle$ can be reached. As for the $S = 1$ case, the local spin can fluctuate by co-tunneling events. By coupling to all triplet states, the singlet state enhances the spin exchange interaction between the dot and the leads, resulting in a higher rate for spin fluctuations. This particular situation yields a strong Kondo effect, which is characterised by an enhanced T_K. This type of Kondo effect has not been considered before, probably because a singlet-triplet degeneracy does not occur in magnetic elements. Recent scaling calculations indeed indicate a strong enhancement of T_K at the singlet-triplet degeneracy[25]. Ref. 25 also argues that the total spin of the many-body Kondo state behaves as in the case of $S = 1$.

Our quantum dot has the external shape of a rectangular pillar (see Fig. 2a,b) and an internal confinement potential close to a two-dimensional ellipse[26]. The tunnel barriers between the quantum dot and the source and drain electrodes are thinner than in our previous devices[9,26] such that co-tunneling processes are enhanced. Figure 2d shows the linear response conductance (dc bias voltage $V_{sd} = 0$) versus gate voltage, V_g, and magnetic field, B. Dark blue regions have low conductance and correspond to the regimes of Coulomb blockade for $N = 3$ to 10. In contrast to previous experiments[10-13] on the Kondo effect, all performed on lateral quantum dots with unknown electron number, here the number of confined electrons is precisely known. Red stripes represent Coulomb peaks as high as $\sim e^2/h$. The B-dependence of the first two lower stripes reflects the ground-state evolution for $N = 3$ and 4. Their similar B-evolution indicates that the 3rd and 4th electron occupy the same orbital state with opposite spin, which is observed also for $N = 1$ and 2 (not shown). This is not the case for $N = 5$ and 6. The $N = 5$ state has $S = 1/2$, and the corresponding stripe shows a smooth evolution with B. Instead, the stripe for $N = 6$ has a kink at $B \approx 0.22$ T. From earlier analyses[26] and from measurements of the excitation spectrum at finite V_{sd} (discussed below) we can identify this kink with a transition in the ground state from a spin triplet to a spin singlet.

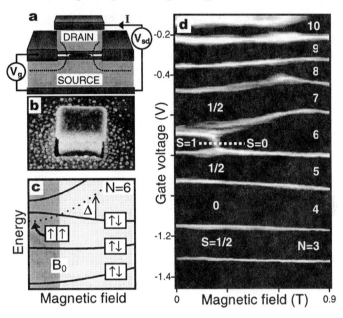

Figure 2. (a) Cross-section of our rectangular quantum dot. The semiconductor material consists of an undoped AlGaAs(7nm)/InGaAs(12nm)/AlGaAs(7nm) double barrier structure sandwiched between n-doped GaAs source and drain electrodes. A gate electrode surrounds the pillar and is used to control the electrostatic confinement in the quantum dot. A dc bias voltage, V_{sd}, is applied between source and drain and current, I, flows vertically through the pillar. In addition to V_{sd}, we apply a modulation with rms amplitude $V_{ac} = 3$ μV at 17.7 Hz for lock-in detection. The gate voltage, V_g, can change the number of confined electrons, N, one-by-one from ~10 at $V_g = 0$ to 0 at $V_g = -1.8$ V. A magnetic field, B, is applied along the vertical axis. Temperature, T, is varied between 14 mK and 1 K. The lowest effective electron temperature is 25±5 mK. (b) Scanning electron micrograph of a quantum dot with dimensions 0.45×0.6 μm^2 and height of ~0.5 μm. (c) Schematic energy spectrum. Solid lines represent the B-evolution of the first four orbital levels in a single-particle model. The dashed line is obtained by subtracting the two-electron exchange coupling from the fourth level. At the crossing between this dashed line and the third orbital level at $B = B_0$ the ground state for $N = 6$ undergoes a triplet-to-singlet transition. $B_0 \approx 0.22$ T with a slight dependence on V_g. We define Δ as the energy difference between the triplet and the singlet states. (d) Gray-scale representation of the linear conductance versus V_g and B. White stripes denote conductance peaks of height $\sim e^2/h$. Dark regions of low conductance indicate Coulomb blockade. The $N = 6$ ground state undergoes a triplet-to-singlet transition at $B_0 \approx 0.22$ T, which results in a conductance anomaly inside the corresponding Coulomb gap.

Strikingly, at the triplet-singlet transition (at $B = B_0$ in Fig. 2c) we observe a strong enhancement of the conductance. In fact, over a narrow range around 0.22 T, the Coulomb gap for $N = 6$ has disappeared completely.

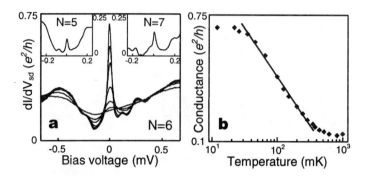

Figure 3. (a) Kondo resonance at the singlet-triplet transition. The dI/dV_{sd} vs V_{sd} curves are taken at $V_g = $ -0.72 V, $B = 0.21$ T and for $T = 14, 65, 100, 200, 350, 520,$ and 810 mK. Insets to (a): Kondo resonances for $N = 5$ (left inset) and $N = 7$ (right inset), measured at $V_g = -0.835$ V and $V_g = -0.625$ V, respectively, and for $B = 0.11$ T and T = 14 mK. (b) Peak height of zero-bias Kondo resonance *vs* T as obtained from (a) (solid diamonds). The line demonstrates a logarithmic T-dependence, which is characteristic for the Kondo effect. The saturation at low T is likely due to electronic noise.

To explore this conductance anomaly, we show in Fig. 3a differential conductance measurements, dI/dV_{sd} vs V_{sd}, taken at $B = B_0$ and V_g corresponding to the dotted line in Fig. 2d. At $T = 14$ mK the narrow resonance around zero bias has a full-width-at-half-maximum, FWHM ≈ 30 μV. This is several times smaller than the lifetime broadening, $\Gamma = \Gamma_R + \Gamma_L \approx 150$ μV, as estimated from the FWHM of the Coulomb peaks. The height of the zero-bias resonance decreases logarithmically with T (see Fig. 3b). These are typical fingerprints of the Kondo effect. From FWHM $\approx k_B T_K$, we estimate $T_K \approx 350$ mK. We note that we can safely neglect the Zeeman spin splitting since $g\mu_B B_0 \approx 5$ μV $\ll k_B T_K$, implying that the spin triplet is in fact three-fold degenerate at $B = B_0$. This condition is essential to the Kondo effect illustrated in Fig. 1b. Alternative schemes have recently been proposed for a Kondo effect where the degeneracy of the triplet state is lifted by a large magnetic field[27,28].

For $N = 6$ we find markedly anomalous T-dependence only when singlet and triplet states are degenerate. Away from degeneracy, the valley conductance increases with T due to thermally activated transport. For $N = 5$ and 7, zero-bias resonances are clearly observed (see insets to Fig. 3a) which are related to the ordinary spin-1/2 Kondo effect. Their height, however, is much smaller than for the singlet-triplet Kondo effect.

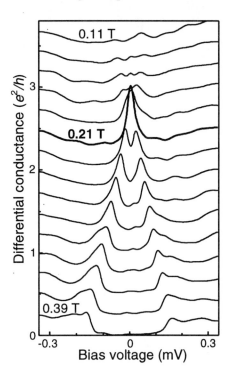

Figure 4. dI/dV_{sd} vs V_{sd} characteristics taken along the dotted line in Fig. 2d ($V_g = -0.72$ V) at equally spaced magnetic fields $B = 0.11, 0.13, ..., 0.39$ T. Curves are offset by 0.25 e^2/h.

We now investigate the effect of lifting the singlet-triplet degeneracy by changing B at a fixed V_g corresponding to the dotted line in Fig. 2d. Near the edges of this line, i.e. away from B_0, the Coulomb gap is well developed as denoted by the dark colours. The dI/dV_{sd} vs V_{sd} traces still exhibit anomalies, however, now at finite V_{sd} (see Fig. 4). For $B = 0.21$ T we observe the singlet-triplet Kondo resonance at $V_{sd} = 0$. At higher B this resonance splits apart showing two peaks at finite V_{sd}. It is important to note that these peaks occur inside the Coulomb gap. They result from "inelastic" co-tunneling events where "inelastic" refers to exchanging energy between quantum dot and electrodes[18,29]. The upper traces in Fig. 4, for $B < 0.21$ T, also show peak structures, although less pronounced.

Inelastic co-tunneling occurs when $eV_{sd} = \pm\Delta$, and this condition is independent of V_g. We believe that for small Δ ($\lesssim k_B T_K$) the split resonance reflects the singlet-triplet Kondo anomaly shifted to finite bias. This resembles the splitting of the Kondo resonance by the Zeeman effect[10,11,30], although on a very different B-scale. In the present case, the splitting occurs between two different multi-particle states and originates from the *B*-dependence of the orbital motion. For increasing Δ, the shift to larger V_{sd} induces spin-decoherence processes, which broaden and suppress the finite-bias peaks[30]. For $B \approx 0.39$ T the peaks have evolved into steps[29] which may indicate that the spin-coherence associated with the Kondo effect has completely vanished.

We acknowledge M. Eto, Yu. V. Nazarov, L. Glazman, K. Maijala, S. M. Cronenwett, J. E. Mooij, and Y. Tokura for valuable discussions.

REFERENCES

1. J. Kondo, Prog. Theor. Phys. **32**, 37 (1964).
2. A. C. Hewson, *The Kondo Problem to Heavy Fermions* (Cambridge University Press, Cambridge, 1993).
3. D. L. Cox and M. B. Maple, Physics Today **48**, 32-40 (1995).
4. G. A. Prinz, Science **282**, 1660-1663 (1998).
5. D. Loss and D. P. DiVincenzo, Phys. Rev. A **57**, 120-126 (1998).
6. V. Madhavan *et al.*, Science **280**, 567-569 (1998).
7. J. Li *et al.*, Phys. Rev. Lett. **80**, 2893-2896 (1998).
8. D. D. Awschalom and D. P. DiVincenzo, Physics Today **48**, 43-48 (1995).
9. S. Tarucha *et al.*, Phys. Rev. Lett. **84**, 2485-2488 (2000).
10. D. Goldhaber-Gordon *et al.*, Nature **391**, 156-159 (1998).
11. S. M. Cronenwett, T. H. Oosterkamp, and L. P. Kouwenhoven, Science **281**, 540-544 (1998).
12. J. Schmid *et al.*, Physica B **256-258**, 182-185 (1998).
13. F. Simmel *et al.*, Phys. Rev. Lett. **83**, 804-807 (1999).
14. L. P. Kouwenhoven *et al.*, in *Mesoscopic Electron Transport*, edited by L. L. Sohn, L. P. Kouwenhoven, and G. Schön, (Kluwer, Series E 345, 1997), p.105-214.
15. L. I. Glazman, and M. E. Raikh, JETP Lett. **47**, 452-455 (1988).
16. T. K. Ng, and P. A. Lee, Phys. Rev. Lett. **61**, 1768-1771 (1988).
17. T. Inoshita *et al.*, Phys. Rev. B **48**, 14725-14728 (1993).
18. D. V. Averin and Y. V. Nazarov, in *Proceedings of a NATO Advanced Study Institute*, Grabert, H. & Devoret, M. H. Eds., Les Houches, France, 5 March to 15 March, 1991 (Series B, **294**, 217, Plenum Press, New York, 1991).
19. D. C. Mattis, Phys. Rev. Lett. **19**, 1478-1481 (1967).
20. P. Nozières and A. Blandin, J. Physique **41**, 193-211 (1980).
21. Y. Wan, P. Phillips, and Q. Li, Phys. Rev. B **51**, 14782-14785 (1995).
22. W. Izumida, O. Sakai, and Y. Shimizu, J. Phys. Soc. Jpn. **67**, 2444-2454 (1998).
23. S. M. Maurer *et al.*, Phys. Rev. Lett. **83**, 1403-1406 (1999).
24. J. Schmid *et al.*, preprint.
25. M. Eto and Yu. V. Nazarov, cond-mat/0002010.
26. D.G. Austing *et al.*, Phys. Rev. B **60**, 11514-11523 (1999).
27. M. Pustilnik, Y. Avishai, and K. Kikoin, Phys. Rev. Lett. **84**, 1756-1759 (2000).
28. D. Giuliano and A. Tagliacozzo, Phys. Rev. Lett. **84**, 4677-4680 (2000).
29. Y. Funabashi *et al.*, Jpn. J. Appl. Phys. **38**, 388-391 (1999).
30. N. S. Wingreen and Y. Meir, Phys. Rev. B **49**, 11040-11052 (1994).

COMPENSATION OF THE SPIN OF A QUANTUM DOT AT COULOMB BLOCKADE

Kondo correlation with the contacts leads to a macroscopic separation of the charge from the spin in the dot

Domenico Giuliano

Department of Physics, Stanford University, Stanford, California 94305

asterion@partenope.stanford.edu

Benoit Jouault

GES, UMR 5650, Université Montpellier II, 34095 Monpellier Cedex 5, France

jouault@ges.univ-montp2.fr

Arturo Tagliacozzo

Dipartimento di Scienze Fisiche, Università di Napoli "Federico II ", Monte S.Angelo via Cintia, I-80126 Napoli, Italy

and Istituto Nazionale di Fisica della Materia (INFM), Unitá di Napoli

arturo@na.infn.it

Keywords: quantum dot, Kondo effect. PACS 71.10.Ay,72.15.Qm,73.23.-b,73.23.Hk,79.60.Jv, 73.61.-r

Abstract

We discuss a new entangled state that has been observed in the conduction across a quantum dot. At Coulomb blockade, electrons from the contacts correlate strongly to those localized in the dot, due to cotunneling processes. Because of the strong Coulomb repulsion on the dot, its electron number is unchanged w.r.to the dot in isolation, but the total spin is fully or partly compensated. In a dot with $N = even$ at the singlet-triplet crossing, which occurs in large magnetic field, Kondo correlations lead to a total spin $S = 1/2$.

Introduction

A very special kind of Macroscopic Quantum Coherence has been recently measured. This is the Kondo anomaly of the conductance across a quantum dot (QD) polarized at Coulomb blockade (CB) [1, 2, 3]. Due to confinement, Coulomb correlations between electrons added to a QD are strong. An appropriate choice of the gate voltage V_g blocks direct sequential tunneling, what is called Coulomb blockade regime[4]. If coupling between the QD and the leads is not weak and the temperature is low enough, an anomaly appears in the differential conductance at zero voltage bias. Kondo correlations are established by cotunneling processes. This leads to a macroscopic ground state (GS) with very peculiar properties, in which the dot and the contacts are entangled.

Kondo conduction is a well known phenomenon occurring in non magnetic metals with diluted magnetic impurities[5]. Typical examples are Fe in Cu or Mn in Ag: a minimum in the resistivity of the metal occurs in lowering the temperature below T_K, before saturating to a residual value corresponding to potential (non magnetic) scattering. Kondo showed that [6] there is a crossover to a correlated state of the system in the neighborhood of T_K, in which exchange interaction between the conduction electrons and the local moment of the impurity leads to scattering events on the microscopic scale, in which the electronic spin is flipped. This gives rise to a term in the impurity contribution to the resistivity that increases with decreasing temperature. Spin flip can only occur if the spin on the local magnetic moment of the impurity atom changes accordingly, to achieve compensation. This shows up in the disappearance of a Curie-Weiss spin susceptibility which becomes constant at low temperature.

To some extent the conductance across a QD in the CB regime can be regarded as the mesoscopic realization of the same physics[7]. This is not so surprising, as QD have always been referred to as artificial atoms and the contacts provide the delocalized conduction electrons of the host metal.

Of course the energy scales are quite different. Energy level separation in a QD is of the order of meV while it is of two or three orders of magnitude larger for the d, f electronic levels in the impurity atom. This implies that the temperature scale (the Kondo temperature T_K) is correspondingly reduced from tens of K to $100mK$ and below. Also, because the anomalous contribution to the conductance stems from cotunneling processes, the internal structure of the QD is of big relevance. Coulomb interaction is dominant in QD, so that the levels involved are many particle levels which strongly depend on the number of electrons added to the dot N by means of a gate voltage V_g.

In this sense one should be cautious when extending the single impurity Anderson model to the dot case[8]. The model describes localized single electron levels with an onsite Coulomb repulsion U, in hybridization with delocalized conduction electrons. It is important that the order of magnitude of U is the same for both the magnetic atom and the QD, so that, in the dot case, we are always in the large U limit, what makes the experimental observation even more striking. In particular, the Kondo peak is observed in a QD in the CB region where conductance is otherwise exponentially small. This is because QDs can be tuned within a very wide range of parameter values and the experiment on Kondo conduction is equivalent to the measurement of one single impurity in the metal host.

As we will show, the investigation of the analogies between the two systems is very fruitful, because new features arise in QD, that are not found when localized moments in diluted alloys are studied.

One example is Kondo conduction in a magnetic field[9]. Because orbital effects are very important in a QD when a magnetic field is applied, the energy scale for the Zeeman spin splitting is not the same as the energy scale for level separation in a QD in magnetic field. This implies that, in diluted alloys, a magnetic field B is always disruptive for Kondo correlations because it lifts the degeneracy of the levels, which is required for the spin flip scattering events. On the dot side, it may even produce crossings of the levels of the dot, which could favour Kondo conductance. We will discuss an example of this in Sect. 2.

The spectroscopy of a QD is usually performed in terms of transport measurements[10]. Two electrodes are attached to the QD and the low-temperature zero-bias conductance is monitored. When the gate voltage V_g changes, the conductance undergoes a series of peaks at zero source-drain voltage V_{sd} each time the increase in V_g matches the chemical potential for adding an extra electron to the dot (Coulomb oscillations). In between two peaks, the number of particles on the QD is fixed (CB regime)(see Fig. 1 for the setup (a) and for a schematic grey-scale drawing of the conductance vs V_g and V_{sd} (b)) . Contributions to the current at CB are fourth order in the transmission amplitude and can be very small (white regions in Fig.1 b)). In these conditions the total spin of the dot S is the only left dynamical variable, as in a magnetic impurity.

Provided this spin is not a singlet, it can become strongly coupled to the spin density of the contacts electrons if $T < T_K$ and the leads are not weakly linked. The correlated state gives rise to non perturbative

differential conduction at zero voltage bias, fairly independent of the
value of V_g within the CB plateaux.

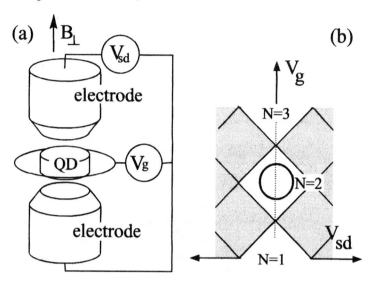

Figure 1 a) Geometry of a vertical QD. b) grey scale picture of the conductance
versus gate voltage V_g and source-drain voltage V_{sd}. In white areas the conductance
is vanishingly small.

The first observations were in dots at CB with an odd number of
electrons [1, 2]. The GS is a doublet and coupling to the contacts in the
CB region generates the Abrikosov-Suhl resonance at the Fermi energy
$\mu \equiv \epsilon_F$ of the contacts which gives a peak in the differential conductance
at zero bias due to cotunneling across the QD.

In Sect. 1 we review the features of the single impurity Anderson
model which are at the basis of the physics involved in the Kondo correlated state for QDs with an odd number of electrons. The level of the
localized impurity is doubly degenerate and located deeply below μ but
double occupancy of this level would cost an energy much larger than
μ. We show that, in the limit of strong onsite repulsion U, in spite of
the coupling to the leads, occupation of the impurity level is frozen. In
addition to the disappearance of the charge degree of freedom, a singlet
state is generated on the impurity, with the help of the conduction electrons. In the symmetrical case, the fixed point GS has $(N = 1, S = 0)$.
Spin-charge separation occurs.

Dots at CB with an even number of electrons are not expected to give
Kondo conduction because the GS is supposed to be a singlet already.

A notable exception is when the GS has higher total spin. In dots with $N = 4$ and $N = 6$ Hund's rule states that the GS is a triplet. This is turned into a singlet by applying a small B (denoted "TS crossing" hereafter).

In Sect. 2 we consider the case of a realistic isolated QD with few electrons using exact diagonalization methods [11]. A magnetic field $B_\perp$, orthogonal to the dot is applied. This produces large orbital changes in the electronic states and, eventually, crossing of levels. The Kondo effect expected for $S = 1$ is strongly enhanced due to this crossing and the anomalous conductance at zero voltage has been recently observed [9]. The very peculiar physics at the TS crossing is under consideration at present[12, 13].

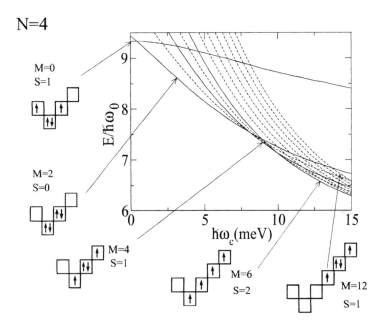

Figure 2 Energy levels of an isolated QD with $N = 4$ electrons *vs.* a magnetic field orthogonal to the dot. The total orbital angular momentum and the total spin of the different GSs are shown, together with the configuration of the Slater determinant which has the highest projection onto the GS.

But there is another possibility of Kondo conductance which has been studied in ref[14]. Increasing B quite substantially, the dot undergoes a new transition from the singlet to triplet state because of orbital effects (denoted ST crossing hereafter), in which the Zeeman spin splitting cannot be ignored. This changes the properties of the GS in a peculiar way, leading to an even charge on the dot with total spin $\frac{1}{2}$.

1.   FROM THE SYMMETRIC ANDERSON MODEL TO THE IMPURITY SPIN DYNAMICS

We refer to a QD connected to the leads in the CB regime, in the limit of the Coulomb interaction being very large. As discussed in the introduction, the dot charge degree of freedom is frozen out, what leaves the dot total spin S as the only dynamical variable. The charge-spin separation emerges. This phenomenon is at the basis of the analogy between this system and that of an impurity local moment in a metal alloy. Here we review the main features of the single impurity spin $\frac{1}{2}$ symmetric Anderson model[8]. The path integral formalism by Anderson,Yuval and Hamann [15, 16] is particularly suited to focus on the spin-charge separation in this model.

The Hilbert space of the spin $\frac{1}{2}$ impurity is conveniently described by two fermionic operators, $d_i^\dagger$, acting on the vacuum $|0\rangle$. The label $i = 1, 2$ is here a spin label, although it could be a generic label for the two degenerate single particle levels located at energy ϵ_d.

The impurity Hamiltonian is:

$$H_d = \epsilon_d \sum_i n_i + U n_1 n_2 \qquad (1.1)$$

where $n_i = d_i^\dagger d_i$ and U is the strength of the onsite repulsion.

We linearize the conduction electron band ϵ_k close to ϵ_F. This allows to write down their Hamiltonian as that of a one-dimensional Fermi gas. To keep contact with the dot picture, we divide the space into a left (L) and a right (R) region:

$$H_{L,R} = \sum_{k\sigma} \epsilon_k b_{(L,R)k\sigma}^\dagger b_{(L,R)k\sigma} \qquad (1.2)$$

If $\epsilon_d = -U/2$ *w.r.to* the chemical potential of the conduction electrons μ (taken as the zero of the single particle energies), the Anderson model which arises is symmetric. In fact, the energies of the empty impurity state,^0E, and that of the doubly occupied impurity state, $^2E = 2\epsilon_d + U$, are both zero, while the singly occupied impurity level has energy $^1E = -U/2$.

Finally, the hybridization Hamiltonian between the conduction electrons and the impurity is:

$$H_t = \frac{1}{\sqrt{2}} \sum_k \{\Gamma^*[(b^\dagger_{Lk\uparrow} + b^\dagger_{Rk\uparrow})d_1 + (b^\dagger_{Lk\downarrow} + b^\dagger_{Rk\downarrow})d_2] + \text{h.c.}\} \qquad (1.3)$$

where all the tunneling amplitudes have been taken equal for simplicity.

We now formulate the quantum dynamics of the system described by eqs.(1.1,1.2,1.3) using the imaginary time Feynman Path Integral formalism. Following [16], we integrate out the conduction electron fields and we obtain the partition function in terms of the Grassman fields d, $d^\dagger$ only ($\hbar = 1$ in the following):

$$\mathcal{Z}(\mu) \propto \int \Pi_i \left(Dd_i Dd^\dagger_i \right) e^{-\mathcal{A}_d} e^{-\beta|\Gamma|^2 \sum_{\omega_m,i} d^\dagger_i K(i\omega_m)d_i} \qquad (1.4)$$

where $i\omega_m$ are Fermionic Matsubara frequencies,

$$\mathcal{A}_d = \int_0^\beta d\tau \left\{ \sum_i \left[d^\dagger_i \frac{\partial}{\partial\tau} d_i + \epsilon_d \, d^\dagger_i d_i \right] + U d^\dagger_1 d_1 d^\dagger_2 d_2 \right\} \qquad (1.5)$$

and $K(i\omega_m)$ is:

$$K(i\omega_m) = \frac{L}{2\pi} \int_{-D}^{D} \frac{dk}{i\omega_m + vk} = \frac{L}{2\pi v_F} \ln\left(-\frac{v_F D + i\omega_m}{v_F D - i\omega_m} \right). \qquad (1.6)$$

Here v_F is the Fermi velocity and D is the band cutoff, symmetrical *w.r.to* ϵ_F.

The average impurity occupation number is calculated from the retarded part of the Green function $G^R(\omega)$:

$$< n_i >= \int_{-\infty}^{\infty} \frac{d\omega}{2\pi} iG^R(\omega) \, .$$

$G^R(\omega)$ is obtained from $G(i\omega_m)$ in the limit to real frequencies: $i\omega_m \to \omega + i0^+$. In the case $U = 0$ we have:

$$G^{-1}_{(0)}(i\omega_m) = i\omega_m - \epsilon_d + |\Gamma|^2 K(i\omega_m) \to \omega - \epsilon_d - \frac{\Delta}{\pi} \ln\left| \frac{v_F D + \omega}{v_F D - \omega} \right| + i\Delta$$

where $\Delta = \pi N(0)|\Gamma|^2$ and $N(0)$ is the density of states at the Fermi energy per spin $L/2\pi\hbar v_F$.

Assuming that $\tilde{\epsilon}_d$ solves the equation $\Re\{(G^0)^{-1}\} = 0$, we get:

$$< n_i >= \int_{-\infty}^{\infty} \frac{d\omega}{2\pi} iG^R_0(\omega) \approx i \int_{-\infty}^{\infty} \frac{d\omega}{2\pi} \frac{1}{\omega - \tilde{\epsilon}_d + i\Delta}$$

The imaginary part is odd and vanishes upon integration. The real part gives:

$$< n_i >= \int_{-\infty}^{\infty} \frac{d\omega}{\pi} \frac{\Delta}{(\omega - \tilde{\epsilon}_d)^2 + \Delta^2} = \frac{1}{2} - \frac{1}{\pi} \arctan \frac{\tilde{\epsilon}_d}{\Delta} \qquad (1.7)$$

Eq.(1.7) describes the Friedel screening around the impurity due to a resonance at $\tilde{\epsilon}_d \approx \epsilon_d$ of width Δ.

We now discuss the role of the onsite Coulomb interaction. In the large-U limit we have:

$$\exp \int_0^\beta d\tau \left\{ -\epsilon_d(n_1 + n_2) - Un_1 n_2 \right\}$$

$$= e^{\frac{\beta}{U} \epsilon_d^2} \delta(n_1 + n_2 + 2\epsilon_d/U) \cdot e^{\frac{U}{4} \int_0^\beta d\tau(n_1 - n_2)^2} , \qquad (1.8)$$

where the delta function implements the constraint of single site occupancy in the symmetric case, $\epsilon_d = -U/2$.

The quartic interaction is decoupled by means of a Hubbard-Stratonovitch boson field $X(\tau)$, according to the identity:

$$e^{\frac{U}{4} \int_0^\beta d\tau(n_1 - n_2)^2} = \int DX e^{-\frac{1}{4U} \int_0^\beta d\tau(X^2(\tau) + 2U(n_1 - n_2)X(\tau))}.$$

Having introduced $X(\tau)$, $\mathcal{Z}(\mu)$ takes the form:

$$\mathcal{Z}(\mu) \propto \int DX e^{-\frac{1}{4U} \int_0^\beta d\tau X^2(\tau)}$$

$$\times \int \Pi_i \left(Dd_i Dd_i^\dagger e^{-\int_0^\beta d\tau d\tau' d_i^\dagger(\tau) G_{(U)}^{-1}(\tau - \tau')d_i(\tau')} \right)$$

$$\times \delta(n_1 + n_2 - 1) \cdot \frac{1}{2} \sum_{j=1,2} e^{(-1)^j \int_0^\beta d\tau[n_j - \frac{1}{2}]X(\tau)} . \qquad (1.9)$$

Note that now the term $\epsilon_d \sum_i n_i$ was included in eq.(1.8), so that $G_{(U)}^{-1}(i\omega_m) = i\omega_m + |\Gamma|^2 K(i\omega_m)$ in this case. At odds with the case $U = 0$ here we have $\tilde{\epsilon}_d \approx 0$. The resonance is at the Fermi level, in spite of the fact that the original localized level is at ϵ_d. This makes eq.(1.7) consistent with the single site occupancy constraint. The partition function in eq.(1.9) describes an effective spin-1/2 coupled to the fluctuating magnetic field $X(\tau)$. Its dynamics is constrained by the requirement that the impurity is singly occupied. This should be implemented at any imaginary time, what is quite hard [17]. In the following it will be accounted for in the average. Indeed, the average over the $(0, \beta)$ interval, of the field configurations we consider, satisfies it.

Therefore, putting aside the delta function, the integration to be performed over the Grassman fields $d_i, d_i^\dagger$ is:

$$\int Dd_i \, Dd_i^\dagger \, e^{\int_0^\beta d\tau d\tau' \, d_i^\dagger(\tau) G_{(U)}^{-1}(\tau-\tau') d_i(\tau') - \int_0^\beta d\tau X(\tau)[n_i(\tau)-\frac{1}{2}]} \qquad (1.10)$$

By including a coupling constant g in front of X we obtain the result:

$$e^{-\int_0^1 dg \int_0^\beta X(\tau)[G(gX,\tau,\tau^+)-\frac{1}{2}]d\tau}$$

Because $G_{(U)}(\tau) \approx -\frac{1}{2\pi\Delta} \mathcal{P}(1/\tau)$, the equation for $G[\xi, \tau, \tau']$:

$$G(\xi,\tau,\tau') = G_{(U)}(\tau-\tau') + \int_0^\beta d\tau'' G_{(U)}(\tau-\tau'')\xi(\tau'')G(\xi,\tau'',\tau') \quad (1.11)$$

is the Mushelishivili equation [16]. G may be split into two contributions, one that is singular as $\tau' \to \tau$, and a second one that is regular. The singular contribution is given by:

$$-\frac{1}{\Delta}\frac{1}{1+\xi^2(\tau)}\left[\frac{1}{\pi}\mathcal{P}\frac{1}{\tau-\tau'} + \xi\Delta\delta(\tau-\tau')\right] \qquad (1.12)$$

where $\xi(\tau) = X(\tau)/\Delta$.

Eq.(1.12) generates an effective potential for the field ξ, $V[\xi]$, given by:

$$V[\xi] = \int_0^\beta d\tau \Delta \left[\frac{\Delta}{4U}\xi^2 - \frac{1}{\pi}\left(\xi\arctan(\xi) - \frac{1}{2}\ln(1+\xi^2)\right)\right] \qquad (1.13)$$

If the condition $\Delta/4U < 1$ is satisfied, V has non-trivial minima at $\xi_0 = \frac{2U}{\pi\Delta}\arctan(\xi_0) \approx \pm U/\Delta$ (for large ξ_0).

The saddle point solution $\xi_{sp}(\tau)$ is either static or, if it is τ−dependent, it is periodic. It represents a series of jumps between the two minima. All these solutions satisfy the constraint on the average within the imaginary time interval $(0,\beta)$, because: $\langle n_i - \frac{1}{2}\rangle \approx -(-1)^i\overline{X_{sp}(\tau)}/2U = 0$.

In the next section we shall map the system onto a 1-dimensional Coulomb Gas (1-d CG) of jumps between the two minima (instantons) and shall derive the low-temperature Kondo physics as condensation of instantons by means of a Renormalization Group (RG) analysis.

1.1. THE EQUIVALENT 1-DIMENSIONAL COULOMB GAS

The mapping between the low-temperature Anderson model and a 1-d CG of instantons has been extensively discussed in the literature [15, 18].

Hence, here we will just skip all the steps leading to the final form of the effective action. The full partition function may be approximated with the sum over the trajectories given by hopping paths and will be given by:

$$Z = \sum_{N=0}^{\infty} \frac{1}{2N!} \int_0^{\beta} d\left(\frac{\tau_1}{\tau_0}\right) \cdots \int_0^{\beta} d\left(\frac{\tau_{2N}}{\tau_0}\right) [e^{\frac{1}{2}\sum_{i\neq j=1}^{2N}(-1)^{i+j}\alpha^2 \ln\left|\frac{\tau_i - \tau_j}{\tau_0}\right|} Y^{2N}]$$

(1.14)

The first relevant parameter is the "fugacity" Y, corresponding to the probability for an istanton to take place within a certain state of the system. It is given by:

$$Y = \tau_0 e^{-\bar{\mathcal{A}}}$$

where $\bar{\mathcal{A}} \sim \tau_0 U$ is the action of the single blip.

The second parameter is provided by the bare "interaction strength" between instantons:

$$\alpha_b = \sqrt{2}(2\arctan(\xi_0)) = \sqrt{2}(1 - 2\Delta/\pi U)$$

(1.15)

The integral over the "centers of the Instantons" has to be understood such that τ_i and τ_j never become closer than τ_0. Here the instantons interact via a logarithmic potential $\ln|(\tau_l - \tau_{l'})/\tau_0|$, so that an ultraviolet cutoff, τ_0 is needed. The requirement that the physics is independent of τ_0 provides the RG equation for the parameters. By rescaling τ_0 to $\tau_0 + d\tau_0$, one obtains the scaling of the fugacity and the renormalization of the coupling constant induced by processes of fusion of charges, which lead to the renormalization group equations [18]:

$$\frac{dY}{d\ln \tau_0} = (1 - \frac{\alpha^2}{2})Y \; ; \; \frac{d\alpha^2}{d\ln \tau_0} = -2Y^2\alpha^2$$

(1.16)

The scaling eq.s (1.16) are discussed in the next subsection. Here it is enough to mention that, according to eq.(1.15) the regime relevant to our analysis has $\alpha^2/2 < 1$. Then, the flow is towards $Y \to \infty$ and $\alpha^2 \to 0$. The corresponding phase is characterized by a "proliferation" of instantons, namely, by a continuous cotunneling between the impurity and the leads. Because of the antiferromagnetic coupling (AF), this produces a continuous flipping of the impurity spin S_{eff} (whose $z-$component is $\langle (n_2 - n_1)(\tau) \rangle$). The latter is screened out, what implies charge-spin separation: the charge on the impurity is 1, while the spin is zero. We estimate now the Kondo temperature, T_K, by using the RG equations (1.16).

1.2. THE KONDO TEMPERATURE

The Kondo temperature T_K is defined as the temperature at which instanton condensation sets in. In order to estimate it, we use the system of equations (1.16), starting the flow from $\Delta/U \ll 1$. If we define $\chi = 1 - \alpha^2/2$, eq.s (1.16) can be approximated as:

$$\frac{dY^2}{d\ln\tau_0} = 2\chi Y^2 \quad ; \quad \frac{d\chi^2}{d\ln\tau_0} = 2\chi Y^2 \tag{1.17}$$

The system (1.17) has a constant of motion, given by $Y^2 - \chi^2/2 = Y_0^2 - \chi_0^2/2 = cnst$. A particular case is when $Y_0 = |\chi_0|/\sqrt{2}$. Its RG trajectory is a straight line in the Y-χ plane, which separates two phases:

i) The phase with $|\chi_0|/\sqrt{2} < Y_0$ and $\chi_0 < 0$, that is attracted by the line ($Y_\infty = 0$, $\chi_\infty = cnst < 0$);

ii) The phase with $|\chi_0|/\sqrt{2} > Y_0$, *or* $\chi_0 > 0$, that is attracted by the point ($Y_\infty = \infty$, $\chi_\infty = \infty$).

It should be kept in mind, however, that eq.s(1.16,1.17) are only valid in the first stages of scaling, so that these flow limits do not strictly hold, as demostrated by the exact solution [19]. In any case, the line $Y_0 = \chi_0/\sqrt{2}$ signals the crossover between the two regimes. The equation for χ on this line becomes: $d\chi/d\ln\tau_0 = \chi^2$. According to this equation $\tau_0 \exp(1/\chi)$ is a scale invariant. Because a decrease of temperature described by $d\beta$ corresponds to a dilation of τ_0 according to $d\tau_0/\tau_0 = d\beta/\beta$, the scale invariant temperature is $T_K = \tau_0^{-1} e^{-1/\chi_0} = \tau_0^{-1} e^{-(\pi U)/(4\Delta)}$.

The prefactor $\tau_0^{-1} \approx (U\Delta)^{\frac{1}{2}}$ can only be obtained by including two loop corrections.

In the language of the Coulomb gas of instantons, T_K can be interpreted as the temperature where the screening of the effective interaction between charges starts. Below this temperature Y flows to infinity and α^2 flows to zero. This corresponds to a fully coherent state, the "instanton condensate", in which the impurity spin is fully compensated.

2. KONDO EFFECT IN MAGNETIC FIELD

In Sect. 1 we reviewed the Anderson model for the electronic state of a localized impurity that hybridizes with a continuum of conduction electrons. This model can be adopted as a paradigm to describe Kondo conduction in a QD, provided some extra features of the QD (which is certainly much a more complex object than a magnetic impurity atom) are accounted for. Clarification of these points also offers the way through to some peculiarities of the Kondo state in QDs that are absent in the conduction of diluted alloys. Indeed, astonishing properties have been experimentally tested of the Kondo effect in a QD in magnetic field [9] and we believe that others will be soon revealed.

Here we quote the most relevant ones.

Because the dot has an internal structure its states are many body states which cannot be obtained by application of fermionic single particle operators d_i as we did in Sect. 1. Only under very special circumstances this is possible, in an approximated way[14]. The role of the excited states should also be considered. One more aspect to be taken care of is the fact that, being the dot not point-like, the symmetry of the coupling to the leads can be important. In this respect a vertical geometry with azimutal symmetry, as the one sketched in Fig. 1 (a), offers some control. In this case the angular momentum is conserved in the tunneling from the contacts to the dot and the theory is effectively one-dimensional, as in Sect.1.

The scenario of Sect.1 can be applied to dots at CB with an odd number of electrons, whose GS is a doublet ($S = \frac{1}{2}$).

In the case of N being even, the GS is expected to be a singlet, what would rule out the possibility of Kondo coupling. Indeed first observations in the absence of magnetic field reported a "parity " effect, by which Kondo behaviour was alternating with increasing N[2].

However, orbital effects are very strong in a QD, when a magnetic field orthogonal to the dot plane, $B_\perp$, is applied. This could produce higher spin states and crossing of levels, which give rise to Kondo conduction also when N is even.

Figure 2 shows the many body energy levels of an isolated dot confined by a parabolic potential in a magnetic field $B_\perp$ ($\omega_c = eB/mc$) for $N = 4$. Quantum numbers are the total angular momentum along the z-direction M, the total spin S and the spin component S_z. The configuration sketched aside represent the filling of single particle orbitals (which are $2d$ harmonic oscillator orbitals: n is an integer, $m = -n, ..n$ is the orbital angular momentum, increasing by steps of two) in the Slater determinant which has largest weight in each state ($M = \sum_{i=1,N} m_i$, $S_z = \sum_{i=1,N} \sigma_i$ where m_i and σ_i are the angular momentum and spin component of each electron along z).

At zero magnetic field Hund's rule applies and the GS has $M = 0$ and $S = 1$. Even a small $B_\perp$ generates large orbital changes in the electronic state. Because of B, Hund's rule breaks down and the GS becomes a singlet, so that triplet and singlet levels cross (the TS crossing). The Kondo effect expected for $S = 1$ is strongly enhanced due to this crossing and the anomalous conductance at zero voltage has been recently observed [9]. The very peculiar physics at the TS crossing has been also theoretically studied [12, 13].

By increasing B further, other level crossings are met (few of which appear in Fig.2). They correspond to an increase of the orbital angular momentum with magnetic field, and possibly of the total spin. The first of these crossings is the one between $(M = 2, S = 0)$ and $(M = 4, S = 1)$ which occurs at a value of the magnetic field which is quite substantial (the ST point).

Occurrence of the ST point is quite generic in dots with $N = even$. In fact, with increasing of B, M increases to take advantage of the Zeeman orbital term and to reduce the Coulomb interaction whose strength increases also. Meanwhile, the total spin increases up to the largest possible value, thus producing a gain in exchange energy. Zeeman spin splitting, being a small correction, is not included.

The prototype of such a crossing is the Singlet-Triplet crossing for $N = 2$, we focus on in the following. The field value is $B_* \sim 4T$, but it can be modulated over a wide range. The GS for $N = 2$ in the absence of $B_\perp$ is a non degenerate spin singlet, $^2S_0^0$. At $B_\perp = B_*$, if Zeeman spin splitting is sizeable, first crossing occurs between $^2S_0^0$ and the spin triplet with total angular momentum $M = 1$, $^2T_1^1$ ($^2T_{S_z}^M : S = 1, M = 1$). It is important that, at $B = B_*$, the total spin of the dot is the only dynamical variable. This can be inferred from Fig. 3, where the charge and spin density are plotted for the $N = 2$ dot at $B = B_*$. While the charge density $\rho(r)$ in the dot is unaffected when B moves across B_*, the spin density jumps dramatically from zero when is $B < B_*$ to $\sigma^z(r) = \frac{1}{2}\rho(r)$ for $B > B_*$.

The dynamics of the dot between the two macroscopic spin states $S = 0, 1$ is induced by the coupling to the contacts. Tunneling to and from the dot is only virtual and occupation of the dot, if tuning is appropriate, is still $N = 2$. As shown by a Schrieffer-Wolff transformation [14], coupling to the leads involves both orbital and spin variables. However, in the special situation here devised, the exchange of angular momentum is locked in with that of the spin in such a way that coupling is overall of AF type.

This Kondo coupling is peculiar. Of the four levels involved, the two crossing levels play the role of the levels of an effective spin $\frac{1}{2}$, S_{eff}, which is acted on by the deviations $\delta B = B - B_*$, only. The other two levels can be attributed to another spin $\frac{1}{2}$, S_r which is fully decoupled. In the dynamics of S_{eff}, the external magnetic field B_* disappears completely, provided conduction electrons of both spin orientations are present at ϵ_F. If the hybridization with the contacts, Δ, is large enough and $T < T_K$, the system flows towards the strongly-coupled fixed point of a standard spin $\frac{1}{2}$ Kondo model, unlike what happens at the TS crossing point. The spin S_{eff} is screened out by the spin density of the delocalized electrons

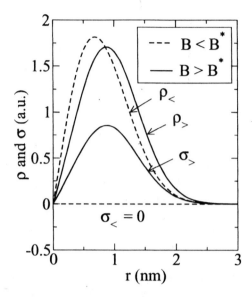

Figure 3 Charge density as a function of the radius r. $B < B_*$ ($B > B_*$) should be understood as B slightly smaller (higher) than B_*. The charge density ρ is only slightly affected when B goes across B_*. The spin density σ is zero when $B < B_*$. It has the profile of the charge density as $B > B_*$ ($\sigma_> = \frac{1}{2}\rho_>$).

and S_r only survives. We end up with charge $N = 2$ on the dot and a dot spin $S_r = 1/2$ with levels splitted in the magnetic field. This is the inverted effect *w.r.to* Kondo coupling for $N = odd$ and leads to fractionalization of the spin in the QD.

3.   SUMMARY

To conclude, Kondo conduction in a QD at CB is the striking realization of a macroscopic entangled state between the dot and the contacts. Alternatively, it can be seen as an extreme condition by which the measuring apparatus is fully invasive. Tunneling across the dot is no longer perturbative and separation of the dot total spin from the dot charge sets in. Because of the internal structure of the dot, new properties of the Kondo conduction arise, which are not present in the conductivity of diluted alloys where the Kondo effect was first discovered. In particular, while the presence of a magnetic field, by lifting the degeneracy of the impurity levels is disruptive in diluted alloys, strong Coulomb interaction in dots can give rise to drastic orbital changes and to level crossing. In these conditions the magnetic field acts in favour of a Kondo coupling and strong conduction anomalies in dots at Coulomb blockade can be measured. The Kondo temperature in these systems is rather low ($T_K \sim 100mK$ and below). This notwithstanding, it is the very discreteness of the levels in the confined dot geometry to support the flow to strong coupling at low temperature, irrespective of the influence of the environment.

Acknowledgments

The authors acknowledge useful discussions with B.Altshuler, S.De Franceschi, B.Kramer, G.Morandi and J.Weis.

Work supported by INFM (Pra97-QTMD) and by EEC with TMR project, contract FMRX-CT98-0180.

References

[1] D.Goldhaber-Gordon, H.Shtrikman, D.Mahalu, D.Abusch-Magder, U.Meirav and M.A.Kastner, *Nature* **391**, 156 (1998)

[2] S.M.Cronenwett,T.H.Oosterkamp and L.P.Kouwenhoven *Science* **281**,540 (1998)

[3] J.Schmid,JWeis,K.Eberl and K.v.Klitzing, Physica B **256-258**,182 (1998)

[4] L.P. Kouwenhowen *et al.*, in "Mesoscopic electron transport", NATO ASI Series E **345**,105; L.Sohn, L.P.Kouwenhoven and G.Schön eds., Kluwer, Dordrecht,Netherlands (1997); L.P. Kouwenhoven *et al.* , *Science* **278**, 1788 (1997); S. Tarucha *et al.*, Phys. Rev. Lett. **77**, 3613 (1996).

[5] A.C. Hewson: "The Kondo Effect to Heavy Fermions" (Cambridge University Press, Cambridge, 1993); G.D.Mahan, "Many-Particle Physics" (New York: Plenum Press, 1990).

[6] J.Kondo, Prog.Theoret. Phys. **32** 37 (1964); Solid State Physics, vol. 23, F.Seitz and Turnbull eds, Academic Press, New York, pg.183 (1969)

[7] L.I.Glazman and M.E.Raikh, Pis'ma Zh.Eksp.Teor.Fiz. **47**,378 (1988) [JETP Lett.**47**, 452 (1988)]; T.K. Ng and P.A.Lee, Phys. Rev. Lett. **61**, 1768 (1988); Y.Meir, N.S.Wingreen and P.A.Lee, Phys.Rev.Lett. **70**, 2601 (1993)

[8] P.W.Anderson, Phys.Rev.**124**, 41 (1961)

[9] S. Sasaki, S. De Franceschi, J.M. Elzerman, W.G. van der Wiel, M. Eto, S. Tarucha and L.P. Kouwenhoven, *Nature* **405**, 764 (2000).

[10] S. Tarucha, D.G. Austing, Y. Tokura, W.G. van der Wiel and L.P. Kouwenhoven, Phys. Rev. Lett. **84**, 2485 (2000).

[11] B. Jouault, G. Santoro and A. Tagliacozzo, Phys. Rev. B **61**, 10242 (2000).

[12] M. Eto and Y. Nazarov, Phys.Rev.Lett.**85**,1306 (2000)

[13] M. Pustilnik and L.I. Glazman, Phys.Rev.Lett.**85**,2993 (2000)

[14] D. Giuliano and A. Tagliacozzo, Phys. Rev. Lett. **84**, 4677 (2000); D. Giuliano, B.Jouault and A. Tagliacozzo, cond-mat/0010054 (2000)

[15] P.W.Anderson, G.Yuval and D.R.Haman *Phys.Rev.* **B11**, 4464

(1970)

[16] D.R.Hamann, Phys.Rev. **B2**,1373 (1970)

[17] M.Di Stasio, G.Morandi and A.Tagliacozzo, Phys.Lett. **A206** ,211 (1995)

[18] B.Nienhuis, "Coulomb Gas Formulation of Two-dimensional Phase Transitions",C. Domb and J.Lebowitz Eds., (Academic Press) (1987)

[19] A.M.Tsvelick and P.B.Wiegmann, Advances in Physics **32**, 453 (1983); P.B.Wiegmann and A.M.Tsvelick, J.Phys.C,2281 (1983), *ibidem*, 2321.

Multi-particle entanglement in quantum computers

Klaus Mølmer and Anders Sørensen

Institute of Physics and Astronomy
University of Aarhus, DK-8000 Århus C, Denmark

Abstract

The superposition principle and entanglement are the central properties for quantum computing, and a fully operational quantum computer
can be used to synthesize any desired entangled state of its constituents.
In this paper we show that some of the most attractive, and potentially
most applicable, multi-particle entangled states are also the ones most
easily prepared, and they are already within reach of current experiments.

1 Introduction

Algorithms exist by which evolution and detection of quantum states provide
computational powers exceeding the ones of classical computers [1, 2], and
many strategies for practical quantum computing have been proposed and investigated [3]. Several research groups now study these proposals, and fundamental gate operations on systems with few bits are underway in very different
systems, presenting candidates for the back bone of future quantum computers. The main goal of making a system which is truly superior over classical
computers will be achieved, when we can operate entangled states of many
systems reliably, but the requirements for carrying out all the millions of steps
in the proposed algorithms on hundreds of two-level quantum systems are so
terrifying, that only few researchers have seriously tried to analyze already at
this stage the parameters for such a complete system [4].

Without precise access to individual qubits and without sufficient elimination of decoherence, we shall not succeed to build a quantum computer. We
have proposed, however, to operate some physical systems as *reduced instruction set quantum* - RISQ - computers and devices [5]. Examples of such RISQ
computers are ones with dynamics taylored to be equivalent to – and thus
to simulate – interaction effects in physically interesting many-body systems.
This is much in the spirit of Feynman's proposal for quantum computing [6],
and the reduced capabilities of RISQ systems may just coincide with natural
restrictions imposed by locality and by symmetries in the physical system of
interest [5, 7], so that "Feynman quantum computing" may quickly become

a reality. RISQ systems of many particles potentially representing qubits in a quantum computer, offer means for multi-particle entanglement which may improve spectroscopic resolution, clocks, and length and frequency standards.

In this paper, we shall focus on current and near-future achievements with the ion trap proposal for quantum computing [8]. In Section 2, we review the results of a recent experiment by the NIST group in Boulder [9], in which four trapped ions were prepared in an entangled state. In Section 3, we present a robust means for two-qubit operations in the ion trap, and in Section 4, we present the theory behind the Boulder experiment, which suggests that it is possible to entangle also larger collections of ions. In Section 5, we briefly comment on the role of various imperfections and we present estimates of the increasing demands for preparation of entangled states with many particles. In Section 6 we conclude the paper, and we summarize a few features that make atomic quantum computers worthwile building, even if the prospects for achieving a fully operational quantum computer turn out to be less realistic, or maybe to be more realistic in other physical systems.

2   Experiment by the NIST group

At low temperatures trapped ions freeze into a crystalline structure, and the vibrations of the ions are strongly coupled due to their Coulomb interaction. One may excite a single collective vibrational mode by tuning a laser to one of the upper or lower sidebands of the ions, *i.e.*, by choosing the frequency of the laser equal to the resonance frequency of an internal transition in an ion plus or minus the vibrational frequency of the mode. Theory, as reviewed in the following sections, suggests that by illuminating an even number, N, of ions in a linear trap with a single light pulse containing two frequencies, symmetrically detuned around the atomic transition frequency in the ions, one can produce the state

$$|\Psi_N\rangle = (|gg...g\rangle + i^{N+1}|ee...e\rangle)/\sqrt{2}. \tag{1}$$

In Eq.(1), g and e denote the two internal states of each ion, which in the experiment [9] are the $2S_{1/2}$, $F = 2, m_F = -2$ and $F = 1, m_F = -1$ hyperfine states in $^9Be^+$ ions. The states are separated by the Bohr frequency $\omega_{eg} = 2\pi \cdot 1.25$ GHz, and the actual transition is a Raman transition via the $2P_{1/2}$ excited state in the ions, and the two laser beams for the Raman transition are chosen with a difference frequency near ω_{eg} and with the wave vector difference along the axis of the linear trap, so that the internal state transitions couple to the vibrational motion of the ions. For technical reasons, the center-of-mass mode of vibration of the ions is strongly influenced (heated) by ambient electric fields, and the experiments make use of a higher vibrational mode. Experiments have been done with 2 and 4 ions, which both have a stretch mode, where all ions experience the same excursions. The stretch mode frequency of vibration in the experiments is 8.8 MHz.

The state (1) has several applications both in fundamental physics and technology. It is a Schrödinger cat superposition of states of mesoscopic separation, and it is ideal for spectroscopic investigations. In current frequency standards the atoms or ions are independent, and when they are interrogated by the same field, the outcome of a measurement fluctuates as the square root of the number of atoms N. The relative frequency uncertainty in samples with many atoms thus behaves like $\frac{1}{\sqrt{N}}$. By binding the ions together as in Eq. (1) we are sensitive to the Bohr frequency between $|gg...g\rangle$ and $|ee...e\rangle$ which is proportional to N, and consequently the frequency uncertainty scales with $\frac{1}{N}$ [10].

The state (1) is prepared following the proposal outlined in the following sections, and subsequent measurements are performed to probe the system. The two states g and e in an ion are easily distinguished by illumination with light resonant with the $2S_{1/2}, F = 2, m_F = -2 \rightarrow 2P_{3/2}, F = 3, m_F = -3$ cycling transition, which produces fluorescence, if the ion is in the state g. During a 200 μs detection period, on average 15 fluorescence photons would be detected from each ion in the state g, and even without the ability to distinguish the fluorescence signal from individual ions in the trap, it is thus possible to determine the number of ions in state g in the state prepared. To acertain that the system is not simply in an incoherent micture of states with different number of excited ions, it is necessary to determine also the coherence between the two components in (1). This is done simply by driving the ions individually on the Raman transition with a $\pi/2$ pulse. The final state occupation of the state g and e depends on the evolution of the phase between the two components in (1), and by varying that phase an interference pattern emerges with a visibility that immediately provides the magnitude of the coherence $\rho_{g^N e^N}$, between these two states. The measured values read

$$P_0 = 0.43, \; P_1 = 0.11, \; P_2 = 0.46, \; \rho_{g^N e^N} = 0.385 \; (2 \; ions)$$
$$P_0 = 0.35, \; P_1 = P_2 = P_3 = 0.10, \; P_4 = 0.35, \; \rho_{g^N e^N} = 0.215 \; (4 \; ions). \qquad (2)$$

According to Eq.(1), P_0 and P_N should both equal one half and all other populations should vanish, and $|\rho_{g^N e^N}|$ should equal one half, so the state is not produced with perfect accuracy. The measured numbers (2) point to an overlap of the state produced with the desired state (1) of 0.83 for N=2, and 0.57 for N=4.

As pointed out in [9], a population of the entangled state (1) above 0.5 is incompatible with a separable state of the ions; the ions must be in an entangled state. But, since simple product states, like $|gg...g\rangle$, have a finite overlap with the state (1), one may imagine a 'conspiracy' so that the system choses among a set of simple product states, with built-in classical correlations which account so well for the data, that only with 14 % probability [9, 11], do we really need to restrict the 4 ions to the state (1): $\rho = 0.14|\Psi_4\rangle\langle\Psi_4| + 0.86\rho_{consp}$. Future experiments may well succeed to control the heating in the trap better, and on the detection side improvement is possible, so there is good reason to expect that the numbers in Eqs.(2) will get closer to the ideal values, and that it will be possible to proceed to larger numbers of ions.

3   Bichromatic excitation of trapped ions

In the original proposal for ion trap quantum computing [8], the collective vibration of the ions is suggested to act like a channel, where the state of the ith ion is encoded and subsequently fed to the jth ion at another location in the string of ions. Our proposal offers the same general access possibilities, but it utilizes the vibrational degree of freedom in a different way, which fortuitously removes the restriction of the original proposal that the vibrations must be in the exact ground state at the beginning of each gate operation.

We treat each ion as a two-level system, and operators acting on the states of the ions are conveniently described by Pauli spin matrices, where the states $|g\rangle_i$ and $|e\rangle_i$ of the ith ion are defined as the eigenstates of $\sigma_{z,i}$. In this framework, a Hamiltonian proportional with $\sigma_{y,i}$ provides a rotation (Rabi-oscillation) between the two states $|g\rangle_i$ and $|e\rangle_i$.

We now illuminate two selected ions with radiation of two different frequencies (with two Raman field pairs of two different frequency differences), $\omega_{1,2} = \omega_{eg} \pm \delta$, where δ is a detuning, not far from the trap frequency ν. In Fig. 1, we illustrate the action of such a bichromatic laser field on the state of the ions. As shown in the figure, the initial and final states $|ggn\rangle$ and $|een\rangle$, separated by $2\omega_{eg} = \omega_1 + \omega_2$ are resonantly coupled, and so are the degenerate states $|egn\rangle$ and $|gen\rangle$, where the first (second) letter denotes the internal state e or g of the ith (j^{th}) ion and n is the quantum number for the relevant vibrational mode of the trap. These resonant couplings lead to an effective Hamiltonian of the form $\sigma_i^+ \sigma_j^+ + \sigma_i^- \sigma_j^- - \sigma_i^+ \sigma_j^- - \sigma_i^- \sigma_j^+ \propto \sigma_{y,i}\sigma_{y,j}$.

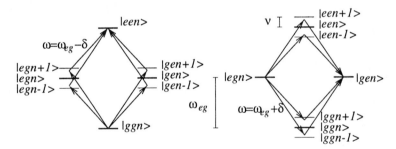

Figure 1: Energy level diagram for two ions with quantized vibrational motion illuminated with bichromatic light. The one photon transitions indicated in the figure are not resonant, $\delta \neq \nu$, so only the two-photon transitions shown from $|ggn\rangle$ to $|een\rangle$ and from $|egn\rangle$ to $|gen\rangle$ are resonant.

We can choose the fields so weak and the detuning from the sideband so large that the intermediate states with $n \pm 1$ vibrational quanta are not populated in the process. In the experiment [9], the ratio between the ground state width and the radiation wave length (the Lamb-Dicke parameter) is $\eta = 0.23/\sqrt{N}$ which is sufficiently small to decribe the system in the Lamb-Dicke approximation. It now turns out [12, 13] that the internal state transition is insensitive to the vibrational quantum number n. This is due to interference between the interaction paths: The transition via an upper sideband excitation $|n+1\rangle$, has a strength of $n + 1$ ($\sqrt{n+1}$ from raising and $\sqrt{n+1}$ from lowering the vibrational quantum number), and the transition via $|n - 1\rangle$ yields a factor of n. Due to opposite signs of the intermediate state energy mismatch, the terms interfere destructively, and the n dependence disappears from the coupling.

The coherent evolution of the internal atomic state is thus insensitive to the vibrational quantum numbers, and it may be observed with ions in any superposition or mixture of vibrational states, even if the ions exchange vibrational energy with a surrounding reservoir. The control of the thermal motion is one of the technical difficulties in ion trap experiments, and the tolerance to vibrations is a major asset of our bichromatic proposal.

Surprisingly, the care, with which we ensure not to populate the intermediate states in the processes in Fig.1 is not needed, if we only make sure that the changes of the vibrational state are cancelled towards the end of the process [13, 14]. This possibility is most easily revealed if we rewrite our bichromatic interaction Hamiltonian in the interaction picture with respect to the atomic and vibrational Hamiltonian

$$H_{\text{int}} = -\sqrt{2}\eta\Omega J_y[x\cos(\nu - \delta)t + p\sin(\nu - \delta)t], \tag{3}$$

where we have introduced the dimensionless position and momentum operators for the centre-of-mass vibrational mode $x = \frac{1}{\sqrt{2}}(a + a^\dagger)$ and $p = \frac{i}{\sqrt{2}}(a^\dagger - a)$, and the collective internal state observable $J_y = \frac{\hbar}{2}(\sigma_{y,i} + \sigma_{y,j})$ of the two ions illuminated. Ω is the Rabi frequency of the field-atom coupling, and η is the Lamb-Dicke parameter.

The exact propagator for the Hamiltonian (3) can be represented by the ansatz

$$U(t) = e^{-iA(t)J_y^2/\hbar^2} e^{-iF(t)J_y x/\hbar} e^{-iG(t)J_y p/\hbar}, \tag{4}$$

where the Schrödinger equation $i\hbar\frac{d}{dt}U(t) = HU(t)$ leads to the expressions $F(t) = -\sqrt{2}\eta\Omega\int_0^t\cos((\nu - \delta)t')dt'$, $G(t) = -\sqrt{2}\eta\Omega\int_0^t\sin((\nu - \delta)t')dt'$, and $A(t) = \sqrt{2}\eta\Omega\int_0^t F(t')\sin((\nu - \delta)t')dt'$.

If the functions $F(t)$ and $G(t)$ both vanish after a period τ, the propagator reduces to $U(\tau) = e^{-iA(\tau)J_y^2/\hbar^2}$ at this instant, and the vibrational motion is returned to its original state as announced. The vibrational state may be the ground state or any vibrationally excited state – the internal state evolution is *independent* of the vibrational state. Note that $(\sigma_{y,i})^2 = 1$ implies that $J_y^2 = \frac{\hbar^2}{4}(2 + 2\sigma_{y,i}\sigma_{y,j})$, yielding precisely the interaction that we need to produce the gate between ions i and j.

4   Multi-particle entangled state

... But what was still more curious, whoever held his finger in the smoke of the kitchen-pot, immediately smelt all the dishes that were cooking on every hearth in the city. (Hans Christian Andersen, *The Swineherd* [15]).

The coupling illustrated in Fig.1 and analyzed in Eq.(4), shows that the bichromatic radiation produces an entangled state of the form (1), when applied to two trapped ions. Let us now consider a situation where we apply a Hamiltonian which acts identically on all ions in a string with N ions because the laser fields extend over the whole ion cloud. It follows that a state of the system, which is initially symmetric under exchange of different ions, will remain symmetric at all later times. A convenient representation of such states is given by the eigenstates of a fictitious total angular momentum, $|JM\rangle$, the so-called Dicke states [16]. (Every single two-level ion is generically described by 2×2 Pauli spin matrices, and the associated fictitious spin 1/2 add up to a total $J = N/2$ angular momentum.) In the Dicke representation, $N = 2J$ is the total number of ions, and M counts the number of excited ions, $N_e = J + M = 0, 1, \ldots N$. A single resonant laser field, which excites all ions with same amplitude acts as the angular momentum raising operator J_+ on the symmetrical states (and the adjoint lowering operator J_-), and effectively it acts as a geometrical rotation of the state vector.

We saw in the above section that a bichromatic field effectively produces an interaction where atoms are excited and deexcited in pairs, and if these fields are applied to all ions in the trap, these processes occur within all possible pairs, and effectively, one produces the collective operator $J_y^2 = (\frac{\hbar}{2} \sum \sigma_{y,i})^2$, where the summation is extended over all ions in the trap. This operator is symmetric under exchange of particles, and hence it only couples the Dicke states among themselves. More precisely, it only couples Dicke states, with M quantum numbers separated by an even number. If the number of ions is itself even, the rotation properties of angular momenta can be used to show [17] that J_y^2 generates directly the maximally entangled state (1) if it is applied to a whole ensemble of ground state ions for the right amount of time. If the number of ions is odd, one can also produce that state, it only requries a joint rotation by J_y of all ions (by a simple resonant laser pulse), followed by the bichromatic pulse [17].

In Fig.2, we show the population of the states $|gg...g\rangle$ and $|ee...e\rangle$ as function of time, or rather as function of the argument $A(\tau)$ in the time evolution operator (4), under the condition that $F(\tau) = G(\tau) = 0$. The left panel shows the analytical results given by (4) for 4, 8, and 20 ions, and the right panel shows the results of a simulation with four ions in a not too slow gate with heating of the vibrational motion. Around $\nu t \sim 1500$, we obtain approximately the state (1), as demonstrated in the NIST experiment [9].

Such entangled states do not have the full communicative powers of the kitchen-pot in the fairy-tale (sold to the princess at a price of 10 kisses). It rather represents a RISQ version, allowing its owner to check if the fire is turned on (state e) or not (state g) in the hearths all over town.

5   Scalability and fidelity of entanglement

We know the exact time evolution operator (4), corresponding to the interaction (3), and this enables a simple analysis of effects, left out of the analysis so far. Extra contributions to the interaction and, e.g., dissipative effects due to coupling to the surroundings may be dealt with by perturbation theory, and their effects on the state of the trapped ions is quite precisely obtained in the interaction picture with respect to the dominant interaction (3). Details of the

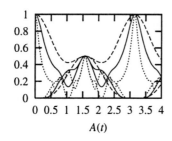 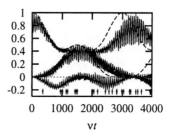

A(t) vt

Figure 2: The left panel shows the time evolution of the population of the joint ionic ground state $|gg...g\rangle = |N/2, -N/2\rangle$ (curves starting from the value of unity at $t = 0$), and of the jexcited state $|ee...e\rangle = |N/2, N/2\rangle$: N=4 (dashed curves), N=8 (solid curves) and N=20 (dotted curves). For $A(\tau) = \pi/2$, the states are both populated 50 %, and a Schrödinger cat state is obtained. The right panel shows the results of a numerical integration of the Schrödinger equation. The populations and coherences are obtained as a trace over the vibrational degrees of freedom; the ions interact with a thermal reservoir, and this gives rise to quantum jumps in the vibrational states at random instants of time, indicated by arrows in the figure.

calculations may be found in Ref.[13]. The fidelity F is here defined as the population of the state (1) after application of the bichromiatic fields for the optimum duration. Imperfections have different causes as listed below, and an estimate of the overall fidelity is obtained as the product of the individual expressions. We wish to emphasize that the deviation of the fidelity from unity depends on the number of ions, N, so that given the experimental parameters, the following expressions can be used directly to estimate how many ions may be reliably entangled in a given experiment.

• *Off-resonant coupling*:
The Hamiltonian (3) is written in a rotating wave approximation which neglects off-resonant excitation of the ions (other intermediate states than the ones connected by arrows in Fig.1). Transitions via states with unchanged vibrational quantum number are suppressed by an energy denominator, but the transition matrix elements are larger, because they do not involve the change of the spatial wave function.

$$F = 1 - \frac{N\Omega^2}{2\delta^2}$$

The reduction in fidelity comes about as an oscillatory evolution of the state vector around the one predicted by (4), and without changing the parameters of the system, the fidelity can be made closer to unity if the laser pulses can be shut off with a timing of better than δ^{-1}.

• *Deviations from Lamb-Dicke regime*:
The atom-field coupling is described in the Lamb-Dicke regime, where the spatial variation of the electric field is expanded to linear order around the

position of the ions, assuming that the ionic excursions are smaller than the transition wavelength. The resulting expression, $\langle n|e^{i\eta(a+a^\dagger)}|n+1\rangle = i\eta\sqrt{n}$, is crucial for the intereference between paths in Fig.1. Going to higher order in the coupling, and allowing a thermal excitation of the center-of-mass mode used for the gate operation [18] with mean vibrational quantum number $\bar{n}_1$, we get

$$F = 1 - \eta^4 \frac{\pi^2 N(N-1)}{8}(\bar{n}_1^2 + \bar{n}_1)$$

- *Spectator vibrational modes*:
N trapped ions share N vibrational modes with low frequency vibrations along the trap axis. The center-of-mass mode has frequency ν, and the other modes have a sequence of higher frequencies. The modes have been identified numerically [19], and the off-resonant coupling to vibrationally excited states lead to a reduction in the fidelity, both due to the population going into these states

$$F = 1 - N\frac{\eta^2\Omega^2}{\nu^2}0.8(\bar{n}_1 + 1),$$

and due to the 'Debye-Waller' effect, causing a spreading of the effective coupling of the desired transitions,

$$F = 1 - \eta^4\frac{\pi^2 N(N-1)}{8}(0.2\bar{n}_1^2 + 0.4\bar{n}_1).$$

In these expressions, we assume that the vibrational motion is thermally excited, and thus fully determined by $\bar{n}_1$ of the center-of-mass mode.
- *Heating of the vibrational motion*:
The preparation of the cat state (1) requires both the vanishing of the functions $F(\tau)$ and $G(\tau)$ in (4), and the appropriate value of the quantity $A(\tau)$ (odd multiple of $\pi/2$). This is fulfilled if $\tau = 2\pi K/(\nu-\delta)$ and $\nu - \delta = 2\sqrt{K}\eta\Omega$, $K = 1, 2, ...$ The fidelity of the gate is reduced, if the vibration relaxes with rate Γn_{th} towards a mean vibrational quantum number n_{th},

$$F = 1 - N\frac{\Gamma(1 + 2n_{th})\tau}{8K}.$$

A small K implies a fast gate, which is only tolerant to heating if the duration of the gate is much shorter than the heating time. A large K, however, implies a slow gate, which is tolerant to heating, even for long gate duration.

6   Outlook

The publication by Sackett *et al*, brought several pieces of good news for the quest to build quantum computers. Most importantly, of course, it showed that an entangled state of several particles can be built utilizing the quantum motion in a man made laboratory device, the ion trap. It is the first example of deterministic entanglement of particles by a strategy which is scalable to any number of particles, and progress with more ions and with higher fidelity of the desired state is expected in the near future.

It is worth mentioning a few central aspects of the experimental procedure, which might guide ideas for implementation of quantum computing in other physical systems:

It is possible to build a quantum computer out of a string of quantum systems if one-bit gates can be carried out on each system, and if two-bit gates can be implemented within any neighbouring pair of systems. The low speed of gates between remote qubits due to the required sequence of nearest neighbour interactions does not alter the conclusions about the effectiveness of quantum computation. The ion trap proposal [8], and proposals involving cavity field modes [20] or superconducting circuits [21], which can act as a 'data-bus' to mediate the interaction among arbitrary qubits, have many advantages, and especially at the current stage of research, the reduced number of gate operations may be important for the obtainment of results.

A priori, it has been assumed that the auxiliary degree of freedom used for the two-qubit operation must be coherently controlled to the same extent as the qubit degrees of freedom. This poses a practical problem, since, by its function, the 'data-bus' is extended over space, and it couples willingly to the qubits but also to noise from the environment. The theory, presented here, and the experiment by Sackett *et al*, shows that it is possible to utilize a quantum degree of freedom to communicate reliably between two quantum systems, even when the state of this degree of freedom is unknown. Eq.(4) expresses this fact by showing that the time evolution of the qubit internal states can be completely disentangled from the other degrees of freedom, and our analysis points to two regimes with quite different explanations for that possibility: *i)* in the 'slow' regime, states of the 'data-bus' degree of freedom are accessed as intermediate, virtual states, and the process proceeds without ever populating other states than the initial=final 'data-bus' state; *ii)* in the 'fast' regime, the 'data-bus' undergoes quite sizable dynamics, entangled with the qubit states, but this entanglement is magically removed towards the end of the gate. The slow gate is not only operational for arbitrary initial states of the 'data-bus', it is also tolerant to changes in these states during operation.

Finally, we wish to comment on the fact that the multi-particle entangled state is created by means of only one single bichromatic light pulse. It is possible that such a collective operation may be useful for quantum computing, but of course in a general quantum computer it has to be suppplemented with operations with individual access. The RISQ operation presented here, however, is relevant in its own right, because the state produced (1) is interesting and technologically useful [10]. Ions and atoms may not be the most suitable systems for quantum computing in the long run, but because an atom or ion is a unique quantity with identical properties in all laboratories, they will forever form the back bone for standardization and metrology technologies, and the gain in precision following from entanglement of ions and atoms makes this branch of quantum information a both promising and a realistic one already for the near future.

References

[1] P. Shor, In *Proceedings of the 35th Annual Symposium on Foundations of Computer Science*, edited by S. Goldwasser (IEE Computer Society, Los Alamos, CA, 1994).

[2] L. K. Grover, Phys. Rev. Lett. **79**, 325 (1997).

[3] For a recent review, see D. DiVincenzo, *The Physical Implementation of Quantum Computation*. To appear in Fortschhritte der Physik, quant-ph/0002077.

[4] A. Steane, Nature **399**, 124 (1999); *Quantum Computing with Trapped Ions, Atoms and Light*, to appear in Fortschritte der Physik, quant-ph/0004053.

[5] K. Mølmer and A. Sørensen, *RISQ – reduced instruction set quantum computers*, to appear in J. Mod. Opt.

[6] R. P. Feynman, Int. J. Theor. Phys. **21**, 467 (1982).

[7] S. Lloyd, Science **273**, 1072 (1996).

[8] J. I. Cirac and P. Zoller, Phys. Rev. Lett. **74**, 4091 (1995).

[9] C. A. Sackett, *et al.*, Nature **404**, 256 (2000).

[10] J. J. Bollinger, *et al.* Phys. Rev. A **54**, 4649 (1996).

[11] Chris Monroe, private communication.

[12] A. Sørensen and K. Mølmer, Phys. Rev. Lett. **82**, 1971 (1999).

[13] A. Sørensen and K. Mølmer, *Entanglement and quantum computation with ions in thermal motion*, to appear in Phys. Rev. A., quant-ph/0002024.

[14] G. Milburn, *Simulating nonlinear spin models in an ion trap*, quant-ph/9908037.

[15] Hans Christian Andersen, *The Swineherd*, (1842). An english translation of the full text can be found on multiple internet sites, *e.g.*, http://underthesun.cc/Classics/Andersen/FairyTales/swineherd.htm.

[16] R. H. Dicke, Phys. Rev. **93**, 99 (1954).

[17] K. Mølmer and A. Sørensen, Phys. Rev. Lett. **82**, 1835 (1999).

[18] Our theoretical proposal requires that all ions experience excursions of same magnitude, which for higher ion numbers only occurs in the centre-of-mass mode, which is therefore the mode applied in our theoretical analysis.

[19] D. V. F. James, Appl. Phys. B. **66**, 181 (1998).

[20] A. Imamoglu *et al*, *Quantum information processing using quantum dot spins and cavity-QED*, quant-ph/9904096.

[21] Y. Makhlin *et al*, *Josephson-Junction Qubits with Controlled Couplings*, cond-mat/9808067.

STABILIZATION
OF QUANTUM INFORMATION:
A UNIFIED DYNAMICAL-ALGEBRAIC
APPROACH

Paolo Zanardi

Institute for Scientific Interchange (ISI) Foundation,

Villa Gualino, Viale Settimio Severo 65, I-10133 Torino, Italy

Istituto Nazionale per la Fisica della Materia (INFM)

zanardi@isiosf.isi.it

Abstract The notion of symmetry is shown to be at the heart of all error correction/avoidance strategies for preserving quantum coherence of an open quantum system S e.g., a quantum computer. The existence of a nontrivial group of symmetries of the dynamical algebra of S provides state-space sectors immune to decoherence. Such noiseless sectors, that can be viewed as a noncommutative version of the pointer basis, are shown to support universal quantum computation and to be robust against perturbations. When the required symmetry is not present one can generate it artificially resorting to active symmetrization procedures.

1.   INTRODUCTION

Stabilizing quantum-information processing against the environmental interactions as well as operation imperfections is a vital goal for any practical application of the protocols of Quantum Information and Quantum Computation theory. To the date three kind of strategies have been devised in order to satisfy such a crucial requirement: a) Error Correcting Codes which, in analogy with classical information theory, stabilize actively quantum information by using redundant encoding and measurements; b) Error Avoiding Codes pursue a passive stabilization by exploiting symmetry properties of the environment-induced noise for suitable redundant encoding; c) Noise suppression schemes in which, exploiting symmetry properties and with no redundant encoding, the unwanted interactions are averaged away by frequently iterated external pulses. We shall show how all these schemes derive conceptually from a unified dynamical-algebraic framework.

Such an underlying dynamical-algebraic structure is provided by the the reducibility of the operator algebra describing the faulty interactions of the coding quantum system. This property in turns amounts to the existence of a non-trivial group of symmetries for the global dynamics. We describe a unified framework which allows us to build systematically new classes of error correcting codes and noiseless subsystems. Moreover we shall argue how by using symmetrization strategies one can artificially produce noiseless subsystems and to perform universal quantum computation within these decoherence-free sectors.

2. NOISELESS SUBSYSTEMS

Suppose that S is quantum system coupled to an environment. Without further assumptions the decoherence induced by this coupling is likely to affect all the states of S. Suppose now that S turns out to be bi-partite, say $S = S_1 + S_2$, and moreover that the environment is actually coupled just with S_1. If this is the case one can encode information is the quantum state of S_2 in a obvious noiseless way.

The idea underlying the algebraic constructions that will follow is conceptually nothing but an extension of the extremely simple example above: *the symmetry of the system plus environment dynamical algebra provides S with an hidden multi-partite structure such that the environment is not able to extract information out of some of these "virtual" subsystems.*

Let S be an open quantum system, with (finite-dimensional) state-space $\mathcal{H}$, and self-Hamiltonian H_S, coupled to its environment through the hamiltonian $H_I = \sum_\alpha S_\alpha \otimes B_\alpha \neq 0$, where the S_α's (B_α's) are system (environment) operators. The unital associative algebra $\mathcal{A}$ closed under hermitian conjugation $S \mapsto S^\dagger$, generated by the S_α's and H_S will be referred to as the interaction algebra. In general $\mathcal{A}$ is a reducible †-closed subalgebra of the algebra $\mathrm{End}(\mathcal{H})$ of all the linear operators over $\mathcal{H}$: it can be written as a direct sum of $d_J \times d_J$ (complex) matrix algebras each one of which appears with a multiplicity n_J

$$\mathcal{A} \cong \oplus_{J \in \mathcal{J}} \mathbf{Id}_{n_J} \otimes M(d_J, \mathbf{C}). \tag{1.1}$$

where $\mathcal{J}$ is suitable finite set labelling the irreducible components of $\mathcal{A}$. The associated state-space decomposition reads

$$\mathcal{H} \cong \oplus_{J \in \mathcal{J}} \mathbf{C}^{n_J} \otimes \mathbf{C}^{d_J}. \tag{1.2}$$

These decompositions encode all information about the possible quantum stabilization strategies.

Knill et al noticed that (1.1) implies that each factor $\mathbf{C}^{n_J}$ in eq. (1.2) corresponds to a sort of effective subsystem of S coupled to the environment in a state independent way. Such subsystems are then referred to as noiseless. In particular one gets a noiseless code i.e., a decoherence-free subspace, $\mathcal{C} \subset \mathcal{H}$ when in equation (1.2) there appear one-dimensional irreps J_0 with multiplicity greater than one $\mathcal{C} \cong \mathbb{C}^{n_{J_0}} \otimes \mathbb{C}$.

The commutant of $\mathcal{A}$ is defined as $\mathcal{A}'$ in $\mathrm{End}(\mathcal{H})$ of $\mathcal{A}$ by $\mathcal{A}' := \{X \mid [X, \mathcal{A}] = 0\}$.

The existence of a NS is equivalent to

$$\mathcal{A}' \cong \oplus_{J \in \mathcal{J}} M(n_J, \mathbb{C}) \otimes \mathbf{Id}_{d_J} \neq \mathbb{C}\,\mathbf{Id} \tag{1.3}$$

The condition $\mathcal{A}' \neq \mathbb{C}\mathbf{Id}$ is amounts to the existence of a non-trivial group of symmetries $\mathcal{G} \subset U\,\mathcal{A}'$. One has that the more symmetric a dynamics, the more likely it supports NSs.

When $\{S_\alpha\}$ is a commuting set of hermitian operators. $\mathcal{A}$ is an abelian algebra and Eq. (1.2) [with $d_J = 1$] is the decomposition of the state-space according the joint eigenspaces of the S_α's. The pointer basis discussed in relation to the so-called environment-induced superselection is nothing but an orthonormal basis associated to the resolution (1.2). Thus *the NS's provide in a sense the natural noncommutative extension of the pointer basis.*

Now we discuss the relation between NSs and error correction. The interaction algebra $\mathcal{A}$ has to to be thought of generated by error operators and it is assumed to satisfy Eq. 1.1. Let $|J\lambda\mu\rangle$ $(J \in \mathcal{J}, \lambda = 1, \ldots, n_J; \mu = 1, \ldots, d_J)$ be an orthonormal basis associated to the decomposition (1.1). Let $\mathcal{H}_\mu^J := \mathrm{span}\{|J\lambda\mu\rangle \mid \lambda = 1, \ldots, n_J\}$, and let $\mathcal{H}_\lambda^J$ be defined analogously. The next proposition shows that to each NS there is associated a family of ECCs called $\mathcal{A}$-*codes*. It is a simple consequence of (1.1) and of the definition of ECCs.

The $\mathcal{H}_\mu^J$'s ($\mathcal{H}_\mu^J$) are $\mathcal{A}$-codes ($\mathcal{A}'$-codes) for any subset E of error operator such that $\forall e_i, e_j \in E \Rightarrow e_i^\dagger e_j \in \mathcal{A}$

The standard stabilizer codes are recovered when one considers a N-partite qubit system, and an abelian subgroup $\mathcal{G}$ of the Pauli group $\mathcal{P}$. Let us consider the state-space decomposition (1.2) associated to $\mathcal{G}$. If $\mathcal{G}$ has $k < N$ generators then $|\mathcal{G}| = 2^k$, whereas from commutativity it follows $d_J = 1$ and $|\mathcal{J}| = |\mathcal{G}|$. Moreover one finds $n_J = 2^{N-k}$: each of the 2^k joint eigenspaces of $\mathcal{G}$ (stabilizer code) encode $N - k$ logical qubits. It follows that

$$\mathcal{H} = \oplus_{J=1}^{2^k} \mathbb{C}^{2^{N-k}} \otimes \mathbb{C} \cong \mathbb{C}^{2^{N-k}} \otimes \mathbb{C}^{2^k}. \tag{1.4}$$

The allowed errors belong to the algebra $\mathcal{A} = \mathbf{Id}_{2^{N-k}} \otimes M(2^k, \mathbb{C})$. In particular errors $e_i, e_j \in \mathcal{A} \cap \mathcal{P}$ are such that $e_i^\dagger e_j$ either belong to $\mathcal{G}$ or anticommute with (at least) one element $\mathcal{G}$. In this latter case one has a a non-trivial action on the $\mathbb{C}^{2^k}$ factor.

Let us finally stress out that the NSs approach described so far works even in the case in which the interaction operators S_α represent, rather than coupling with external degrees of freedom, internal unwanted *internal* interactions i.e., $H'_S = H_S + \sum_\alpha S_\alpha$.

3.   COLLECTIVE DECOHERENCE

Collective decoherence arises when a multi-partite quantum system, is coupled symmetrically with a common environment. This is the paradigmatic case for the emergenge of noiseless subspaces and NS's as well. More specifically one has a N-qubit system $\mathcal{H}_N := (\mathbb{C}^2)^{\otimes N}$ and the relevant interaction algebra $\mathcal{A}_N$ coincides with the algebra of completely symmetric operators over $\mathcal{H}_N$. The commutant $\mathcal{A}'_N$ is the group algebra $\nu(\mathbb{C}\mathcal{S}_N)$, where ν is the natural representation of the symmetric group $\mathcal{S}_N$ over $\mathcal{H}_N : \nu(\pi) \otimes_{j=1}^N |j\rangle = \otimes_{j=1}^N |\pi(j)\rangle$, $(\pi \in \mathcal{S}_N)$. Uing elementary $su(2)$ representation theory one finds:

$\mathcal{A}_N$ supports NS with dimensions

$$n_J = \frac{(2J+1)\,N!}{(N/2+J+1)!\,(N/2-J)!} \qquad (1.5)$$

where J runs from 0 (1/2) for N even (odd). If in the above $\mathcal{A}_N$ is replaced by its commutant, the above result holds with $n_J = 2J+1$. Of course collective decoherence allows for $\mathcal{A}_N$-codes as well.

In order to illustrate the general ideas let us consider r $N = 3$. One has $(\mathbb{C}^2)^{\otimes 3} \cong \mathbb{C} \otimes \mathbb{C}^4 + \mathbb{C}^2 \otimes \mathbb{C}^2$. The last term can be written as $\mathrm{span}\{|\psi_\beta^\alpha\rangle\}_{\alpha\beta=1}^2$ where

$$
\begin{aligned}
|\psi_1^1\rangle &= 2^{-1/2}(|010\rangle - |100\rangle),\ |\psi_2^1\rangle = 2^{-1/2}(|011\rangle - |101\rangle) \\
|\psi_1^2\rangle &= 2/\sqrt{6}\,[1/2(|010\rangle + |100\rangle) - |001\rangle], \\
|\psi_2^2\rangle &= 2/\sqrt{6}\,[|110\rangle - 1/2(|011\rangle + |101\rangle)].
\end{aligned} \qquad (1.6)
$$

The vectors $|\psi_\beta^1\rangle$ and $|\psi_\beta^2\rangle$ ($|\psi_1^\alpha\rangle$ and $|\psi_2^\alpha\rangle$) span a two-dimensional $\mathcal{A}_3$-code ($\mathcal{A}'_3$-code). Taking the trace with respect to the index α (β) one gets the $\mathcal{A}'_3$ ($\mathcal{A}_3$) NS's. Moreover the first term corresponds to a trivial four-dimensional $\mathcal{A}'_3$ code.

4.   NS SYNTHESIS BY SYMMETRIZATION

A typical situation in which NSs could arise is when the interaction algebra is contained in some reducible group representation ρ of a finite order (or compact) group $\mathcal{G}$ Suppose that the irrep decomposition of $\mathcal{H}$ associated to r ρ has the form of Eq. (1.2) in which the $\mathcal{J}$ labels a set

of $\mathcal{G}$-irreps ρ_J (dim $\rho_J = d_J$). ρ by extends linearly to the group algebra $\mathbb{C}\mathcal{G} := \oplus_{g \in \mathcal{G}} \mathbb{C}|g\rangle$ giving rise to a decomposition as in Eq. 1.1. It follows
If $\mathcal{A} \subset \rho(\mathbb{C}\mathcal{G})$ then the dynamics supports (at least) $|\mathcal{J}|$ NS's with dimensions $\{n_J(\rho)\}_{J \in \mathcal{J}}$.

The non-trivial assumption in the above statement is the reducibility of ρ in that any subalgebra of operators belongs to a group-algebra. As already stressed this is equivalent to a symmetry assumption. When this required symmetry is lacking one can *artificially* generate it by resorting to the so-called bang-bang techniques. These are physical procedures, involving iterated external ultra-fast "pulses" $\{\rho_g\}_{g \in \mathcal{G}}$, whereby one can synthesize, from a dynamics generated by the S_α's, to a dynamics generated by $\pi_\rho(S_\alpha)$'s where

$$\pi_\rho \colon X \to \pi_\rho(X) := \frac{1}{|\mathcal{G}|} \sum_{g \in \mathcal{G}} \rho_g \, X \, \rho_g^\dagger \in \rho(\mathbb{C}\mathcal{G})'. \qquad (1.7)$$

projects any operator X over the commutant of the algebra $\rho(\mathbb{C}\mathcal{G})$ generated by the bang-bang operations. If one will, preserving the system self-dynamics, to get rid of unwanted interactions $\S_\alpha$ with the environmnet he has to look for a group $\mathcal{G} \subset U(\mathcal{H})$, such that i) $H_S \in \mathbb{C}\mathcal{G}'$, ii) the S_α's transform according to non-trivial irreps under the (adjoint) action of $\mathcal{G}$. Then it can be shown that $\pi_\mathcal{G}(S_\alpha) = 0$: the effective dynamics of S is unitary.

To understand how this strategies might be useful for artificial NSs synthesis is sufficient to notice that Prop. 2 holds even for the commutant by replacing the n_J's with the d_J's. Since the $\mathcal{G}$-symmetrization of an operator belongs to $\rho(\mathbb{C}\mathcal{G})'$, one immediately finds that:
$\mathcal{G}$-symmetrization of $\mathcal{A}$ supports (at least) $|\mathcal{J}|$ NS's with dimensions $\{d_J(\rho)\}_{J \in \mathcal{J}}$.

It is remarkable that NSs do not allow just for safe encoding of quantum information but even for its manipulation. Form the mathematical point of view this result stems once again quite easily from the basic Eq. 1.1which shows that the elements of $\mathcal{A}'$ have non-trivial action over the $\mathbb{C}^{n_J}$ factors. Therefore:
If an experimenter has at disposal unitaries in $U\mathcal{A}'$ universal QC is realizable within the NSs. When such gates are not available from the outset they can be obtained through a $\mathcal{G}$-symmetrization of a couple of generic hamiltonians, where $\mathcal{G} := U\mathcal{A}$.

5. NS: ROBUSTENESS

In this section we prove a robustness result for NSs extending analogous ones obtained for decoherence-free subspaces. Let $\mathcal{H}_S$ ($\mathcal{H}_B$) denote the system state (environment) state-space. Here S represents the NS and the environment includes both the coupled factor in Eq (1.2) and the external degrees of freedom.

The evolution of the subsystem S is given by $\mathcal{E}_\varepsilon^t(\rho) := \mathrm{tr}_B[e^{t\mathcal{L}_\varepsilon}(\rho \otimes \sigma)]$, where $\rho \in \mathcal{S}(\mathcal{H}_S)$, $\sigma \in \mathcal{S}(\mathcal{H}_B)$. and the Liouvillian operator is given by $\mathcal{L}_\varepsilon := \mathcal{L}_0 + \varepsilon \, \mathcal{L}_1$, where $e^{t\mathcal{L}_0}$ acts trivially over S i.e., $e^{t\mathcal{L}_0}(\rho \otimes \sigma) = \rho \otimes \sigma'_t$. In particular $\rho \otimes \mathbf{I}_B$ is a fixed point.

The fidelity is defined as

$$F_\varepsilon(t) := \mathrm{tr}_S[\rho \, \mathcal{E}_\varepsilon^t(\rho)] = < \rho \otimes \mathbf{I}_B, \, e^{t\mathcal{L}}(\rho \otimes \sigma) > = \sum_{n=0}^\infty \varepsilon^n f_n(t). \qquad (1.8)$$

One has $e^{t\mathcal{L}} = e^{t\mathcal{L}_0} \, \mathcal{E}_t^\varepsilon \, e^{-t\mathcal{L}_0}$, in which, by defining $\mathcal{L}_\varepsilon(\tau) := e^{-\tau \mathcal{L}_0} \, \mathcal{L}_1 \, e^{\tau \mathcal{L}_0}$

$$\begin{aligned}
\mathcal{E}_\varepsilon^t &:= e^{-t\mathcal{L}_0} \, e^{t\mathcal{L}} \, e^{t\mathcal{L}_0} = \mathbf{T} \exp\left(\int_0^t \mathcal{L}_1(\tau)d\tau \right) \\
&= \sum_{n=0}^\infty \varepsilon^n \int_0^t d\tau_1 \cdots \int_0^{\tau_{n-1}} d\tau_n \, \mathcal{L}_1^{\tau_1} \cdots \mathcal{L}_1^{\tau_n}. \qquad (1.9)
\end{aligned}$$

$$\begin{aligned}
F_\varepsilon(t) &= < \rho \otimes \mathbf{I}_B, \, e^{t\mathcal{L}}(\rho \otimes \sigma) > = < \rho \otimes \mathbf{I}_B, \, e^{t\mathcal{L}_0} \mathcal{E}_\varepsilon^t(\rho \otimes \sigma'_t) > \\
&= < e^{t\mathcal{L}_0^\dagger}(\rho \otimes \mathbf{I}_B), \, \mathcal{E}_\varepsilon^t(\rho \otimes \sigma'_t) > = < \rho \otimes \mathbf{I}_B, \, \mathcal{E}_\varepsilon^t(\rho \otimes \sigma'_t) > \\
&= \sum_{n=0}^\infty \varepsilon^n \int_0^t d\tau_1 \cdots \int_0^{\tau_{n-1}} d\tau_n < \rho \otimes \mathbf{I}, \, \mathcal{L}_1^{\tau_1} \cdots \mathcal{L}_1^{\tau_n}(\rho \otimes \sigma'_t) > \\
&= 1 + \varepsilon \int_0^t d\tau < \rho \otimes \mathbf{I}_B, \, \mathcal{L}_1^\tau(\rho \otimes \sigma'_t) > + o(\varepsilon^2) \\
&= 1 + \varepsilon \int_0^t d\tau < \rho \otimes \mathbf{I}_B, \, \mathcal{L}_1(\rho \otimes \sigma''_\tau) > + o(\varepsilon^2) = 1 + o(\varepsilon^2)
\end{aligned}$$

Where, for obtaining the last equality, we assumed

$$< \rho \otimes \mathbf{I}_B, \, \mathcal{L}_1(\rho \otimes \sigma''_\tau) > = 0. \qquad (1.10)$$

We assume that the infinitesimal generator $\mathcal{L}$ of the dynamical semi-group has the following standard (Lindblad) form that holds for Markovian evolutions

$$\mathcal{L}(\rho) := \frac{1}{2} \sum_\mu ([L_\mu \rho, \, L_\mu^\dagger] + [L_\mu, \, \rho L_\mu^\dagger]). \qquad (1.11)$$

Perturbing the Lindblad operators $L_\mu \mapsto L_\mu + \varepsilon \, \delta L_\mu$, one gets $\mathcal{L} \mapsto \mathcal{L} + \varepsilon [\mathcal{L}_1 + \varepsilon \, \mathcal{L}_2]$, where

$$\mathcal{L}_1(\omega) := \frac{1}{2} \sum_\mu ([\delta L_\mu \omega, \, L_\mu^\dagger] + [\delta L_\mu, \, \omega L_\mu^\dagger] + [L_\mu \omega, \, \delta L_\mu^\dagger] + [L_\mu, \, \omega \delta L_\mu^\dagger]) \qquad (1.12)$$

$$\mathcal{L}_2(\omega) := \frac{1}{2}\sum_\mu([\delta L_\mu\omega,\ \delta L_\mu^\dagger] + [\delta L_\mu,\ \omega\delta L_\mu^\dagger]). \tag{1.13}$$

Moreover $L_\mu := \mathbf{I}_S \otimes B_\mu$, and $\delta L_\mu := X_\mu \otimes A_\mu$. Let us consider the first two terms of Eq. (1.12) for a given μ and with $\omega = \rho \otimes \sigma$

$$2\,\delta L_\mu\,(\rho\otimes\sigma)\,L_\mu^\dagger - L_\mu^\dagger\delta L_\mu\,(\rho\otimes\sigma) - (\rho\otimes\sigma)\,L_\mu^\dagger\delta L_\mu$$

$$= 2\,X_\mu\rho\otimes A_\mu\sigma L_\mu^\dagger - X_\mu\rho\otimes L_\mu^\dagger A_\mu\sigma - \rho X_\mu\otimes\sigma L_\mu^\dagger A_\mu \tag{1.14}$$

multiplying by $\rho \otimes \mathbf{I}_B$ and taking the trace

$$\mathrm{tr}_S(\rho X_\mu\rho)\,\mathrm{tr}_B(2\,A_\mu\sigma L_\mu^\dagger - L_\mu^\dagger A_\mu\sigma - \sigma L_\mu^\dagger A_\mu) = 0. \tag{1.15}$$

Reasoning in the very same way, even the last terms of Eq. (1.12) give a vaninshing contribution. This show that relation (1.10) is fulfilled by $\mathcal{L}_1$.

6. CONCLUSIONS

The possibility of noiseless enconding and processing of quantum information is traced back to the existence of an underlying multi-partite structure. The origin of such hidden structure is purely algebraic and it is dictated by the interactions between the systems and the environmentr: When the latter admits non trivial symmetry group then *noiseless subsystems* allowing for universal quantum computation exist. This NSs approach, introduced by Knill et al as a generalization of decoherence-free subspaces, is robust against perturbations and provides an analog of the pointer basis in the noncommutative realm. The notion of NS has been shown to be crucial for a unified understanding and designing of e error correction/avoidance strategies. In particular we argued how one can use decoupling/symmetrization techniniques for artificial synthesis of systems supporting NSs, and how to perform on such NSs non-trivial computations.

References

E. Knill, R. Laflamme and L. Viola, Phys. Rev. Lett.**84**,2525 (2000);
P. Zanardi, LANL e-print ArXiv quant-ph/9910016
L. Viola, E. Knill, S. Lloyd, LANL e-print ArXiv quant-ph/0002072

1/F NOISE DURING MANIPULATION OF JOSEPHSON CHARGE QUBITS

[1,2,3]E. Paladino, [2,4]L. Faoro, [1,2,5]G. Falci, and [1,2]Rosario Fazio
[1]*Dipartimento di Metodologie Fisiche e Chimiche (DMFCI), Università di Catania (Italy).* [2]*Istituto Nazionale per la Fisica della Materia, Unitá di Catania e di Torino (Italy).* [3]*Consorzio Ennese Universitario, Cittadella degli Studi, Enna (Italy).* [4]*Dipartimento di Fisica del Politecnico di Torino (Italy).* [5]*Laboratoire Etudes Proprietès Electroniques des Solides, C.N.R.S., Grenoble (France).*

Abstract   Josephson nanocircuits have been recently proposed as a physical realization of Quantum Bits. In this contribution we analyze the decoherence in Josephson qubits in the charge regime. In particular we study the effect of background charges in the substrate responsible for the 1/f noise.

Keywords: Quantum Computation, Dissipative Quantum Mechanics, Josephson Effect

1.   INTRODUCTION

The study of quantum dynamics of nanofabricated devices has recently attracted a substantial interest in connection with development of the theory of quantum information processing [1]. The quest for large scale integrability and for tunability of the components finds in superconducting nanocircuits [4, 5, 6, 7, 8] natural candidates for the implementation of a quantum computer. The indirect observation of superpositions of charge states in a Josephson devices [9, 10] and of Rabi oscillation in SQUID's [11, 12], and the direct observation of controlled coherent time evolution of quantum states [13] in a Cooper pair box (see Fig.1.1) are the first important steps towards the implementation of a solid state quantum computer based on Josephson devices.

A quantum computer requires low dephasing on a large integration scale. One should be able to prepare a given input multiqubit state, to perform a controlled unitary manipulation and to perform efficient readout of the output state [3]. The interaction of the computer with the environment is a source of error in the above three steps. We will discuss dephasing effects during unitary manipulation which cause loss of memory of the input, focusing on the effect of $1/f$ noise.

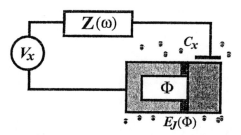

Figure 1.1 The tunable Cooper pair box which implements the JC Qubit: a superconducting island is connected to a voltage source V_x via a capacitor (C_x) and the two arms of a SQUID (C is the sum of the two capacitances).

We consider the Josephson-Charge (JC) Qubit [5, 13]. The device is Cooper pair box (see Fig.1.1) formed by a SQUID pierced by a magnetic flux Φ_x and with an applied gate voltage V_x. It is described by the Hamiltonian [14]

$$\mathcal{H}_{Box} = E_C \left(n - \frac{C_x V_x}{2e} \right)^2 - E_J(\Phi) \cos \theta \qquad (1.1)$$

where the couplings are the charging energy for a Cooper pair $E_C = 2e^2/(C + C_x)$, whose effect is tuned by V_x, and the Josephson coupling, $E_J(\Phi) \cos(\pi \Phi_x/\Phi_0)$, ($E_J$ is the Josephson energy and Φ_0 is the flux quantum [14]), which is tuned by Φ_x. The phase θ and the number n of extra Cooper pairs in the box are canonically conjugate variables, $[\theta, n] = i$. The device operates in the regime $E_C \gg E_J$ so at temperatures $T \ll E_C$ only two charge states are important [15].

By projecting eq.(1.1) on the computational two-state space, one obtains the Qubit Hamiltonian $\mathcal{H}_Q$. It is convenient to use a spin representation, where the charge basis $\{|0\rangle, |1\rangle\}$ is represented by the eigenstates $\{|\uparrow\rangle, |\downarrow\rangle\}$ of σ_z and

$$\mathcal{H}_Q = -\frac{1}{2} \vec{B}_0 \cdot \vec{\sigma} \qquad ; \qquad \vec{B}_0 \equiv (E_J, 0, 1 - C_x V_x/e)$$

is the Hamiltonian of a spin (σ_i, where $i = x, y, z$, are the Pauli matrices) in a fictious magnetic field $\vec{B}$.

The interaction of the computer with the environment may cause relaxational dynamics and dephasing [17, 18] with associated time scales, T_1 and T_2. Of course one cannot expect that in general Bloch equations follow from an arbitrary environment, but this happens for weak coupling and if $t > \tau_c$, where τ_c is the correlation time of the environment [19, 20]. For a source of $1/f$ noise these conditions are not strictly met. In single electron (SET) devices $1/f$ noise is routinely observed [21] and it is attributed to mobile charges trapped in the substrate.

In fact the environment has a quantum nature and this has striking consequences for the long time behavior [18, 22]. Noise coming from the *linear* electromagnetic environment can be studied fully quantum mechanically by a mapping in the spin-boson problem. [18].

We will present a quantum mechanical model for a Qubit coupled to a set of background charges. This will be first analyzed by weak-coupling evolution equations for the Reduced Density Matrix (RDM) of the Qubit which allow to express the dephasing rate in terms of the power spectrum of the environment. The picture which emerges and other possible issues can be tested using a different approach, where the dynamics of a Qubit coupled with a finite set of charges is studied.

2. MODEL

Mobile background charges in the substrate polarize the Qubit. A simple Hamiltonian which captures this feature has been introduced by Bauernschmitt and Nazarov [23]. For a single background charge it reads

$$\mathcal{H} = -\frac{1}{2} \vec{B} \cdot \vec{\sigma} - \frac{V}{2} \sigma_z b^\dagger b + \varepsilon_c b^\dagger b + \sum_k [T_k c_k^\dagger b + T_k^* b^\dagger c_k] + \sum_k \varepsilon_k c_k^\dagger c_k \tag{1.2}$$

where the background charge is an electron (b) which may occupy a localized level (ε_c) close to the junction. In these conditions it induces an extra voltage drop on the Qubit, which generates the interaction V and other terms which renormalize the parameters. The localized electron may tunnel to a continous band (field operator c_k, energies ε_k), these processes being responsible for dissipation. Other microscopic pictures can be considered to describe the background charge, however the model eq.(1.2) is a convenient simple choice. For many independent background charges Hamiltonian eq.(1.2) can be generalized in a straightforward way. Then the standard road [24] to obtain $1/f$ noise is to consider a suitable distribution of the proper rates γ_i of the background charges.

Following Lax [25] we derive an equation for the RDM $\rho(t)$ of the Qubit, treating the background charges and the band as an environment. It is convenient to define the operator $E = V/2 [b^\dagger b - \langle b^\dagger b \rangle_0]$, where the average is taken with respect to the environment. The interaction term becomes $-E\,\sigma_z$. One seeks a solution of the partial averaged Neumann equation for $W(t)$ of the form $W(t) = \rho(t) \otimes W_0^E + \Delta W(t)$, where W_0^E is the equilibrium density matrix of the environment alone. We start from a factorized initial state, $W(0) = \rho(0) \otimes W_0^E$. Physically we are assuming that the environment stays very close to equilibrium. The resulting equations are

$$\frac{\partial \rho}{\partial t} = \frac{i}{2} [\vec{B} \cdot \vec{\sigma}, \rho] - \int_0^t dt' \, \pi \, S_E(t - t') \, [\sigma_z, [\tilde{\sigma}_z, \tilde{\rho}]]$$
$$- \frac{i}{2} \int_0^\infty dt' \, G_R(t - t') \, [\sigma_z, [\tilde{\sigma}_z, \tilde{\rho}]_+] \qquad (1.3)$$

The first term in the r.h.s. describes free propagation. The second and the third term contain the effect of the environment, which enters via the exact spectral function $S_E(t) = 1/(2\pi) \langle [\hat{E}(t), \hat{E}(0)]_+ \rangle_0$ and the susceptibility $\chi_E(t) = -i\Theta(t) \langle [\hat{E}(t), \hat{E}(0)] \rangle_0$. The operators $\tilde{\sigma}_z = \mathcal{U}_Q(t - t') \, \sigma_z \, \mathcal{U}_Q^\dagger(t - t')$ and $\tilde{\rho} = \mathcal{U}_Q(t - t') \, \rho(t') \, \mathcal{U}_Q^\dagger(t - t')$ (here $\mathcal{U}_Q(t)$ is the time evolution operator of the Qubit alone) appear.

For a short time correlated environment one could perform the Markov approximation $\tilde{\rho} = \rho(t)$. In our case we work out the anticommutators appearing in eq.(1.2) and obtain equations where possible memory effects are retained.

It is convenient to perform a rotation of the reference frame which brings the $\hat{x}$ axis on $\vec{B}$. We then expect the dynamics of the new $\langle \sigma_x(t) \rangle$ to show relaxation (T_1) and that of $\langle \sigma_y(t) \rangle$ and $\langle \sigma_z(t) \rangle$ to show dephasing (T_2). The free evolution term simplifies at the expenses of having a more complicated structure for the other terms, for instance $S_E [\sigma_z, [\tilde{\sigma}_z, \tilde{\rho}]] \to \sum_{ij} S^{ij} [\sigma_i, [\tilde{\sigma}_j, \tilde{\rho}]]$, where $i, j = x, y, z$. Explicitly $S_E^{ij}(t) = S_E(t) \beta_i \beta_j$ where $\{\beta_i\} \equiv (B_z/B, 0, B_x/B)$, B_i being the components of $\vec{B}$ in the original reference frame. One finds eventually that the term in $\chi_E(t)$ does not depend on $\rho(t)$ whereas the term in $S_E(t)$ contains information about the relaxation and dephasing rates. These are calculated in the Wigner-Weisskopf approximation. A single real pole is involved in the dynamics of $\langle \sigma_x(t) \rangle$, and the relaxation rate reads

$$\frac{1}{T_1} = 2\pi \, \tilde{S}_E(B) \left(\frac{B_x}{B} \right)^2 \qquad (1.4)$$

where the Fourier transform $\tilde{S}_E(\omega)$ is the power spectrum of the noise. Two complex conjugated poles are involved in the dynamics of $\langle \sigma_y(t) \rangle$

and $\langle \sigma_z(t) \rangle$, which give

$$\frac{1}{T_2} = \pi \tilde{S}_E(B) \left(\frac{B_x}{B}\right)^2 + 2\pi \tilde{S}_E(0) \left(\frac{B_z}{B}\right)^2 \qquad (1.5)$$

For $B_z = 0$ (charging splitting, $1 - C_x V_x/e = 0$) the noise acts on the axis perpendicular to $\vec{B}$ and produces broadening of the Qubit's levels. This leads to both relaxation and dephasing. Rates are proportional to $\tilde{S}_E(B) = \tilde{S}_E(E_J)$. On the other hand if some charging splitting is present ($B_z \neq 0$), the additional term $\propto \tilde{S}_E(0)$ appears. It is the noise "parallel" to $\vec{B}$, i.e. random fluctuations of the energy splitting which produce errors in the accumulated phases. This term is analogous to the secular broadening in NMR [20] and may be relevant in our case $\tilde{S}_E(\omega) \propto 1/\omega$.

For the spin-boson model a weak coupling systematic resummation of diagrams [18] yields formally the same equations (1.4,1.5). Notice that the derivation we presented here works, under weak coupling conditions, for a generic environment, including quantum harmonic oscillators. This is consistent with the analysis in Ref. [18], where it was argued that a weakly coupled environment is similar to a bath of oscillators, so the validity of the results for the spin-boson model can be extended to more general situations. We mention on passing that the hamiltonian eq.(1.2) is not a spin bath [26] because of the coupling of the individual background charges with the band.

For a weakly coupled environment T_1 and T_2 respectively, are calculated by the Redfield approach [19, 18] used in NMR [20], i.e. by treating the environment as a classical noise source with short τ_c. T_1 and T_2 are expressed in terms of the power spectra of the noise. Of course one cannot expect that in general Bloch equations follow from an arbitrary environment.

3. DISCUSSION

Equations (1.4, 1.5) can be used to calculate decay times due to broadening from a known power spectrum. This has to be in general related to other kind of measurement, for instance the impedance of the circuit for an oscillator environment. In our case the underlying model eq.(1.2) implies that the noise spectrum is given by

$$S_E(\omega) = \sum_i \left[\frac{V_i}{4\cosh(\beta\varepsilon_{ci}/2)}\right]^2 \frac{\gamma_i}{\gamma_i^2 + \omega^2}$$

and may have certain simple features, as being $S_E(\omega) = 2\pi\alpha^2/\omega$, where α can be inferred from other experiments. The operator E produces fluctuations of the charge Q_x polarizing the box, $E \leftrightarrow E_C \, \delta Q_x / e$, and recognize $S_E(\omega) = E_C^2 \, S_Q(\omega) / (\pi e^2)$. So we need the $1/f$ part of $S_Q(\omega)$ at $\omega = E_J \sim GHz$. Noise measurements in SET devices [21] show $1/f$ behavior up to kHZ frequencies, because at higher frequencies intrinsic shot noise prevails. We then extrapolate the observed $1/f$ spectrum to GHz frequency. Using the results of Ref. [21] for parameters of the experiment [13] we get $T_1 \approx 130 \, ns$. We remark that the extrapolation of $1/f$ behavior up to GHZ is neither accurate nor easily justifiable assuming equilibrium at mK temperatures. However this procedure seems to explain recent observations in electron pumps [27]. We conclude that background charges may be in general an active source of dephasing in CJ Qubits. We notice that actual values of T_1 are dependent both on fabrication parameters and on operations. In particular implementation of operations which require low values of E_J (and in general low energy splittings) will be surely affected by $1/f$ noise.

We remark that our analysis covers only few of the aspects related to possible effects of background charges on the dynamics of the CJ qubit. For instance single background charges inside the junction or correlated clusters in the substrate may be more stongly coupled and generate telegraph noise. Moreover slow and out of equilibrium background charges prevent to prepare repeatedly a given (generic) Qubit. This effect may determine a reduction of the amplitude of the output in the operation scheme used in the experiment [13].

Another interesting issue is the effect of nonequilibrium distributions of background charges on the coherent time evolution. The Lax equation approach discussed so far cannot be used to study this problem, because it assumes that the environment is practically at equilibrium. Instead one can use a different strategy, namely take the band in eq.(1.2) as the environment and study the reduced dynamics of the system composed by the Qubit and a finite number of background charges. In this case the system is charachterized by a large number of poles and in principle may show a dynamics characterized by many time scales. We studied dephasing at charge degeneracy ($B_z = 0$) and the results show that for set of charges giving a $\sim 1/f$ spectrum in some decade around $\omega \sim E_J$ with parameters as the ones discussed above, the dynamics is still well described by a single time scale $\propto S_E(E_J)$. Charges in nonequilibrium further broaden the spectral line, even if they switch at rates smaller than the resonance.

Finally we discuss the effect of parallel noise which, according to eq. (1.5), contributes to the dephasing rate with the unfriendly term $\propto \tilde{S}_E(0)$. One may argue that for frequencies smaller than some cutoff ω_{min} the $1/f$ spectrum somehow flattens and very slow charges act as a constant background which does not produce dephasing. However many background charges with $\gamma_i \leq \omega_{min}$ contribute to $\tilde{S}_E(0)$. Calculation of this (unmeasurable) contribution would require detailed knowledge of the environment. We remark that at low frequencies $1/f$ noise may be so large (and so slow) to invalid the weak coupling picture. Let consider a model where only "parallel" noise is present. For an oscillator environment the dynamics can be solved exactly [17] for a generic spectral density. In the ohmic case a superposition of $|\vec{B}\pm\rangle$ states will dephase according to a power law, instead of exponentially. The exponential decay of coherences is recovered in the thermal dominated regime. In practice there is initially a window of times where saturation occurs. In this window the effect of parallel noise is a slow reduction of the amplitude, which cannot be described by a rate. For later times thermal processes prevail and the decay becomes exponential. Notice that if a non vanishing σ_x field (or equivalently some "orthogonal" noise) is present the dynamics of the above model is in general not exactly solvable.

Acknowledgments

We warmly acknowledge D. Averin, D. Esteve, L. Glazman, F. Hekking, P. Lafarge, R. Leoni, Y. Nazarov, G.M. Palma, M. Rasetti, U. Weiss and A. Zorin for very useful discussions. This work has been supported by the European Community (grants TMR-FMRX-CT-97-0143, IST-SQUBIT, IST-EQUIP) and by INFM-PRA-SSQI.

References

[1] R. Cleve, A. Ekert, L. Henderson, C. Macchiavello, M. Mosca *On quantum algorithms* Complexity Vol.4, No.1, pp.33-42 (1998); quant-phys/9903061 Reprinted in *Quantum computation and quantum information theory* C.Macchiavello, G.M.Palma & A.Zeilinger Ed.s, World Scientific, (Singapore, 2000). A. Ekert and R. Jozsa, Rev. Mod. Phys. **68**, 733 (1996). Bennett, C. H., Oct. 1995, *Physics Today.*, p.24; Di Vincenzo, D., 1995, *Science* **270**, 255; Steane, A., 1998, *Rept. Prog. Phys.* **61**, 117.

[2] Leggett, A.J., 1986, in *Directions in Condensed Matter Physics*, (G. Grinstein and G. Mazenko Eds.), World Scientific, Singapore, vol 1, p. 187.

[3] See the contribution of Y. Makhlin et al. in this volume.

[4] D.A. Averin, Sol. State Comm. **105** 659 (1998).

[5] Y. Makhlin, G. Schön and A. Shnirman, Nature **398**, 305 (1999); A. Shnirman, G. Schön and Z. Hermon, Phys. Rev. Lett. **79**, 2371 (1997).

[6] L.B. Ioffe, V.B. Geshkenbein, M.V. Feigelman, A.L. Faucher, and G. Blatter, Nature **398**, 679 (1999).

[7] J.E. Mooij, T.P. Orlando, L. Tian, C. van der Wal, L. Levitov, S. Lloyd, and J.J. Mazo, Science **285**, 1036 (1999).

[8] G. Falci, R. Fazio, G.M. Palma, J. Siewert, and V. Vedral, Nature **407**, 355 (2000).

[9] M. Matters, W. Elion, and J.E. Mooij, Phys. Rev. Lett. **75**, 721 (1995).

[10] V. Bouchiat, D. Vion, P. Joyez, D. Esteve, and M. Devoret, Physica Scripta **T76**, 165 (1998).

[11] J.R. Friedman, V. Patel, W. Chen, S.K. Tolpygo, J.K. Lukens, Nature, **406**, 43 (2000).

[12] See the contribution of C.H. van der Wal et al. in this volume.

[13] Y. Nakamura, Yu.A. Pashkin, J.S. Tsai, Nature **398**, 786 (1999).

[14] M. Tinkham, *Introduction to Superconductivity*, 2nd ed, (McGraw-Hill, New York 1996).

[15] The influence of higher charge states has been studied in R. Fazio, G.M. Palma, and J. Siewert, Phys. Rev. Lett. **83**, 5385 (1999).

[16] J. Jones, V. Vedral, A.K. Ekert, C. Castagnoli, Nature, **403**, 869 (2000).

[17] G.M. Palma, K. Suominen, A.K. Ekert, Proc. Roy. Soc., London Ser. **A 452**, 567 (1996).

[18] U. Weiss, *Quantum Dissipative Systems*, 2nd ed (World Scientific, Singapore 1999).

[19] A.G. Redfield, IBM J. Research Develop, **1**, 19 (1957).

[20] A. Abragam, *The Principles of Nuclear Magnetism* (Claredon Press, Oxford 1961).

[21] A.B. Zorin, F.-J Ahlers, J. Niemeyer, T. Weimann, H. Wolf, V.A. Krupenin, S.V. Lotkhov, Phys. Rev. B, **53**, 13682 (1996).

[22] a simple example is the Fano-Anderson model, see M.O. Scully and M.S. Zubairy, *Quantum Optics*, (Cambridge University Press, 1997).

[23] R. Bauernschmitt, Y.V. Nazarov, Phys. Rev. B, **47**, 9997 (1993).

[24] M.B. Weissman, Rev. Mod. Phys. **60**, 537 (1988).

[25] M. Lax, Phys. Rev. **145**, 110 (1966).

[26] N. Prokof'ev, P. Stamp, Rep. Prog. Phys. **63(4)**, 669 (2000).

[27] M. Covington, M.W Keller, R.L. Kautz and J.M. Martinis, Phys. Rev. Lett., **84**, 5192 (2000).

Non-Markovian dynamics in continuous–wave atom lasers

H. P. Breuer[1], D. Faller[1], B. Kappler[1] and F. Petruccione[1,2]

[1] Albert-Ludwigs-Universität, Fakultät für Physik.
Hermann-Herder-Straße 3, D-79104 Freiburg im Breisgau, Federal Republic of Germany
[2] Istituto Italiano per gli Studi Filosofici, Palazzo Serra di Cassano
Via Monte di Dio 14, I-80132 Napoli, Italy

Abstract. The non-Markovian dynamics of a continuous-wave atom laser model including gravitational effects and interactions inside the Bose-Einstein condensate is studied. A generalization of the time-convolutionless projection operator technique presented in this article enables a correct modeling of memory effects in the atom laser which takes into account correlations in the initial state of the combined atom-reservoir system. This generalization allows the computation of both the occupation number of the condensate and the spectrum of the atoms coupled out of the condensate. The non-Markovian evolution is shown to yield substantial deviations from results obtained in the Born-Markov approximation.

1. Introduction

Shortly after the first experimental realization of a Bose–Einstein condensate (BEC) in the laboratory [1, 7], an atom laser, a source of a coherent atomic beam, has also been realized. The integral part of an atom laser is the output coupling mechanism which coherently transfers the atoms out of the trap. The experimental realization of this output coupling mechanism is based on a state change of the trapped atoms. This state change is caused either by a rf pulse [2, 18], or through a two-photon Raman transition [11], see Refs. [8, 17] for a short survey of the experimental situation.

Usually the analysis of atom laser models is based on the Born-Markov approximation [10, 12, 19, 25–27] which is widely used in quantum optics, e.g. to describe the dynamics of optical lasers. However, the investigation of very simple atom laser models, which can be solved exactly, reveals that in the realistic parameter regime non-Markovian effects dominate and, thus, the Born-Markov approximation fails [22]. As shown in Ref. [3] the non-Markovian dynamics of realistic atom laser models may be studied with the help of the time-convolutionless (TCL) projection operator technique [4, 6, 24].

The TCL projection operator technique is based on an expansion of the exact equation of motion of the density operator of the combined system in powers of the coupling strength between system and reservoir. The resulting equation of motion for the system density operator is local in time and the second order expansion has a form similar to the Born–Markov master equation [4]. By taking into account higher orders of the expansion it is thus possible to study systematically non-Markovian effects. It is important to note that the TCL projection operator technique is based on an expansion of the exact equation of motion and not on an expansion of its solution. Hence the results obtained are also valid in the long time limit if one does not leave the parameter regime of moderate coupling between system and reservoir, see

Ref. [4]. In order to determine the spectrum of the atoms coupled out of the atomic trap, the TCL method was generalized to take into account the effect of non-factorizing initial conditions between system and reservoir [9]. Thus, there is no need to make use of the Quantum–Regression theorem which is based on the Born–Markov approximation in order to determine the spectrum of the atoms coupled out of the atomic trap.

The atom laser model investigated in this article has been originally proposed in [12,20,25]. While these papers make use of the Born–Markov approximation, we will use the TCL projection operator technique to investigate the influence of a non-Markovian output coupling mechanism based on a two-photon Raman transition. This output coupling proposed in Ref. [13, 21–23] was experimentally realized recently [11]. In addition to previous studies of this atom laser model our more realistic model includes interactions inside the BEC and accounts for the effects of gravity towards the atoms coupled out of the atomic trap.

2. Model of a continuous–wave atom laser

Our study is based on the binary-collision atom laser model [12, 20, 25]. This simplified model takes into account only three atomic modes of the trap which are described by a harmonic potential. As shown in Fig. 1 the pump mode 1 of the trap is coupled to a thermal reservoir from which atoms are allowed to enter or leave the trap. Through collisions with other atoms in the pump mode, atoms are distributed to the source mode 0 which contains the BEC and to mode 2 from which atoms are strongly coupled out to enable evaporative cooling which is needed to achieve the necessary low temperatures. The atoms in the source mode 0 are coupled out of the trap with the help of a two-photon Raman transition. This output coupling mechanism which was originally proposed in Ref. [13, 21–23] leads to non-Markovian effects which will be investigated in this paper.

In order to obtain a BEC in the source mode 0 the rate κ_1 at which atoms enter the trap must be much larger than the rate κ_0 at which the atoms from the BEC are coupled out of the trap. To enable the evaporative cooling process in mode 2 the corresponding coupling rate κ_2 must be much larger than all other rates. This yields the condition

$$\kappa_2 \gg \kappa_1 \gg \kappa_0. \tag{1}$$

In our study only the coupling of the atoms out of the BEC is described taking into account the non-Markovian dynamics, whereas the coupling to the other two reservoirs will be treated within the Born–Markov approximation. Furthermore mode 2 can be adiabatically eliminated, as shown in Ref. [12,25]. It is only the output coupling of atoms from the source mode 0 which requires a non-Markovian treatment.

The Hamiltonian H for this binary collision atom laser model consists of the system Hamiltonian H_{sys}, the Hamiltonian V describing the interactions

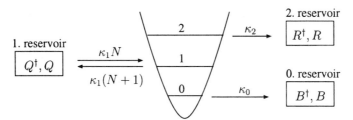

Figure 1. Model of a continuous–wave atom laser consisting of three modes. Atoms entering the trap at the pump mode 1 interact with each other and are distributed inside the trap to the source mode 0 forming the BEC and to mode 2 from which the atoms are coupled out strongly to enable evaporative cooling.

inside the atomic trap, and the three interaction Hamiltonians $H_{I,i}$, $i = 0, 1, 2$, which describe the coupling of the atomic modes to the corresponding reservoirs [3]

$$H = H_{\text{sys}} + V + H_{I,0} + H_{I,1} + H_{I,2} .\qquad(2)$$

The interaction Hamiltonian V does not include interactions which are energetically unfavored at low densities such as the annihilation of two atoms in the source mode and the creation of two atoms in mode 2. After the adiabatic elimination of mode 2 the Hamiltonian V describing the interaction between the atoms takes the form [27],

$$V = \left\{ g_{0000} a_0^{\dagger 2} a_0^2 + g_{0101} a_0^\dagger a_0 a_1^\dagger a_1 + g_{1111} a_1^{\dagger 2} a_1^2 \right\} .\qquad(3)$$

Here, the $a_i^\dagger$, a_i are the creation and annihilation operators for the atoms in modes $i = 0, 1$, and the corresponding coupling constants are denoted by g. The adiabatic elimination of mode 2 introduces an effective reservoir which describes the annihilation of two atoms in the pump mode 1 and the creation of an atom in the source mode 0 and in mode 2 which is coupled out immediately. The coupling to this reservoir is described by the Hamiltonian H_{eff} given below.

Since V only affects off-diagonal elements of the system's density operator it only influences the spectrum of the out-coupled atoms. In [27] the first two terms have been analyzed separately and were both shown to lead to a broadening of the spectrum, while the last term in V does not affect the spectrum of this two-mode model at all. Hence, for the sake of simplicity we may restrict the following discussion to the first term describing interactions inside the BEC. Thus in the interaction picture the Hamiltonian $H(t)$ is given by

$$H(t) = H_{I,0}(t) + H_{I,1}(t) + H_{\text{eff}}(t) + g_{0000}\, a_0^\dagger a_0^\dagger a_0 a_0 ,\qquad(4)$$

with the interaction Hamiltonians

$$H_{I,0} = -i \left(B(t) a_0^\dagger e^{i\omega_0 t} - a_0 B^\dagger(t) e^{-i\omega_0 t} \right), \tag{5a}$$

$$H_{I,1} = \sqrt{\kappa_1} \left(Q(t) a_1^\dagger e^{i\omega_1 t} + a_1 Q^\dagger(t) e^{-i\omega_1 t} \right), \tag{5b}$$

$$H_{\text{eff}} = \sqrt{\Omega} \left(R^\dagger(t) a_0^\dagger a_1^2 e^{i(\omega_0 - 2\omega_1)t} + a_0 a_1^{\dagger 2} R(t) e^{-i(\omega_0 - 2\omega_1)t} \right). \tag{5c}$$

The operators $B, B^\dagger, Q, Q^\dagger$ and $R, R^\dagger$ are the creation and annihilation operators of the different reservoirs shown in Fig. 1.

Gravitational forces can be accounted for by an appropriate modification of the reservoir correlation function. In the following we shall denote by $f(\tau)$ and $f_{\text{grav}}(\tau)$ the reservoir correlation functions without and with gravity, respectively. In the case of the coupling to the vacuum $f(\tau)$ and $f_{\text{grav}}(\tau)$ can be determined analytically [14–16]

$$f(\tau) = \text{Tr}_{\text{res}} \left\{ B(\tau) B^\dagger(0) \rho_{\text{res}}(0) \right\} e^{i\omega_0 \tau} = \frac{\Gamma}{\sqrt{1 + i\alpha\tau}} e^{i\omega_0 \tau} \tag{6a}$$

$$f_{\text{grav}}(\tau) = f(\tau) \exp \left(-\frac{\tilde{g}^2 \tau^2}{32\lambda^2 \sigma_k^2} - \frac{i\tilde{g}^2 \tau^3}{48\lambda} \right). \tag{6b}$$

Here α is defined by $\alpha = \hbar \sigma_k^2/(2m)$, σ_k denotes the width of the coupling in k-space and $\lambda = \hbar/(2m)$. The constant $\tilde{g}$ is the effective gravitational constant, $\tilde{g} = g \sin(\theta)$, where θ denotes the angle between the horizontal direction and the optical waveguide, which bounds the motion of the out-coupled atoms of our one-dimensional atom laser model in the other two space directions. Just as in [14, 15] we have chosen $\theta = \pi/20$.

As discussed in [3] the reservoir correlation function exhibits a slow algebraic asymptotic decay, $f(\tau) \sim 1/\sqrt{\tau}$. Thus, a strong non-Markovian behavior of the atom laser is to be expected which is caused by the quadratic dispersion relation for massive particles.

3. Time-convolutionless projection operator technique

Usually one assumes that a non-Markovian dynamics implies an equation of motion containing a time-convolution kernel. However, the time-convolutionless (TCL) projection operator technique enables the derivation of an equation of motion for the system's density operator which is local in time, that is, which does not involve a memory kernel. Starting from the Liouville-von Neumann equation in the interaction picture for the density operator $\rho_{\text{tot}}(t)$ of the combined system,

$$\frac{\partial}{\partial t} \rho_{\text{tot}}(t) = -i\alpha[H(t), \rho_{\text{tot}}(t)] = \alpha L(t) \rho_{\text{tot}}(t), \tag{7}$$

the TCL projection operator technique leads to an exact equation of motion for the atomic density operator $\rho_{\text{sys}}(t)$ which takes the form [4, 6, 24]

$$\frac{\partial}{\partial t} \mathcal{P} \rho_{\text{tot}}(t) = K(t) \mathcal{P} \rho_{\text{tot}}(t) + I(t) \mathcal{Q} \rho_{\text{tot}}(0). \tag{8}$$

Here α denotes the strength of the coupling between system and reservoir and the projection operators $\mathcal{P}$ defined by $\mathcal{P}\rho_{\text{tot}}(t) = \rho_{\text{sys}}(t) \otimes \rho_{\text{res}}(0)$ and $\mathcal{Q} = 1 - \mathcal{P}$ have been introduced. The generator $K(t)$ and the inhomogeneity $I(t)$ are defined by

$$K(t) = \alpha \mathcal{P}L(t)\left[1 - \Sigma(t)\right]^{-1}\mathcal{P}, \tag{9a}$$

$$I(t) = \alpha \mathcal{P}L(t)\left[1 - \Sigma(t)\right]^{-1}\mathcal{G}(t,0). \tag{9b}$$

Detailed derivations of the above equations and the definitions of the super-operators $\Sigma(t)$ and $\mathcal{G}(t,0)$ can be found in Refs. [3,4].

If we presume factorizing initial conditions for the density operator $\rho_{\text{tot}}(t)$ of the combined system, then the inhomogeneity $I(t)$ in Eq. (8) vanishes. Unfortunately the solution of the master equation (8) is as difficult as the direct solution of the Liouville-von Neumann equation. In order to approximate the above master equation we expand the generator $K(t)$ and the inhomogeneity $I(t)$ in powers of the coupling strength α. This expansion enables the systematic investigation of the non-Markovian dynamics.

3.1. Expansion of the Generator $K(t)$

If the inverse operator in Eqs. (9) exists, which is always the case for short times or in a moderate coupling regime, see Ref. [4], it is possible to derive a general expression for the expansion of the generator $K(t)$, see Ref. [6,24]

$$K(t) = \sum_{n=1}^{\infty} \alpha^n K^{(n)}(t). \tag{10}$$

One obtains

$$K^{(n)}(t) = \int_0^t dt_1 \int_0^{t_1} dt_2 \cdots \int_0^{t_{n-2}} dt_{n-1} k_n(t, t_1, \ldots, t_{n-1}), \tag{11a}$$

$$k_n(t, t_1, \ldots, t_{n-1}) = \sum (-1)^q \mathcal{P}L(t) \ldots L(t_i)\mathcal{P}L(t_j) \ldots L(t_l)\mathcal{P} \ldots . \tag{11b}$$

The sum in the above equation extends over all possible insertions of the projection operator $\mathcal{P}$ in between the superoperators $L(t_i)$, preserving the chronological time ordering between successive insertions of two projection operators $\mathcal{P}$, that is, $t_j > \cdots > t_l$. The sign of the resulting expression is determined by the number q of inserted operators $\mathcal{P}$. For the expansion up to third order one finds

$$k_1(t) = \mathcal{P}L(t)\mathcal{P}, \tag{12a}$$

$$k_2(t, t_1) = \mathcal{P}L(t)\mathcal{Q}L(t_1)\mathcal{P}, \tag{12b}$$

$$k_3(t, t_1, t_2) = \mathcal{P}L(t)\mathcal{Q}L(t_1)\mathcal{Q}L(t_2)\mathcal{P} - \mathcal{P}L(t)\mathcal{Q}L(t_2)\mathcal{P}L(t_1)\mathcal{P}. \tag{12c}$$

This expansion is considerably simplified if one assumes that the trace Tr_{res} taken over odd moments of the interaction Hamiltonian $H_{\text{int}}(t)$ vanishes

$$\text{Tr}_{\text{res}}\left\{H_{\text{int}}(t_1) \ldots H_{\text{int}}(t_{2k+1})\rho_{\text{res}}(0)\right\} = 0, \tag{13}$$

which would be the case if we do not include the interaction Hamiltonian V. Then all terms containing an odd number of superoperators $L(t_i)$ between two insertions of $\mathcal{P}$ disappear and, furthermore, all expansion terms containing an odd power of the coupling strength α vanish.

3.2. Expansion of the Inhomogeneity $I(t)$

The inhomogeneous part $I(t)$ of the master equation (8) expresses the effect of non-factorizing initial conditions between the system and the reservoir. In order to approximate the master equation (8) the superoperator $I(t)$ has to be expanded in powers of the coupling strength α. In addition one must find a suitable expression for the term $\mathcal{Q}\rho_{\text{tot}}(0)$.

The expansion of the superoperator $I(t)$ can be performed in a way similar to the expansion of the generator $K(t)$, see Ref. [5],

$$I(t) = \sum_{n=1}^{\infty} \alpha^n I^{(n)}(t). \tag{14}$$

Interestingly up to the fourth order one obtains the same results as in the expansion of $K(t)$ in Eqs. (12) if one replaces in every term the last projection operator $\mathcal{P}$ with $\mathcal{Q}$.

In order to eliminate the initial density operator $\rho_{\text{tot}}(0)$ of the combined system in the master equation (8) we assume that $\rho_{\text{tot}}(0)$ can be written as

$$\rho_{\text{tot}}(0) = \mathcal{A}\rho_{\text{cor}}(0), \tag{15}$$

where $\mathcal{A}$ is an arbitrary superoperator acting on the Hilbert space of the system. The only restriction for the density operator $\rho_{\text{cor}}(0)$ is, that it was in a factorizing initial state at an earlier time $t = -T$, i.e., $Q\rho_{\text{cor}}(-T) = 0$. Such conditions are complied with, e. g. if one calculates correlation functions of observables of a system which was in a factorizing state at some instant in the past.

In order to eliminate $\rho_{\text{tot}}(0)$ it suffices to determine $\mathcal{Q}\rho_{\text{cor}}(0)$ because the superoperator $\mathcal{A}$ acts only in the Hilbert space of the system and thus commutes with $\mathcal{Q}$. Applying the TCL projection operator technique to this combined system which was in a factorizing state at time $t = -T$, one obtains a connection between $\mathcal{Q}\rho_{\text{cor}}(0)$ and $\mathcal{P}\rho_{\text{cor}}(0)$, namely

$$\mathcal{Q}\rho_{\text{cor}}(0) = R\mathcal{P}\rho_{\text{cor}}(0). \tag{16}$$

Here the superoperators R and $\Sigma_T(0)$ are defined by

$$R = [1 - \Sigma_T(0)]^{-1}\Sigma_T(0), \tag{17a}$$

$$\Sigma_T(0) = \alpha \int_{-T}^{0} ds \, \mathcal{G}(t,s)\mathcal{Q}L(s)\mathcal{P}G(t,s). \tag{17b}$$

The definitions of the propagators $G(t,s)$ and $\mathcal{G}(t,s)$ can be found in Ref. [4].

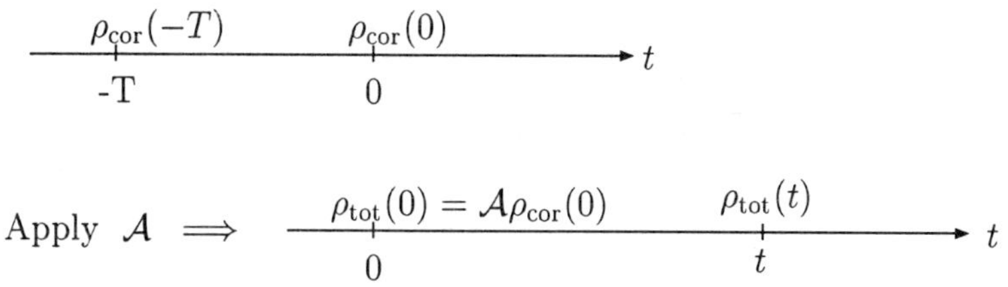

Figure 2. Connection between the different density operators of the combined system. At $t = -T$ the system is described by the factorizing density operator $\rho_{\text{cor}}(-T)$. The superoperator $\mathcal{A}$ is applied at $t = 0$ and one obtains the initial density operator of the system $\rho_{\text{tot}}(0)$.

Just as the operators $K(t)$ and $I(t)$ one can expand R in powers of the coupling strength α

$$R = \sum_{n=1}^{\infty} \alpha^n R^{(n)}.$$ (18)

For this expansion one obtains to third order

$$R^{(1)} = \int_{-T}^{0} dt_1 \; \mathcal{Q}L(t_1)\mathcal{P},$$ (19a)

$$R^{(2)} = \int_{-T}^{0} dt_1 \int_{-T}^{t_1} dt_2 \left\{ \mathcal{Q}L(t_1)\mathcal{Q}L(t_2)\mathcal{P} - \mathcal{Q}L(t_2)\mathcal{P}L(t_1) \right\},$$ (19b)

$$R^{(3)} = \int_{-T}^{0} dt_1 \int_{-T}^{t_1} dt_2 \int_{-T}^{t_2} dt_3 \{ \mathcal{Q}L(t_1)\mathcal{Q}L(t_2)\mathcal{Q}L(t_3)\mathcal{P}$$
$$- \mathcal{Q}L(t_2)\mathcal{Q}L(t_3)\mathcal{P}L(t_1) - \mathcal{Q}L(t_1)\mathcal{Q}L(t_3)\mathcal{P}L(t_2)$$
$$+ \mathcal{Q}L(t_3)\mathcal{P}L(t_2)\mathcal{P}L(t_1) - \mathcal{Q}L(t_2)\mathcal{P}L(t_1)\mathcal{Q}L(t_3)\mathcal{P}$$
$$- \mathcal{Q}L(t_3)\mathcal{P}L(t_1)\mathcal{Q}L(t_2)\mathcal{P} \}.$$ (19c)

Having determined the expansions of $I(t)$ and R we can put the results together. We managed to expand the inhomogeneity in the TCL master equation (8) and to eliminate the term $\mathcal{Q}\rho_{\text{tot}}(0)$ which involves initial correlations between the reservoirs and the system's degrees of freedom. Summarizing we obtain to fourth order

$$I(t)\mathcal{Q}\rho_{\text{tot}}(0) = I(t)\mathcal{A}R\mathcal{P}\rho_{\text{cor}}(0)$$ (20)
$$= \left[\alpha^2 I^{(1)}(t)\mathcal{A}R^{(1)} + \alpha^3 \left(I^{(1)}(t)\mathcal{A}R^{(2)} + I^{(2)}(t)\mathcal{A}R^{(1)} \right) \right.$$
$$\left. + \alpha^4 \left(I^{(1)}(t)\mathcal{A}R^{(3)} + I^{(2)}(t)\mathcal{A}R^{(2)} + I^{(3)}(t)\mathcal{A}R^{(1)} \right) \right] \mathcal{P}\rho_{\text{cor}}(0)$$
$$+ O(\alpha^5).$$

3.3. COMPUTATION OF CORRELATION FUNCTIONS

The TCL projection operator technique and the expansions of various operators presented above enable the computation of correlation functions without
assuming factorizing initial conditions for the density operator of the combined system. There is thus no need to make use of the Quantum–Regression
theorem, which is normally used to compute correlation functions. The only
restriction was that the system has been in a factorizing initial state $\rho_{cor}(-T)$
at some instant in the past, see Eq. (15).

This condition is complied with, e. g., if one calculates the correlation func-
tion $c(t) \equiv \langle a_0^\dagger(t) a_0(0) \rangle$. We assume, that at $t = -T$ the trapping potential
and the Raman lasers are switched on. Thus the system and the reservoir
coupling to mode 0 are independent and the state $\rho_{cor}(-T)$ factorizes, and
hence $\mathcal{Q}\rho_{cor}(-T) = 0$. Thus the initial dynamics can be described by a homogeneous master equation. However, at time $t = 0$ the density operator of the
combined system will not factorize anymore, and the inhomogeneous master
equation must be used for the computation of the further time evolution. In
order to compute $c(t)$ the general superoperator $\mathcal{A}$ is chosen to be a_0 acting
from the left.

Fig. (2) summarizes how the correlation functions $c(t)$ can be calculated. At
time $t = -T$ the combined system has been in the factorizing state $\rho_{cor}(-T)$.
The corresponding system density operator is now propagated according to
Eq. (8) until $t = 0$, where the annihilation operator a_0 is applied from the left.
This propagation takes place without the inhomogeneous part of the master
equation. From now on the further evolution of the system density operator is
described by the full master equation (8) including the inhomogeneous part.
At time t the creation operator $a_0^\dagger$ is applied and the trace over the system
can be computed.

4. Derivation of the quantum master equation

With the help of the TCL projection operator technique presented in the
previous section we are now able to derive an exact equation of motion for
the continuous–wave atom laser model from section 2, which is local in time.
In order to approximate this master equation it has to be expanded in powers
of the coupling strength α between system and reservoir. It is important to
note that we do not expand the solution of an exact equation of motion, we
expand the equation of motion. Thus, if we do not leave the parameter regime
in which this method is valid, see section 3.1, the results obtained will also be
valid in the long time limit.

The resulting master equation consists of two parts

$$\frac{\partial}{\partial t}\rho_{sys}(t) = \mathcal{L}_{hom}(t)\rho_{sys}(t) + \mathcal{L}_{in}(t)\rho_{cor}(0), \tag{21}$$

the expansion of the homogeneous part $\mathcal{L}_{hom}(t)$ and the expansion of the
inhomogeneous part $\mathcal{L}_{in}(t)$. At the beginning the atomic trap is assumed to

be empty and the initial density operator $\rho_{\text{tot}}(-T)$ factorizes, see Fig. 2. Hence, the inhomogeneous part $\mathcal{L}_{\text{in}}(t)$ can be neglected if one calculates, e.g., the occupation number of the BEC. But in order to calculate correlation functions, e.g. $c(t) = \langle a_0^\dagger(t)a_0(0)\rangle$, at time $t = 0$ the inhomogeneous part has to be considered, as explained in section 3.3.

The following discussion is based on two assumptions concerning the parameter regime in which the atom laser operates. We assume that we are in the weak collision regime, where the collision rate in the source mode Ω is much smaller than the rate κ_0 at which atoms are coupled out of the trap, that is, $\Omega \ll \kappa_0$. This parameter regime would be the generic operating regime for an atom laser, see Ref. [25]. In addition we assume that the interaction inside the BEC, given by g_{0000} is much smaller than all other rates

$$g_{0000} \ll \Omega \ll \kappa_0. \tag{22}$$

Because of this relation we can neglect all terms of order $\mathcal{O}(\alpha\, g_{0000})$ in the expansions presented below. This is a technical assumption which greatly simplifies the following calculations but could be dropped in principle.

4.1. EXPANSION OF THE HOMOGENEOUS PART OF THE MASTER EQUATION

To determine the expansion of the homogeneous part $\mathcal{L}_{\text{hom}}$ one has to insert the interaction Hamiltonian $H_{\text{int}}(t)$ from Eq. (4) into the expansion of the generator $K(t)$, see Eqs. (12). One obtains to fourth order

$$\mathcal{L}_{\text{hom}}\, \rho_{\text{sys}}(t) = -ig_{0000}\left[a_0^\dagger a_0^\dagger a_0 a_0, \rho_{\text{sys}}(t)\right] \tag{23}$$

$$+ \left(\mathcal{L}_0^{(2)} + \mathcal{L}_{\text{out}}^{(2)} + \mathcal{L}_{\text{col}} + \mathcal{L}_{\text{in}}\right)\rho_{\text{sys}}(t) + \left(\mathcal{L}_0^{(4)} + \mathcal{L}_{\text{out}}^{(4)} + \mathcal{L}_{\text{oc}}\right)\rho_{\text{sys}}(t).$$

Here the superoperators $\mathcal{L}_0^{(n)}$ and $\mathcal{L}_{\text{out}}^{(4)}$ describe the Lamb shift and the output coupling of the atoms inside the BEC, while $\mathcal{L}_{\text{in}}$ and $\mathcal{L}_{\text{coll}}$ are, respectively, responsible for the coupling to the thermal reservoir from which atoms enter and leave the trap and for the redistribution of atoms inside the trap. One obtains

$$\mathcal{L}_0^{(n)}\, \rho = -\frac{i}{2}S^{(n)}(t)\left[a_0^\dagger a_0, \rho\right], \tag{24a}$$

$$\mathcal{L}_{\text{out}}^{(n)}\, \rho = \gamma^{(n)}(t)\, \mathcal{D}[a_0]\rho, \tag{24b}$$

$$\mathcal{L}_{\text{in}}\, \rho = N\kappa_1 \mathcal{D}[a_1^\dagger]\rho + (1 + N)\kappa_1 \mathcal{D}[a_1]\rho, \tag{24c}$$

$$\mathcal{L}_{\text{coll}}\, \rho = \Omega\, \mathcal{D}[a_1^{\dagger 2} a_0]\rho, \tag{24d}$$

where $\mathcal{D}$ is the Lindblad superoperator, defined by

$$\mathcal{D}[a]\rho = a\rho a^\dagger - \frac{1}{2}(a^\dagger a\rho - \rho a^\dagger a). \tag{25}$$

The time-dependent rates $S^{(n)}(t)$ and $\gamma^{(n)}(t)$ can easily be evaluated as simple integrals over certain combinations of the reservoir correlation function $f(\tau)$.

The exact definitions of these functions can be found in Ref. [3], for example in second order one finds

$$\gamma^{(2)}(t) = \int_0^t ds\, \Re(f(s))\,, \qquad S^{(2)}(t) = \int_0^t ds\, \Im(f(s)). \qquad (26)$$

If one extends the integrals in the second order functions presented above to infinity, then the master equation (23) in second order coincides with the Born–Markov master equation, see Ref. [3].

4.2. EXPANSION OF THE INHOMOGENEOUS PART OF THE MASTER EQUATION

The expansion of the inhomogeneous part of the master equation is similar to the expansion of the generator $K(t)$. One gets

$$\mathcal{L}_{\text{in}}\, \rho_{\text{cor}}(0) = \left(\tilde{\mathcal{L}}_0^{(2)} + \tilde{\mathcal{L}}_{\text{out}}^{(2)}\right) a_0 \tilde{\rho}_{\text{cor}}(0) + \left(\tilde{\mathcal{L}}_0^{(4)} + \tilde{\mathcal{L}}_{\text{out}}^{(4)} + \tilde{\mathcal{L}}_{\text{oc}}^{(4)}\right) a_0 \tilde{\rho}_{\text{cor}}(0)$$

$$+ \tilde{\mathcal{L}}_{\text{oc},2}^{(4)}\, \tilde{\rho}_{\text{cor}}(0). \qquad (27)$$

The inhomogeneous part of the master equation (21) depends on the initial system density operator $\tilde{\rho}_{\text{cor}}(0) = \text{Tr}_{\text{res}}\{\rho_{\text{cor}}(0)\}$. The expansion of the generator $K(t)$ and that of the inhomogeneity have a very similar structure. The superoperators $\tilde{\mathcal{L}}_0^{(n)}, \tilde{\mathcal{L}}_{\text{out}}^{(n)}, \tilde{\mathcal{L}}_{\text{in}}$, and $\tilde{\mathcal{L}}_{\text{oc}}^{(4)}$ can be obtained from the definitions of the corresponding superoperators in the expansion of $K(t)$ just by replacing the transition rates. For instance, $\tilde{\mathcal{L}}_{\text{out}}^{(2)}$ is given by

$$\tilde{\mathcal{L}}_{\text{out}}^{(2)}\rho = \tilde{\gamma}^{(2)}(t)\left\{ a_0 \rho a_0^\dagger - \frac{1}{2} a_0^\dagger a_0 \rho - \frac{1}{2} \rho a_0^\dagger a_0 \right\}. \qquad (28)$$

This is exactly the definition of the superoperator $\mathcal{L}_{\text{out}}^{(2)}$ with the original transition rate $\gamma^{(2)}(t)$ replaced by

$$\tilde{\gamma}^{(2)}(t) = \gamma^{(2)}(t+T) - \gamma^{(2)}(t), \qquad (29a)$$

$$\tilde{S}^{(2)}(t) = S^{(2)}(t+T) - S^{(2)}(t). \qquad (29b)$$

As expected, these equations clearly show that the influence of the inhomogeneity vanishes for long times, because the decay rate $\gamma^{(2)}(t)$ approaches asymptotically the constant Born-Markovian decay rate γ_{M} [3]. The other transition rates can be evaluated as integrals over appropriate combinations of the real and imaginary parts of $f_{\text{grav}}(\tau)$,

$$\tilde{\gamma}^{(4)}(t) + i\, \tilde{S}^{(4)}(t) = 2 \int_0^t dt_1 \int_0^{t_1} dt_2 \int_{-T}^{t_2} dt_3\, f^*(t-t_2) f^*(t_1-t_3) \qquad (29c)$$

$$+ 2 \int_{-T}^0 dt_1 \int_{-T}^{t_1} dt_2 \int_{-T}^{t_2} dt_3\, \{f(t-t_2)f(t_1-t_3) + f(t-t_3)f(t_1-t_2)\}.$$

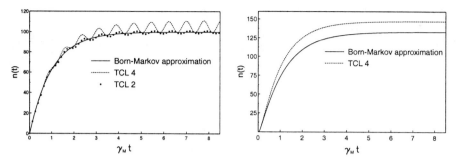

Figure 3. The occupation number $n(t)$ of the BEC without (left) and with gravity (right). The figures show the results of the TCL master equation in 2nd and 4th order and the corresponding Born-Markov approximation in the weak collision regime.

Only the superoperator $\tilde{\mathcal{L}}^{(4)}_{oc,2}$ given by

$$\tilde{\mathcal{L}}^{(4)}_{oc,2}\,\rho = -\tilde{r}(t)\,a_1 a_1\,[a_0,\rho]\,a_1^\dagger a_1^\dagger, \qquad (30)$$

has no counterpart in the expansion of the homogeneous part. The time dependent rate $\tilde{r}(t)$ is defined by

$$\tilde{r}(t) = \Omega \int_{-T}^{0} dt_1 \int_{-T}^{t_1} dt_2\, f(t - t_2). \qquad (31)$$

Having derived the master equation (21) for the system density operator $\rho_{\text{sys}}(t)$, we are now able to investigate the non–Markovian dynamics of the atom laser model. To this end the resulting master equation is solved numerically and the results are compared with the results obtained using the Born–Markov approximation.

5. Results

As pointed out, e.g. in [25], the generic operation regime of an atom laser would be the weak collision regime, where the interaction rate Ω of atoms in mode 1 is much smaller than the transition rate κ_1 at which atoms enter the trap. Hence we have chosen $\Omega = 0.1\,\kappa_0$ and $\kappa_1 = 10\,\kappa_0$. All other parameters are chosen according to [3]. Because of our previous assumption $g_{0000} \ll \Omega$ we have chosen $g_{0000} = 0.01\,\kappa_0$. Hence we can neglect all terms of order $\mathcal{O}(\alpha\,g_{0000})$ in Eqs. (23) and (27).

Fig. 3 shows the occupation number of the BEC with and without including gravity into the atom laser model. In both cases there are substantial deviations between the results of a 4th order expansion of the TCL master equation and the results obtained with the help of the Born–Markov approximation. The increase of the occupation number due to non-Markovian effects if gravity is not included in the model was already seen in [3]. This increased occupation number is also observed if gravity is included in the model. But, including gravity the oscillations of the occupation number disappear. As discussed in

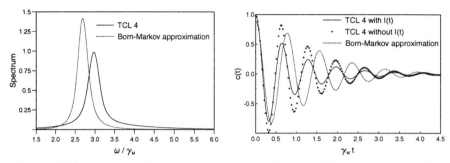

Figure 4. The spectrum of the atoms coupled out of the Bose–Einstein condensate (left Figure) and a correlation function $c(t) = \langle a_0^\dagger(t)a_0(0)\rangle$. Plotted are the results obtained from an expansion of the TCL master equation to 4th order (Eq. (23)), together with the corresponding Born-Markov approximation.

Ref. [3] these oscillations are mainly caused by the quadratic dispersion relation of massive particles which lead to a slow algebraic decay of the reservoir correlation function $f(\tau)$. Gravitational forces lead to an exponential decay of $f_{\text{grav}}(\tau)$ and, hence, the oscillations in the transition rates vanish much faster and are no longer observable in the occupation number.

In Fig. 4 we depict the spectrum of the atoms coupled out of the trap. The non-Markovian spectrum is shifted to higher frequencies and broadened in comparison to the results of the Born–Markov approximation. The frequency shift is caused by the increased occupation number of the BEC, see Fig. 3. Since the atom laser model includes interactions inside the Bose–Einstein condensate an increased occupation number leads to a higher potential energy and thus to a frequency shift towards higher frequencies.

The broadening of the spectrum on the other hand has two reasons. As shown in Ref. [27] a higher occupation number of the BEC leads to a broadening of the spectrum. Furthermore the initial correlations between system and reservoir contribute to the broadening of the spectrum of the atoms coupled out of the trap. This can be seen in Fig. 4, where the correlation functions obtained within the TCL projection operator method with and without including the initial correlations, described by the inhomogeneity $I(t)$ in Eq. (23), as well as the results of the Born–Markov approximation are plotted. The non-Markovian correlation function which includes the non-factorizing initial conditions has the same oscillation frequency as the corresponding correlation function which does not include the superoperator $I(t)$, but decays much faster. Hence, the initial correlations between system and reservoir cause an additional broadening of the spectrum.

Summarizing we can draw the following important conclusions. If we include gravity in the atom laser model then the slowly decaying oscillations of the occupation number, see Ref. [3], of the BEC vanish, because the gravitational forces lead to a reservoir correlation function $f_{\text{grav}}(\tau)$ showing an exponential decay. But the mean occupation number is still substantially larger than that of the corresponding Born-Markov approximation. The generalization of the TCL projection operator technique presented in this article which allows a correct modeling of non-factorizing initial conditions and thus the computation of correlation functions enables one to compute the spectrum of the atoms coupled out of the atomic trap without using the Quantum–Regression theorem. The non–Markovian spectrum is substantially shifted towards higher frequencies and broadened. The frequency shift could be traced back to the increased occupation number of the BEC and the broadening of the spectrum has two main reasons. It is partly caused by the increased occupation number, but also the non-factorizing initial conditions between system and reservoir contribute to the spectral broadening. The observed changes in the occupation number and the spectrum represent a clear signature of non-Markovian effects for which an experimental verification could perhaps be feasible in the future.

References

1. Anderson, M. H., J. R. Ensher, M. R. Matthews, C. E. Wiemann, and E. A. Cornell: 1995, 'Observation of Bose-Einstein Condenation in a Dilute Atomic Vapor'. *Science* **269**, 198.
2. Bloch, I., T. W. Hänsch, and T. Esslinger: 1999, 'An Atom Laser with a cw Output Coupler'. *Phys. Rev. Lett.* **82**, 3008.
3. Breuer, H. P., D. Faller, B. Kappler, and F. Petruccione: 1999a, 'Non–Markovian dynamics in pulsed- and continuous-wave atom lasers'. *Phys. Rev A* **60**, 3188–3198.
4. Breuer, H. P., B. Kappler, and F. Petruccione: 1999b, 'Stochastic wave function method for non-Markovian quantum master equations'. *Phys. Rev. A* **59**, 1633.
5. Chang, T.-M. and J. L. Skinner: 1993, 'Non-Markovian population and phase relaxation and absorption lineshape for a two-level system strongly coupled to a harmonic quantum bath'. *Physica A* **193**, 483–539.
6. Chaturvedi, S. and F. Shibata: 1979, 'Time-Convolutionless Projection Operator Formalism for elimination of fast variables and applications to Brownian Motion'. *Journal of the Physical Society of Japan* **35**, 297.
7. Davis, K. B., M.-O. Mewes, M. R. Andrews, N. J. van Druten, D. S. Durfee, D. M. Kurn, and W. Ketterle: 1995, 'Bose-Einstein Condensation in a Gas of Sodium atoms'. *Phys. Rev. Lett.* **75**, 3969.
8. Durfee, D. S. and W. Ketterle: 1998, 'Experimental studies of Bose-Einstein condensation'. *Optics Express* **2**, 299.
9. Faller, D.: 1999, 'Nicht-Markovsche Dynamik im gepulsten und kontinuierlichen Atomlaser'. Diploma-Thesis, Universität Freiburg. http://webber.physik.uni-freiburg.de/~fallerd.
10. Guzman, A. M., M. G. Moore, and P. Meystre: 1996, 'Theory of a coherent atomic-beam generator'. *Phys. Rev. A* **53**, 977.
11. Hagley, E. W., L. Deng, M. Kozuma, J. Wen, K. Helmerson, S. L. Rolston, and W. D. Phillips: 1999, 'A Well-Collimated Quasi-Continuous Atom Laser'. *Science* **283**, 1706.
12. Holland, M., K. Burnett, C. Gardiner, J. I. Cirac, and P. Zoller: 1996, 'Theory of an atom laser'. *Phys. Rev. A* **54**, 1757.
13. Hope, J. J.: 1997, 'Theory of input and output from an atomic trap'. *Phys. Rev. A* **55**, 2531.
14. Hope, J. J., G. M. Moy, M. J. Collett, and C. M. Savage, 'The linewidth of a non-Markovian atom laser'. Preprint.
15. Hope, J. J., G. M. Moy, M. J. Collett, and C. M. Savage: 2000, 'The steady state quantum statistics of a non-Markovian atom laser'. *Phys. Rev. A* **61**, 023603-1.
16. Jack, M. W., M. Naraschewski, M. J. Collett, and D. F. Walls: 1999, 'The Markov approximation for the atomic output coupler'. *Phys. Rev. A* **59**, 2962.
17. Lubkin, G. B.: 1999, 'New atom lasers eject atoms or run cw'. *Physics Today* **52**, 17.

18. Mewes, M.-O., M. R. Andrews, D. M. Kurn, D. S. Durfee, C. G. Townsend, and W. Ketterle: 1997, 'Output Coupler for Bose-Einstein Condensed Atoms'. *Phys. Rev. Lett.* **78**, 582.
19. Moore, M. G. and P. Meystre: 1997a, 'Dipole-Dipole selection rules for an atom laser cavity'. *Phys. Rev. A* **56**, 2989.
20. Moore, M. G. and P. Meystre: 1997b, 'Monte Carlo investigation of an atom laser with a modulated quasi-one-dimensional cavity'. *Journal of Modern Optics* **44**, 1815–1825.
21. Moy, G. M., J. J. Hope, and C. M. Savage: 1997, 'Atom laser based on Raman transitions'. *Phys. Rev. A* **55**, 3631.
22. Moy, G. M., J. J. Hope, and C. M. Savage: 1999, 'The Born and Markov approximations for atom lasers'. *Phys. Rev. A* **59**, 667.
23. Moy, G. M. and C. M. Savage: 1997, 'Output coupling for an atom laser by state change'. *Phys. Rev. A* **56**, 1087.
24. Shibata, F. and T. Arimitsu: 1980, 'Expansion Formulas in Nonequilibrium Statistical Mechanics'. *Journal of the Physical Society of Japan* **49**, 891.
25. Wiseman, H., A. Martins, and D. Walls: 1996, 'An atom laser based on evaporative cooling'. *Quantum Semiclass. Opt.* **8**, 737–753.
26. Wiseman, H. M. and M. J. Collett: 1995, 'An atom laser based on dark-state cooling'. *Phys. Lett. A* **202**, 246–252.
27. Zobay, O. and P. Meystre: 1998, 'Phase dynamics in a binary-collisions atom laser scheme'. *Phys. Rev. A* **57**, 4710–4719.

TIME-CORRELATED MACROSCOPIC QUANTUM TUNNELING OF SOLITONS IN DENSITY WAVES AND LONG JOSEPHSON JUNCTIONS

John H. Miller, Jr.
Department of Physics and Texas Center for Superconductivity
University of Houston
Houston, Texas 77204-5932 USA

Abstract. A number of many-body systems, including charge and spin density waves, Wigner crystals, etc., exhibit a sharp threshold electric field for inducing nonlinear transport, without showing the requisite polarization below threshold that one would expect classically. Similarly, many Josephson junctions, especially in the cuprates, have $I_c R_n$ products up to three orders of magnitude lower than predicted by the Ambegaokar-Baratov formula. These observations suggest that the observed critical field or current is a pair creation threshold for nucleating soliton-antisoliton pairs – essentially a macroscopic Coulomb blockade effect. This paper discusses the analogy to time-correlated single electron tunneling and introduces the concept of a quantum dynamical phase transition.

Key words: quantum tunneling, soliton, density wave, Josephson junction

1. INTRODUCTION

A charge density wave (CDW) is an example of spontaneous symmetry breaking, in which the electronic charge density becomes modulated, $\rho(x) = \rho_0(x,t) + \rho_1\cos[2k_F x - \phi(x,t)]$, where $\rho_0(x,t)$ contains a uniform background charge and an excess or deficiency of charge $\propto \pm \partial\phi/\partial x$. A spin density wave (SDW) has a modulated spin density, $\Delta S(x) = \Delta S_0\cos[2k_F x - \phi(x,t)]$, and is equivalent to two out-of-phase CDWs for the spin-up and spin-down subbands. Although pinned by impurities, a density wave (DW) can transport a current when an applied field exceeds a threshold value E_T. John Bardeen, [1] proposed a quantum tunneling (QT) model, in which condensed, dressed electrons Zener tunnel through a "pinning gap." This model was motivated by the fact that the current-field (J-E) curves fit a Zener characteristic $J \sim [E-E_T]\exp[-E_0/E]$. [2,3] The Zener activation field E_0 is roughly independent of the system size (i.e. scale invariant) and appears to be derived from a Euclidean action for each microscopic degree of freedom.

Density waves and long Josephson junctions (JJs) are described by a position-dependent phase (or phase difference), $\phi(x,t)$, in a sine-Gordon potential. They are dual in that the roles of charge and flux are interchanged, as well as those of current and voltage (see Table I). For example, the *current* per chain in a density wave is given by $I = (e^*/2\pi)\partial\phi/\partial t$, where $e^* = 2e$. By contrast, the *voltage* in a Josephson junction is proportional to the phase-difference across the junction, i.e. $V = (\Phi_0/2\pi)\partial\phi/\partial t$, where $\Phi_0 = h/2e$. Quantum nucleation of soliton-antisoliton (S-S') pairs has been proposed for both density waves [4,5] and long JJs [6]. The energy per spin-chain of a 2π-soliton in a density wave is $2\mu^{1/2}\hbar\omega_0/\pi$, where ω_0 is the pinning frequency and $\mu = M_F/m_e$ is the Fröhlich mass ratio. The soliton width is $\lambda_0 = c_0/\omega_0$, where $c_0 = \mu^{1/2}v_F$ is the phason velocity, for a density wave, and is equal to the Josephson penetration length λ_J for a long JJ.

Table I. Charge-flux duality between density waves and Josephson junctions

	Density wave	**Josephson junction**
Topological soliton or antisoliton	Kink w/ charge $\pm 2e$ per chain	Josephson vortex w/ flux $\pm\Phi_0$
Type of threshold	Threshold field E_T	Critical current I_c
Transport characteristic	J vs. E. $J \propto \partial\phi/\partial t$.	V vs. I. $V \propto \partial\phi/\partial t$.

In the early and mid-1980's, the author, and others, carried out experiments on the *ac* response of CDWs in TaS$_3$ [7] and NbSe$_3$ [8]. These experiments, which included *ac* admittance, direct mixing, harmonic mixing, and n^{th} harmonic generation, yielded three major results. First, *all* small-signal *ac* responses were found to be independent of *dc* bias below threshold. This indicated that the CDW does not polarize significantly for $E < E_T$, and contradicted classical predictions that the dielectric response should increase, or even diverge, near threshold. Second, a scaling was observed between the field-dependent *dc* conductivity $\sigma(E)$ and the real part of the frequency-dependent *ac* conductivity $\sigma(\omega)$ at zero *dc* bias. Bardeen interpreted this scaling by applying the theory of photon-assisted tunneling (PAT). Third, we found that the frequency- and bias-dependence of the complex admittance, direct mixing response, harmonic mixing response etc. were in good quantitative agreement with PAT theory predictions, with no adjustable parameters.

NMR [9] and x-ray scattering [10] experiments provide further evidence that the density wave does not displace significantly below threshold, suggesting that the measured threshold is far below the classical depinning field. Moreover, Aharonov-Bohm oscillations in the conductance of CDWs with columnar defects [11] provide strong evidence for the quantum nature of DW transport. The periodicity of these oscillations is $\Phi_0 = h/2e$, rather than $h/2Ne$ as predicted by Bogachev et al. [12], where N is the number of coupled parallel chains. Thus, one should think of DW transport as occurring via phase-coherent tunneling of many microscopic degrees of freedom, each of charge $2e$, within the condensate.

The quantum interpretation of the threshold is based on S. Coleman's elegant paper [13] on soliton pair-creation in the massive Schwinger model. An S-S' pair in a DW is analogous to a capacitor of separation L and area A per chain, and produces an internal field of magnitude $E^* = e^*/\varepsilon A$. When an external field E is applied, the difference between the electrostatic energy of a state with a pair and that of the "vacuum" is $\Delta U = \frac{1}{2} \varepsilon AL[(E \pm E^*)^2 - E^2] = e^*L[\frac{1}{2} E^* \pm E]$, which is positive when $|E| < \frac{1}{2} E^*$. Conservation of energy thus forbids the vacuum to produce a pair for fields less than $E_T \equiv \frac{1}{2}E^* = e^*/2\varepsilon A$. The threshold voltage across a region of length L is just the Coulomb blockade voltage $V_T = e^*/2C$, where $C = \varepsilon A/L$. We have extended this model [14], using the analogy to time-correlated single-electron tunneling (SET) to explain the coherent voltage oscillations and mode-locking phenomena above threshold. Numerous experiments on cuprate Josephson junctions show I_cR_n products well below the predictions of the Ambegaokar-Baratov formula, usually by a factor of 10-1000. This suggests that, here too, the critical current may be a soliton (i.e. Josephson vortex) pair creation threshold resulting from the magnetic energy of the interacting Josephson vortices and antivortices.

2. MODEL HAMILTONIAN

The massive Schwinger model can be employed to represent a pinned DW or long Josephson junction, with the Hamiltonian [14]:

$$H = \int dx \left\{ \frac{\Pi^2}{2D} + \frac{1}{2} Dc_0^2 \left(\frac{\partial \phi}{\partial x}\right)^2 + D\omega_0^2 [1 - \cos\phi] + u(\phi - \theta)^2 \right\}, \quad (1)$$

where $D = \mu\hbar/4\pi$ v_F for a density wave, the canonical momentum density is $\Pi = D\partial_t\phi$, $\theta = \pi E/E_T$ ($\theta = \pi I/I_T$ for a JJ), and $u = \frac{1}{2}\varepsilon A(E^*/2\pi)^2$ for a DW (and $u \sim I_T^2$ for a JJ). The quadratic term in Eq. (1) is the electrostatic or inductive energy per unit length, due to the applied field or current and the internal energy generated by phase displacements ϕ with respect to its value as $x \to \pm\infty$. This term will prevent instanton (tunneling) transitions from occurring when $\theta < \pi$, by conservation of energy.

Treating ϕ as a scalar quantum field, we performed [14] a variational calculation that takes the trial states to be coherent states. The energy density was found to be given by, in rescaled units:

$$E = \frac{1}{\gamma^2 \lambda_0^2}\left[1 - \left(\frac{m^2}{m_0^2}\right)^{\gamma^2/8\pi}\cos\xi\right] + \frac{m_0^2}{2\gamma^2}(\xi - \theta)^2 - \frac{m_0^2}{8\pi}\ln\frac{m^2}{m_0^2} + \frac{m^2 - m_0^2}{8\pi}, \qquad (2)$$

where $\xi \equiv \langle\phi\rangle$, $m_0 \equiv [2u/Dc_0^2]^{1/2}$, having units of 1/(length), is a "bare mass" proportional to the energy of an excitation in the absence of the s-G pinning potential, and $\gamma = \sqrt{\hbar/Dc_0} = \sqrt{4\pi/\mu^{1/2}}$. The "mass" m is proportional to the energy per chain $\sim\hbar\omega_0$ of a low-lying excitation in the full potential, including the s-G pinning potential. Minimizing E with respect to m and ξ yields the following self-consistency equations:

$$\frac{m^2}{m_0^2} = \frac{1}{(m_0\lambda_0)^2}\left(\frac{m^2}{m_0^2}\right)^{\gamma^2/8\pi}\cos\xi + 1,$$

$$\qquad (3)$$

$$\frac{1}{(m_0\lambda_0)^2}\left(\frac{m^2}{m_0^2}\right)^{\gamma^2/8\pi}\sin\xi + \xi - \theta = 0.$$

Using Eqs. (2) and (3), we found that the phase displacement $\langle\phi\rangle$ can be much smaller, below threshold, than predicted by classical models when the electrostatic (inductive) energy is much smaller than the sine-Gordon (pinning or Josephson coupling) energy, i.e. when $\tau \equiv 2u/D\omega_0^2 \ll 1$.

3. TIME-CORRELATED S-S' PAIR CREATION

When τ is small, the energy E will be minimized when $\xi \sim 2\pi n$, so $E(\theta) - E(0) \sim (m_0^2/2\gamma^2)[\theta - 2\pi n]^2$ (see Fig. 4 of ref. 14). This is a family of parabolas that intersect at odd integral multiples of π, similar to those (i.e. $[Q - ne]^2/2C$) encountered in time-correlated SET. Suppose a current is applied to a long sample with two contacts placed close together to measure the field, $E - \phi E^*/2\pi \propto \theta' \equiv \theta - \phi$, between the solitons and antisolitons. The current per chain in this region is $I = \sigma_n \theta' + \varepsilon' \partial \theta'/\partial t + (e^*/2\pi)\partial \langle \phi \rangle/\partial t$, where σ_n and ε' are the (suitably rescaled) conductivity of the normal electrons and dielectric constant, respectively. In time-correlated SET, the corresponding expression is $I = G_n V + C dV/dt + e dn/dt$, where the first term is the normal current, the second is the Maxwell displacement current, and the last term is due to tunneling.

If the system starts out in its "vacuum state" ($\langle \phi \rangle = 0$, $\theta = 0$) and the current I is increased then θ will increase with time. When θ exceeds π, what was formerly the "true vacuum" now becomes the "false vacuum," which then decays into the next true vacuum by nucleating an S-S' pair, while the phase ϕ tunnels from 0 to 2π between the soliton and antisoliton. This process repeats itself, leading to a characteristic sawtooth waveform in a plot of $\theta'(t)$. Such waveforms have been observed in coherent oscillations above threshold. In the case of a long JJ, the roles of current and voltage (and of capacitance and inductance) are reversed, but similar arguments apply when describing time-correlated Josephson vortex nucleation.

When $\theta = \pi$ (the quantum pair creation threshold), the potential in Eq. (1) has equivalent minima at $\phi \sim 0$ and $\phi \sim 2\pi$. The coupling between the states ψ_L and $\psi_R(\phi)$, centered at the left and right minima, will lift their energy degeneracy, thus opening energy gaps at $\phi = n\pi$ (where n is odd). Each quantum degree of freedom will then be in a superposition of states $a\psi_L + b\psi_R$. The state of the complete system, incorporating N quantum degrees of freedom, is more appropriately described as a highly entangled (and highly robust) coherent state of the form $(a\psi_L + b\psi_R)^N$, rather than a Schrödinger's catlike state, $a\psi_L^N + b\psi_R^N$. Such a phase-coherent many-body phenomenon could occur at all temperatures below the phase transition temperature T_c. If the system indeed consists of many coupled quantum degrees of freedom, then the Hamiltonian ought to include these interactions. Moreover, the analogy to thermodynamic phase transitions suggests the possibility of a quantum critical point at $\theta = \pi$.

4. THE QUANTUM DYNAMICAL PHASE TRANSITION

The idea of treating the quantum degrees of freedom individually can be discussed in terms of the BCS ground state. Suppose we change the phase ϕ_k of one pair relative to the phase ϕ of the overall wavefunction:

$$|\Psi\rangle = \left(u_k + v_k e^{i\phi_k} c_{k\uparrow}^+ c_{-k\downarrow}^+\right) \prod_{k' \neq k} \left\{u_{k'} + v_{k'} e^{i\phi} c_{k'\uparrow}^+ c_{-k'\downarrow}^+\right\}|0\rangle \qquad (4)$$

One would expect, based on the sign change for the coefficients of the Bogoliubon operators, that changing the phase ϕ_k of one pair by π relative to the ground state would cost an energy of about 2Δ per pair. Similar arguments would apply to a CDW, except that 2Δ would represent the Peierls gap rather than the BCS energy gap.

This suggests writing down a Hamiltonian that describes the coupling between these quantum degrees of freedom:

$$H = \sum_n \left[\frac{\Pi_n^{\,2}}{2D_1} + E_1\left[1 - \cos\phi_n\right] + E_2\left(\phi_n - \theta\right)^2 + \Delta\left[1 - \cos\left(\phi_n - \phi_{n-1}\right)\right] \right] \quad (5)$$

where $\Pi_n = (\hbar/i)\partial/\partial\phi_n$, the energies E_1 and E_2 are small since they correspond to a single degree of freedom, and $\Delta \gg E_1 \gg E_2$ when $\tau \ll 1$. The effective Euclidean action for tunneling should be independent of the size of the system in order to be consistent with experiment. Moreover, its magnitude will not be nearly as large as it would be if one naively calculated the WKB exponent simply by treating the entire system as one macroscopic "object." (The Euclidean action would then scale with the total number of particles.)

An interesting feature of the Hamiltonian in Eq. (5) is that the driving force θ is the same for all of the degrees of freedom ϕ_n. Note that the entire system becomes unstable when θ is slowly increased above π, as illustrated in Fig. 1. This suggests an analogy to thermodynamic phase transitions, in which θ is treated as a parameter somewhat analogous to temperature in thermodynamics. Moreover, tunneling is an example of quantum field theory in Euclidean space, which is formally identical to statistical mechanics. Near a thermodynamic phase transition, the effects of thermal fluctuations become greatly magnified, so that all sizes of fluctuations are observed at the critical temperature. Similarly, right at the quantum critical point, $\theta = \pi$, the effects of quantum fluctuations might become greatly magnified, thus leading to scaling laws similar to those observed in thermodynamic phase transitions.

The quantum phase transition here is going from a metastable state (but with random quantum fluctuations) into a current-carrying state (or finite voltage state for a JJ). Thus, it makes sense to define a current-current (voltage-voltage for a JJ) correlation function as follows:

$$\left\langle \dot{\phi}_n \dot{\phi}_{n'} \right\rangle \sim \exp\left(-|n - n'| / \xi\right) \quad (6)$$

where ξ represents a dimensionless correlation "length." As we approach the quantum critical point, this correlation length is expected to diverge and,

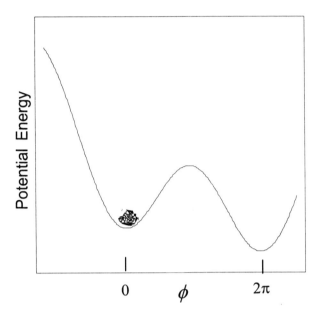

Figure 1. Potential energy for the coupled phases ϕ_n, when $\theta > \pi$. Note the quantum instability of the interacting "particles" in the left-hand well.

right *at* the critical point, $\theta = \pi$, the current-current (voltage-voltage) correlation function would decay in a power law (for large $|n - n'|$):

$$\left\langle \dot{\phi}_n \, \dot{\phi}_{n'} \right\rangle \sim 1/|n - n'|^d \tag{7}$$

where d represents a critical exponent. This suggests the use of renormalization group (RG) theory to better understand the behavior of the system near the quantum critical point. RG theory should provide an explanation for the observed scale invariance, and such work will be the topic of future investigations. In addition, many samples, at certain temperatures, show switching and hysteresis behavior in their *I-V* or *V-I* curves. The system appears, in such cases, to persist in a metastable state (i.e. $\langle \phi \rangle \sim 0$) well above the quantum critical point (when $\theta > \pi$) and may be a manifestation of a first-order quantum phase transition. This will also be a subject of future studies.

ACKNOWLEDGMENTS

The author would like to acknowledge the valuable contributions of E. Prodan and C. Ordonez. This work was supported, in part, by the State of Texas through the Texas Center for Superconductivity and the Texas Higher Education Coordinating Board Advanced Research Program (ARP), by the Robert A. Welch Foundation (E-1221).

REFERENCES

[1] John Bardeen, *Phys. Rev. Lett.* **42**, 1498 (1979); *ibid* **45**, 1978 (1980); *ibid* **55**, 1002 (1985).

[2] P. Monceau, N. P. Ong, A. M. Portis, A. Meerschaut, and J. Rouxel, *Phys. Rev. Lett.* **37**, 602 (1976); N. P. Ong and P. Monceau, *Phys. Rev.* B**16**, 3443 (1977).

[3] R. E. Thorne, J. H. Miller, Jr., W. G. Lyons, J. W. Lyding, and J. R. Tucker, *Phys. Rev. Lett.* **55**, 1006 (1985).

[4] K. Maki, *Phys. Rev. Lett.* **39**, 46 (1977); *Phys. Rev.* B**18**, 1641 (1978).

[5] A. Maiti and J.H. Miller Jr., *Phys. Rev.* B**43**, 12205 (1991).

[6] A. Widom and Y. Srivastava, *Phys. Lett.* **114A**, 337 (1986).

[7] J. H. Miller, Jr., R. E. Thorne, W. G. Lyons, J. R. Tucker, and John Bardeen, *Phys. Rev.* B**31**, 5229 (1985); J. H. Miller, Jr., Ph.D. Thesis, University of Illinois at Urbana-Champaign, 1985 (unpublished).

[8] A. Zettl and G. Grüner, *Phys. Rev.* B**29**, 755 (1989).

[9] J. H. Ross, Jr., Z. Wang, and C. P. Slichter, *Phys. Rev. Lett.* **56**, 663 (1986); *Phys. Rev.* B**41**, 2722 (1990).

[10] H. Requart, R. Ya. Nad, P. Monceau, R. Currat, J. E. Lorenzo, S. Brazovskii, N. Kirova, G. Grubel, and Ch. Vettier, *Phys. Rev. Lett.* **80**, 5631 (1998).

[11] Yu I. Latyshev, O. Laborde, P. Monceau, and S. Klaumünzer, *Phys. Rev. Lett.* **78**, 919 (1997).

[12] E. N. Bogachek, I. V. Krive, I. O. Kulik, and A. S. Rozhavsky, *Phys. Rev.* B**42**, 7614 (1990).

[13] S. Coleman, *Annals of Phys.* **101**, 239 (1976).

[14] J. H. Miller, Jr., C. Ordóñez, and E. Prodan , *Phys. Rev. Lett.* **84**, 1555 (2000).

THE NON-PERTURBATIVE QUANTUM BEHAVIOUR OF A SQUID RING IN A STRONG ELECTROMAGNETIC FIELD

M. J. Everitt,[1] P. Stiffell,[1] J. F. Ralph,[2] T. D. Clark,[1] A. Vourdas,[2] H. Prance,[1] and R. J. Prance[1]

(1) Physical Electronics Group, School of Engineering, University of Sussex, Brighton, East Sussex, BN1 9QT, UK. (2) Department of Electrical Engineering and Electronics, University of Liverpool, Liverpool, L69 3GJ, UK

Key words: Quantum mechanical transitions, Josephson devices, Floquet theory.

Abstract: We investigate the use of a time-dependent electromagnetic field to stimulate quantum transitions in a macroscopic superconducting circuit. Using two different models for the electromagnetic field, we show that the time-dependent Schrödinger equation predicts complex non-perturbative behaviour and multi-photon transitions. Moreover, we show that the strong coupling between the SQUID ring and the applied field means that this non-perturbative behaviour is relatively easy to produce.

1.   INTRODUCTION

In this paper we consider the non-perturbative effect of a strong electromagnetic field applied to a macroscopic quantum mechanical circuit: a thick superconducting ring containing a single Josephson (weak link) junction. This macroscopic quantum circuit, a radio frequency (rf-) SQUID ring, has been the subject of a great deal of interest over the last twenty years (for review articles and recent experimental work on macroscopic quantum behaviour in SQUID rings see references 1-8). It is often considered to be one of the best systems for investigating the quantum mechanical behaviour of macroscopic systems. Indeed, the most recent experimental data that could indicate the superposition of macroscopically distinct states has been reported in the electronic press in the last few months[7]. This work, and previous related work[5,6], relies on the application of a microwave field to induce transitions between the energy states. Our interest in this area is motivated by other experimental work on the appearance of non-adiabatic behaviour in small capacitance SQUID rings subject to a time-dependent field[8]. This non-adiabatic behaviour can be interpreted as the appearance of non-perturbative quantum transitions in the SQUID ring, as predicted by the time-dependent Schrödinger equation (TDSE). This means that the same analysis should be applicable to both types of experiment, albeit for different regions of parameter space. The appearance of these non-perturbative microwave-induced transitions is the subject of this paper.

We begin by summarising previous work on the non-perturbative effects of an applied electromagnetic field in the semi-classical limit[9]. We then introduce a new fully quantum mechanical model that reproduces the behaviour of the semi-classical model in all major respects. Notably, we still see the emergence of strongly non-perturbative behaviour that corresponds to transitions between the (time-averaged) energy levels for the SQUID ring. As with the semi-classical model, these transitions appear at field amplitudes and frequencies far below those predicted by conventional perturbative methods, based on transitions between energy eigenstates of the time-independent Schrödinger equation (TISE) for the SQUID ring. We conclude by discussing the relevance of these results to experimental SQUID systems and to the proposed use of superconducting circuits and Josephson junction devices in quantum computer systems[10,11].

2.   TIME DEPENDENT HAMILTONIAN

Although the rf-SQUID ring has been the subject of extensive theoretical and experimental investigations over the last twenty years, most (if not all) previous work has concentrated on solutions to the time-independent Schrödinger equation, calculating the energy eigenstates[1,4] and looking at dissipative processes[2,3] or perturbative transitions between energy eigenstates[4-6]. Recently, the some of the current authors (and others)[9] have considered the natural extension to this approach, looking at the behaviour of the TDSE directly using techniques familiar to the quantum optics community. As we shall see, even though the underlying techniques are familiar, the behaviour that the TDSE predicts for the SQUID ring is not usually seen in quantum optical systems because the coupling between the electromagnetic field and atomic or molecular systems is very weak compared to the coupling between the field and the SQUID ring. This means that only a comparatively weak field is needed to begin to produce multi-photon excitations[9]. (However, where the field is sufficiently strong, similar non-perturbative effects are predicted, see the example of the ammonia molecule given in reference 9).

For a time varying electromagnetic field, described by a time-dependent magnetic flux, which is inductively coupled to the SQUID ring, the time-dependent Hamiltonian for the SQUID ring is given by[9],

$$H(t) = \frac{Q^2}{2C} + \frac{\Phi^2}{2\Lambda} - \hbar v \cos\left(\frac{2\pi}{\Phi_0}(\Phi + \Phi_x(t))\right) - Q\frac{\partial \Phi_x(t)}{\partial t}$$

$$= H_S(t) - Q\frac{\partial \Phi_x(t)}{\partial t}$$

where the time dependent flux $\Phi_x(t) = \Phi_{xstat} + \mu\Phi_{em}\cdot\sin(\omega_{em}t + \delta)$ has a static and time periodic component (μ is the coupling between the applied field and the SQUID ring), Q is the electric displacement flux across the weak link (which has units of charge), Φ is the magnetic flux enclosed in the ring (noting that Q and Φ are quantum mechanical operators that do not commute, $[Q,\Phi]=i\hbar$), C is the effective capacitance of the weak link, Λ is the geometric inductance of the ring, v is the (angular) frequency for pairs tunnelling across the weak link (the critical current of the weak link is $I_c=qv$, where q is the pair charge $q=2e$), and $\Phi_0=(h/q)$ is the superconducting flux quantum. Where the applied magnetic flux is static, or varying sufficiently slowly so that any changes are adiabatic, the coupling term between Q and the time derivative of $\Phi_x(t)$ vanishes and this Hamiltonian reduces to the usual time-independent one.

The TDSE is solved by calculating the coefficients of superposition for the SQUID wavefunction, using the instantaneous eigenvectors of the Hamiltonian for each time t as basis functions[9]. The eigenvectors are found from solving,

$$H(t)|\kappa(\Phi_x(t))\rangle = E_\kappa(\Phi_x(t))|\kappa(\Phi_x(t))\rangle$$

at each instant in time, and the wavefunction for the SQUID is written as,

$$|\psi\rangle = \sum_{\kappa=0}^{\infty} b_\kappa(t)|\kappa(\Phi_x(t))\rangle$$

where $b_\kappa(t)$ are the time dependent coefficients in the eigenvector basis. The wavefunction therefore has time dependent contributions from both the coefficients and the basis vectors.

Substituting this expression into the TDSE produces a set of coupled differential equations for the coefficients, which can be solved using standard numerical methods for any given initial conditions $b_\kappa(0)$. These equations are given by,

$$\frac{db_\kappa(t)}{dt} = -i\frac{b_\kappa(t)}{\hbar} E_\kappa(\Phi_x(t)) + \frac{\partial\Phi_x(t)}{\partial t}\sum_{\gamma=0}^\infty b_\gamma(t)L_{\kappa\gamma}(\Phi_x(t))$$

where the summation is truncated to a large but finite integer for the purposes of calculation, and $L_{\kappa\gamma}(\Phi_x(t))$ is expressed in terms of the coefficients of the (static) energy eigenvectors of the TISE using the harmonic oscillator energy eigenstates of the static Hamiltonian without the Josephson cosine term as a basis,

$$L_{\kappa\gamma}(\Phi_x(t)) = \sum_{m=0}^\infty \upsilon_\kappa^m\left(\upsilon_\gamma^{m+1}\sqrt{\frac{m+1}{2\hbar\omega_0\Lambda}} - \upsilon_\gamma^{m-1}\sqrt{\frac{m}{2\hbar\omega_0\Lambda}} - \frac{d\upsilon_\gamma^m}{d\Phi_x}\right)$$

where the υ_κ^m are the coefficients found in the solution of the TISE and their dependence on $\Phi_x(t)$ has been suppressed for clarity[9],

$$|\kappa(\Phi_x)\rangle = \sum_{n=0}^\infty \upsilon_\kappa^n(\Phi_x)|n\rangle$$

where the summations in both expressions are again truncated for computational reasons, and the $|n\rangle$ are the harmonic oscillator states.

3.   SEMI-CLASSICAL MODEL

Thus it is possible to obtain a solution to the TDSE for the SQUID ring in a time-dependent electromagnetic field. To make use of these solutions we need to relate these solutions back to a potentially useful physical quantity. To do this, we define a time evolution operator (S) such that,

$$S|\psi\rangle = |\psi(t+T_{em})\rangle$$

where $T_{em}=1/f_{em}=2\pi/\omega_{em}$ is the period of oscillation of the electromagnetic field. The eigenvectors of this operator give a special set of initial conditions, for which the behaviour of the system is periodic (period $= T_{em}$). This allows us to calculate averages for the energy expectation values for the SQUID ring which correspond to the Floquet (quasi-)energies,

$$\langle\langle E(\Phi_{xstat})\rangle\rangle_\kappa = \frac{\omega_{em}}{2\pi}\int_0^{2\pi/\omega_{em}}\langle E(\Phi_{em}(t))\rangle_\kappa dt$$

that are used in atomic and quantum optical calculations[12]. In Fig.1 we show the first three time-averaged quasi-energies over one period of the static magnetic flux (period=Φ_0) for a SQUID ring whose parameters are chosen to be similar to those given in reference 8. The quasi-energies show a great deal of structure. The sharp 'spikes' correspond to transition regions where there is a resonance between the electromagnetic field and the SQUID ring. It is noted that most of the transitions shown in Fig.1 correspond to the absorption/emission of a large number of photons. However, it is also possible to reproduce the perturbative limit for single and two-photon transitions using the same approach (an example is given in reference 9). The fact that these non-perturbative transitions can be accessed relatively easily (the curves shown in Fig.1, the amplitude of the applied field and the coupling to the SQUID ring are relatively weak) is an interesting one, and one that has no analogue in quantum optics.

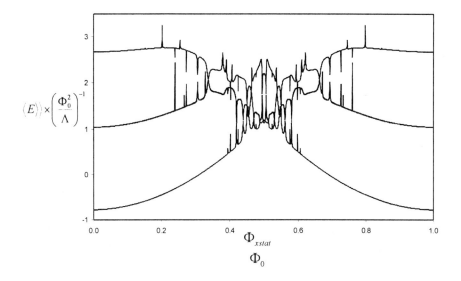

Figure 1. Semi-classical TDSE calculation of time averaged quasi-energies for κ=0,1,2 states of the rf-SQUID ring when a periodic field is applied. (Circuit parameters are $C=10^{-16}$ F, $\Lambda=3\times10^{-10}$ H, $\hbar\nu=0.075 \times \Phi_0^2/\Lambda$, $f_{em}=183.776$ GHz, $\Phi_{em}=2.9325\times \Phi_0$ Wb, $\mu=0.02$).

In addition to calculating the quasi-energies for the time-periodic eigenstates of S, it is also informative to calculate the average energy of the SQUID ring when the 'static' magnetic flux is linearly ramped, as is done in the experiment described in reference 8. Fig.2 shows the average energy (bold line) against the quasi-energies from Fig.1 (faint lines). Here we see very large excursions in energy either side of $\Phi_{xstat}=0.5\Phi_0$, although it is also noticeable that the excursions are somewhat smaller at $\Phi_{xstat}=0.5\Phi_0$ itself. For $\Phi_{xstat}>0.6\Phi_0$ little seems to occur. Past this point, the state of the system is left in a (time-varying) superposition of the κ=0 and κ=1 states (see note below on dissipation).

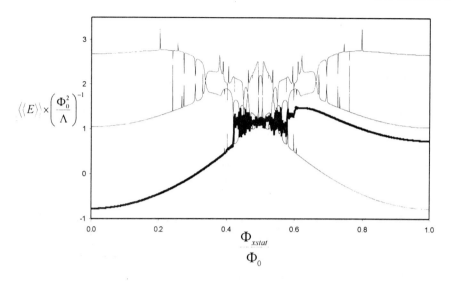

Figure 2. Semi-classical TDSE calculation of time averaged quasi-energy as the mean external flux is swept, the time averaged quasi-energies for κ=0,1,2 states are shown as a reference. (Circuit parameters are the same as those in Fig.1, ramp rate = 184 Φ_0/sec).

4.   FULLY QUANTUM MECHANICAL MODEL

Given the remarkably rich structure found in the semi-classical limit, the next logical step is to compare the predictions of this model against one where the applied field itself is quantum mechanical. For this, we take a new Hamiltonian,

$$H = \frac{Q_{em}^2}{2C_{em}} + \frac{\Phi_{em}^2}{2L_{em}} + \frac{Q^2}{2C} + \frac{(\Phi - \Phi_{xstat})^2}{2\Lambda} - \hbar v \cos\left(\frac{2\pi\Phi}{\Phi_0}\right) - \frac{\mu}{\Lambda}(\Phi - \Phi_{xstat})\Phi_{em}$$

where the electromagnetic field is represented by a harmonic oscillator ($[Q_{em},\Phi_{em}]=i\hbar$), that is linearly coupled to the magnetic flux in the SQUID ring. To provide a comparison with the semi-classical model, the initial state of the coupled system is chosen to be a product of a SQUID energy eigenstate and a coherent state for the electromagnetic field at each point in Φ_{xstat} (the α parameter for the coherent state being fixed by the desired amplitude/energy of the electromagnetic field oscillations). Given this initial state, we can solve the TDSE for the two-mode (linear em-field mode and nonlinear SQUID mode) system using an evolution operator in the usual way. An analogue of the quasi-energies can then be obtained by averaging the instantaneous expectation value of the SQUID part of the Hamiltonian over many cycles of the em-field oscillator. The resultant time-averaged energies are shown in Fig.3.

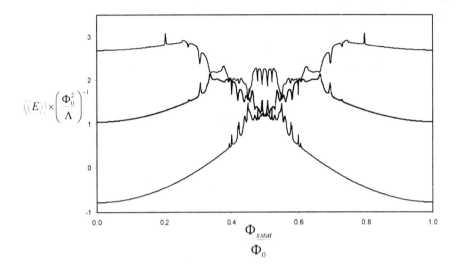

Figure 3. Fully Quantum TDSE calculation of time averaged energies for κ=0,1,2 states of the rf-SQUID ring. (Circuit parameters are the same as those in Fig.1).

There are small differences between the quasi-energy levels shown in Figs. 1 and 3. Some of the very weak transitions toward the outer edges of the curves appear to be suppressed, and the shapes of some of the features toward the middle of the graph are different (particularly for the (κ=2) state). Since the convergence of each solution was carefully checked so that any numerical artefacts were much smaller than the features shown, there are two physical explanations for these differences. The first is that the energy in the electromagnetic mode for the fully quantum system is limited from below, whereas there is no limit to the number of photons available from the electromagnetic field in the semi-classical model. The second reason for the differences is the choice of the initial state, which will have some effect on the energy in the system. These are the most likely origin of the differences between the two sets of curves.

There is no explicit dissipation added to either of the models described above. The fully quantum mechanical model is coherent and, whilst the semi-classical model does not contain explicit decoherence, the electromagnetic field in the first model is classical so that it cannot be said to be truly coherent in the quantum mechanical sense. There are two main reasons for this omission. Firstly, extremely large amounts of computer time are required to generate the numerical solutions, making the introduction of further levels of complexity difficult. Secondly, there are a number of possible sources of dissipation that could be important for any particular experimental set-up, which means that any dynamics that are dominated by decoherent processes are likely to be specific to a particular model. The approach adopted here is pragmatic. As long as any extraneous environmental degrees of freedom that might give rise to dissipation or decoherence are very weakly coupled to the SQUID ring, so that their effect on the behaviour of the SQUID ring can be neglected, then the semi-classical or fully quantum mechanical models should describe the system reasonably well (see the appendix in reference 9).

5.   DISCUSSION AND CONCLUSIONS

We have seen how non-perturbative behaviour appears in the numerical solutions of the TDSE for the rf-SQUID ring coupled to an electromagnetic field mode, in both the semi-classical and fully quantum mechanical limits. We have related this behaviour back to transitions between the time-averaged energy levels for the SQUID ring, which are the familiar Floquet (quasi-) energies used in quantum optics. The next step is to ask how these transitions would manifest themselves in an experimental system? Clearly, an additional device is required to perform some sort of measurement of, or collect a classical record of, the behaviour of the SQUID ring.

In the case of the experiments described in reference 7, this device is another SQUID ring (a dc-SQUID), which performs a projective measurement on the magnetic flux contained in the rf-SQUID ring. If non-perturbative transitions were to occur in this system, as the amplitude of the applied microwaves is increased, it might be possible to compare the experimental measurements with the predictions of the TDSE as the microwave amplitude is increased. An abrupt change in the behaviour of the system might indicate the appearance of higher-order transition processes. However, the analysis of this system is complicated by the fact the measurement process itself is time-dependent and quantifying the amplitude of the applied field might be difficult if the couplings between the microwave mode and the SQUID ring structure are not well characterised.

Another approach to the problem is taken in reference 8, this paper describes an experiment where the classical record is provided by a resonant circuit (the 'tank' circuit, familiar from standard rf-SQUID magnetometer systems). The natural frequency of the tank circuit is around 20 MHz and it is assumed to be entirely classical. This low frequency classical oscillator couples inductively to supercurrents in the rf-SQUID ring, which in turn couple back to the tank circuit, producing a nonlinear response that is periodic in the static external magnetic flux[13]. The experimental procedure is to track the (classical) resonance of the tank circuit, in both frequency and amplitude, as a function of Φ_{ex} for increasing levels of the applied microwave field. The rf-SQUID ring and the coupling between the SQUID ring and the microwave field are characterised by matching the behaviour of the resonance in the adiabatic (weak field) limit[13] (the weak link is made in situ, which makes the effective capacitance difficult to measure directly). Once the ring is characterised, the amplitude of the microwaves is increased incrementally. When the amplitude reaches a certain level, the response of the SQUID ring begins to differ from the adiabatic modulation predicted in reference 13. The amplitude of the classical rf-resonance grows, exhibiting a very large power gain (which can be up to ≈ 20 dB)[8]. This power gain is seen in precisely the regions of static magnetic flux and microwave power where the TDSE predicts large non-perturbative effects, implying that the SQUID ring is down converting the applied microwave energy to radio frequencies. Another reason for believing that these non-adiabatic processes are quantum mechanical in origin is that the solutions to the (classical) capacitive resistively shunted junction (RSJ+C) model[14] for the SQUID show no evidence for the features found in these experiments (for references to the solutions to the classical nonlinear equations see 15).

Lastly we consider the relevance of these results to the proposed use of Josephson circuits for quantum computation[11,12]. Two of the criteria required of a useable qubit are that its Hamiltonian should be carefully characterised and that it should have inherently low decoherence[16]. Both of these criteria may be difficult to achieve if the system cannot be adequately described by a two-state system. As we have shown in the paper, simple perturbative models for the interaction of a SQUID ring with the electromagnetic field do not necessarily give an accurate prediction for the behaviour of the full TDSE. If the predictions of the TDSE are correct, then great care will be needed in ensuring that the system is in fact describable by a two-state model, and that the effect of the environmental and other degrees of freedom is sufficiently weak that no non-perturbative behaviour is seen. This is particularly true for schemes that aim to reduce the effect of decoherence by the introduction of a time-dependent perturbation[17].

The authors would like to thank NESTA for its generous funding of this work. We would also like to thank Dr.A.Sobolev for interesting and informative discussions.

REFERENCES

1. A.Widom, *J. Low Temp. Phys.* **37**, 449 (1979); A.Widom, Y.Srivastava, *Phys. Rep.* **148**, 1 (1987).
2. A.O.Caldeira, A.J.Leggett, *Ann. Phys.* **149**, 374 (1983); A.J.Leggett, S.Chakravarty, A.T.Dorsey, M.P.A.Fischer, A.Garg, W.Zweger, *Rev. Mod. Phys.* **59**, 1 (1987).
3. G.Schön, A.D.Zaikin, *Phys. Rep.* **198**, 237 (1990).
4. T.P.Spiller, T.D.Clark, R.J.Prance, H.Prance, *Prog. Low Temp. Phys.* **XIII**, 219 (Elsevier, 1992); R.J.Prance, T.D.Clark, R.Whiteman, J.Diggins, H.Prance, J.F.Ralph, T.P. Spiller, A. Widom, Y. Srivastava, *Physica* **B203**, 381 (1994).
5. R.Rouse, S,Han, J.E.Lukens, *Phys. Rev. Lett.* **75**, 1614 (1995); S,Han, R.Rouse, J.E.Lukens, *Phys. Rev. Lett.* **76**, 3404 (1996).
6. P.Silvestrini, B.Ruggiero, Y.N.Ovchinnikov, A.Esposito, A.Barone, *Phys. Lett. A* **212**, 347 (1996); P.Silvestrini, B.Ruggiero, C.Granata, E.Esposito, *Phys. Lett. A* (in press).
7. J.R.Friedman, V.Patel, W.Chen, S.K.Tolpygo, J.E.Lukens, 'Detection of a Schrödinger's cat state in an rf-SQUID' cond-mat/0004293.
8. R.Whiteman, T.D.Clark, R.J.Prance, H.Prance, V.Schöllmann, J.F.Ralph, M.Everitt, J.Diggins, *J. Mod. Optics* **45**, 1175 (1998).
9. T.D.Clark, J.Diggins, J.F.Ralph, M.J.Everitt, R.J.Prance, H.Prance, R.Whiteman, *Ann. Phys.* **268**, 1 (1998); J.Diggins, R.Whiteman, T.D.Clark, R.J.Prance, H.Prance, J.F.Ralph, A.Widom, Y.Srivastava, *Physica B* **233**, 8 (1997).
10. T.P.Orlando, J.E.Mooji, L.Tian, C.H. van der Wal, L.S.Levitov, S.Lloyd, J.J.Mazo, *Phys. Rev. B* **60**, 15398 (1999).
11. T.P.Spiller, 'Superconducting circuits for quantum computing' (to be published in Fortschr. der Phys.).
12. E.g. C.Cohen-Tannoudji, J.Dupont-Roc, G.Grynberg, 'Atom-Photon Interactions' (Wiley, New York, 1992) Ch. III and Ch. IV; J.Shirley, *Phys. Rev. B* **138**, 979 (1965); S.I.Chu, *Adv. Atomic Molec. Phys.* **21**, 197 (1985).
13. R.Whiteman, V.Schöllmann, M.Everitt, T.D.Clark, R.J.Prance, H.Prance, J.Diggins, G.Buckling, J.F.Ralph, *J. Phys.: Cond. Matt.* **10**, 9951 (1998).
14. E.g. K.K.Likharev *'Dynamics of Josephson Junctions and Circuits'* (Gordon and Breach, Sidney, 1986).
15. T.D.Clark, J.F.Ralph, R.J.Prance, H.Prance, J.Diggins, R.Whiteman, *Phys. Rev. E* **57**, 4035 (1998); W.C.Shieve, A.R.Bulsara, E.W.Jacobs, *Phys. Rev. A* **37**, 3541 (1988); M.P.Soerensen, M.Barchelli, P.L.Christinasen, A.R.Bishop, *Phys. Lett.* **109A**, 347 (1985).
16. D.P.DiVincenzo, in *'Mesoscopic Electron Transport'*, eds. L.Sohn, L.Kouwenhoven, G.Schön, Vol.345, NATO ASI Series E (Kluwer, 1997) p.657.
17. L.Viola, E.Knill, S.Lloyd, *Phys. Rev. Lett.* **82**, 2417 (1999).

Noise properties of the SET transistor in the co-tunneling regime

D.V. Averin

Department of Physics and Astronomy, SUNY, Stony Brook
NY 11794-3800

Abstract. Zero-frequency spectral densities of current noise, charge noise, and their cross-correlation are calculated for the SET transistor in the co-tunneling regime. The current noise has a form expected for the uncorrelated co-tunneling events. Charge noise is created by the co-tunneling and also by the second-order transitions in a single junction. Calculated spectral densities determine transistor characteristics as quantum detector.

1. INTRODUCTION

Single-electron-tunneling (SET) transistor [1,2,3] is the natural measuring device for the potential quantum logic circuits based on the charge states of mesoscopic Josephson junctions [4,5,6]. This fact, together with the general interest to the problem of quantum measurement, motivates current discussions of the SET transistors as quantum detectors – see, e.g., [7]. Detector characteristics of the transistor are determined by its noise properties, and have been studied so far [8,9] in the regime of classical electron tunneling. The aim of this brief note is to calculate the transistor noise properties in the Coulomb blockade regime, when the current flows in it by the process of co-tunneling. An expected, and confirmed below, advantage of co-tunneling for quantum signal detection is a weaker back-action noise on the measured system produced by the SET transistor.

2. CO-TUNNELING IN THE SET TRANSISTOR

SET transistor [1] is a small conductor, typically a small metallic island, placed between two bulk external electrodes, that forms two tunnel junctions with these electrodes. Due to Coulomb charging of the island by tunneling electrons, the current I through the structure depends on the island electrostatic potential controlled by an external gate voltage V_g. Sensitivity of the current I to variations of the voltage V_g makes it possible to measure this voltage, and is the basis for transistor operation as the detector.

When the bias voltage V across the transistor is smaller than the Coulomb blockade threshold (dependent on V_g), the tunneling is suppressed by the Coulomb charging energy required to transfer an electron in or out of the central electrode. In this regime, the current I flows through the transistor only by the co-tunneling process that consists of two electron jumps across the two transistor junctions in the same direction (for review, see [10]). Energy diagram of the co-tunneling transitions is shown in Fig. 1. Transitions go via two virtual intermediate charge states with $n = \pm 1$ extra electrons on the island and large charging energies $E_{1,2}$. The energies $E_{1,2}$ are equal to the change of electrostatic energy due to the first electron jump in the first or the second junction, and depend on the voltages V, V_g, junction capacitances $C_{1,2}$, and capacitance C_g that couples the gate voltage V_g:

$$E_1 = E_C(\frac{1}{2} + q_0 - (C_1 + \frac{C_g}{2})\frac{V}{e}), \quad E_2 = E_C(\frac{1}{2} - q_0 - (C_2 + \frac{C_g}{2})\frac{V}{e}). \quad (1)$$

Here E_C is the characteristic charging energy of the transistor $E_C = e^2/C_\Sigma$, $C_\Sigma \equiv C_1 + C_2 + C_g$, and q_0 is the charge (in units of electron charge e) induced by the gate voltage, $q_0 \equiv C_g V_g/e$. Equations (1) assume that the voltage V_g is applied to the transistor symmetrically.

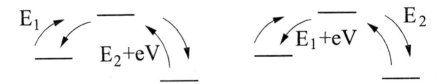

Figure 1: Energy diagrams of the four co-tunneling transitions in the SET transistor. The arrows indicate electron jumps. The left and right diagrams show, respectively, the transitions through the charge states with $n = +1$ and $n = -1$ extra electrons on the central electrode of the transistor. The energy labels are the changes of electrostatic energy in the first electron jump of each co-tunneling process.

The co-tunneling regime is realized for values of V and V_g that make the energies E_j, $j = 1, 2$, sufficiently large, $E_j \gg T, \hbar G_{1,2}/C_\Sigma$, where T is the temperature and $G_{1,2}$ are the tunnel conductances of the transistor junctions. The conductances $G_{1,2}$ are assumed to be small, $g_{1,2} \equiv \hbar G_{1,2}/2\pi e^2 \ll 1$. In this regime, one can neglect thermally-induced "classical" electron transitions of the first order in conductances g_j, and also neglect quantum broadening of the charge states by tunneling [11,12]. Transport characteristics of the transistor are determined then by the co-tunneling transitions of the second order in g_j.

The gate voltage V_g controls the co-tunneling current I in the transistor through the dependence of the energies E_j on the induced charge q_0. Viewed as the detector for measurement of small variations of V_g, the transistor is characterized by: (1) the linear response coefficient $\lambda = \delta\langle I\rangle/\delta V_g$, where $\delta\langle I\rangle$ is the change of the average current due to variation of V_g; (2) the spectral density S_I of the current noise (output noise of the detector); (3) the spectral density S_Q of the fluctuations of the charge $Q = en$ on the central electrode of the transistor ("back-action" noise); and (4) cross-correlation S_{IQ} between the current and charge noise. In general, the back-action noise is the fundamental property of a quantum detector responsible for localization of the measured system in the eigenstates of the measured observable. For the SET transistor, the measured observable is the voltage V_g, and the back-action noise originates from fluctuations of the charge Q in the process of electron transfer through the transistor. The voltage V_g is coupled to Q in the transistor energy as QV_gC_g/C_Σ, where $Q = 0, \pm e$ in the co-tunneling regime. The fluctuations of Q produce random force that acts on the system creating V_g, and lead to mutual decoherence of the states of this system with different values of V_g.

In the next section, the noise spectral densities are calculated in the zero-frequency limit for the SET transistor in the co-tunneling regime. The zero-frequency results determine the detector characteristics of the transistor in the frequency range below the characteristic frequency of electron tunneling.

3. NOISE CALCULATION

Transport properties of the transistor related to co-tunneling can be calculated by straightforward perturbation theory in tunneling in the second order in tunnel conductances g_j. The tunneling part of the transistor Hamiltonian is

$$H_T = \sum_{j=1,2} H_j^{\pm},$$

where $H_j^{\pm}$ describe forward and backward electron transfer in the jth junction. The junction electrodes are assumed to have quasicontinuous density of states. The only nonvanishing free correlators of $H_j^{\pm}$ can be expressed in terms of the conductances g_j (see, e.g., [10]):

$$\langle H_j^{\pm}(t)H_j^{\mp}(t')\rangle = g_j \int \frac{d\omega\,\omega\,e^{i\omega(t'-t)}}{1 - e^{-\omega/T}}. \tag{2}$$

The zero-frequency spectral densities of current and charge noise are given by the standard expressions. For instance,

$$S_I = \frac{1}{\pi} \text{Re} \int_{-\infty}^{t} dt' \text{Tr}\{S^{\dagger}(t)I(t)S(t,t')I(t')S(t')\rho_0\}, \tag{3}$$

where $S(t,t') = \mathcal{T}\exp\{-i\int_{t'}^{t} d\tau H_T(\tau)\}$ is the evolution operator of the transistor due to tunneling, ρ_0 is its unperturbed equilibrium density matrix, $S(t) \equiv S(t,-\infty)$, and the current I can be calculated in either of the two junctions, e.g., $I = ie(H_1^+ - H_1^-)$. The time dependence of all operators H_T is now due to both internal energies of the junction electrodes and the electrostatic charging energy of the transistor. The term $-\langle I \rangle^2$ is omitted in (3), since the average co-tunneling current $\langle I \rangle$ is of the same order in g_j as S_I. Therefore, $\langle I \rangle^2$ is of higher order than S_I, and can be neglected in the perturbative calculation adequate for the co-tunneling regime. The same considerations apply to the spectral densities S_Q and S_{IQ} discussed below.

Expanding all evolution operators in (3) up to the second order in $H_j^{\pm}$, and evaluating the averages with the help of Eq. (2), we find that S_I is given by the standard expression characteristic for uncorrelated tunneling:

$$S_I = \frac{e^2}{2\pi}(\gamma^+ + \gamma^-). \tag{4}$$

Here $\gamma^{\pm}$ are the rates of forward and backward co-tunneling:

$$\gamma^+ = \frac{2\pi g_1 g_2}{\hbar} \int \frac{d\omega_1 \omega_1}{1 - e^{-\omega_1/T}} \frac{d\omega_2 \omega_2}{1 - e^{-\omega_2/T}} \delta(\omega_1 + \omega_2 - eV) \left(\frac{1}{E_1 + \omega_1} + \frac{1}{E_2 + \omega_2}\right)^2,$$

and γ^- is given by the same expression with eV and E_j changed into $-eV$ and $E_j + eV$. The rates $\gamma^{\pm}$ also determine the average co-tunneling current, $\langle I \rangle = e(\gamma^+ - \gamma^-)$.

Expression for the spectral density S_Q of charge fluctuations is obtained from Eq. (3) by replacing the current operators with the charge operators $Q = en$. To find S_Q in the co-tunneling regime one needs to expand the evolution operators in this expression up to the fourth order in the tunneling terms H_T. The non-vanishing contribution to S_Q comes then from one particular choice of orders of expansion of different evolution operators:

$$S_Q = -\frac{e^2}{\pi} \text{Re} \int_{-\infty}^{t} dt_1 \int_{-\infty}^{t} dt_2 \int^{t_2} dt_3 \int^{t_3} dt' \int^{t'} dt_4$$
$$\text{Tr}\{H_T(t_1)nH_T(t_2)H_T(t_3)nH_T(t_4)\rho_0\}. \tag{5}$$

Taking the trace in Eq. (5) we get the two types of contributions to S_Q. In one of them, the pairings of operators H_T belong to the two different junctions, while in the other, all H_T's belong to one junction. The first contribution describes fluctuations of the charge in the process of co-tunneling, and can be split into the two terms, $S^{\pm}$, associated with the forward and backward transitions. The second contribution $\bar{S}$ is created by the back-and-forth tunneling within the same junction. Accordingly, S_Q can be written as

$$S_Q = \bar{S} + S^{\pm}, \tag{6}$$

where the different terms are found from Eq. (5) to be

$$\bar{S} = \sum_j (2\pi e g_j)^2 \frac{\hbar T^3}{6} \left(\frac{1}{E_j^2} - \frac{1}{(E_{j'} + eV)^2} \right)^2, \quad j, j' = 1, 2, \; j' \neq j,$$

$$S^+ = \hbar e^2 g_1 g_2 \int \frac{d\omega_1 \omega_1}{1 - e^{-\omega_1/T}} \frac{d\omega_2 \omega_2}{1 - e^{-\omega_2/T}} \delta(\omega_1 + \omega_2 - eV) \cdot$$
$$\left(\frac{1}{(E_1 + \omega_1)^2} - \frac{1}{(E_2 + \omega_2)^2} \right)^2.$$

The last term S^- is given by the same expression as S^+ with eV and $E_{1,2}$ replaced, respectively, by $-eV$ and $E_{1,2} + eV$.

Finally, we calculate the correlator S_{IQ} between the charge and current noise:

$$S_{IQ} = \frac{1}{2\pi} \int dt' \text{Tr}\{S^{\dagger}(t)I(t)S(t,t')Q(t')S(t')\rho_0\}. \tag{7}$$

Similarly to Eq. (3), it is convenient to break the integral in (7) into the two parts, $t' < t$ and $t' > t$. Expanding then the evolution operators up to the third order in H_T, and evaluating averages using Eq. (2), we find:

$$S_{IQ} = ie^2 g_1 g_2 \int \frac{d\omega_1 \omega_1}{1 - e^{-\omega_1/T}} \frac{d\omega_2 \omega_2}{1 - e^{-\omega_2/T}} \left\{ \delta(\omega_1 + \omega_2 - eV) \cdot \right.$$
$$\left(\frac{1}{E_1 + \omega_1} - \frac{1}{E_2 + \omega_2} \right) \left(\frac{1}{E_1 + \omega_1} + \frac{1}{E_2 + \omega_2} \right)^2 + \delta(\omega_1 + \omega_2 + eV) \cdot \tag{8}$$
$$\left. \left(\frac{1}{E_2 + eV + \omega_2} - \frac{1}{E_1 + eV + \omega_1} \right) \left(\frac{1}{E_1 + eV + \omega_1} + \frac{1}{E_2 + eV + \omega_2} \right)^2 \right\}.$$

The correlator S_{IQ} (8) is purely imaginary. From the perspective of the general theory of quantum linear detection [13] this fact means that the SET transistor in the co-tunneling regime is "symmetric" detector with the output noise and back-action noise uncorrelated at the classical level, since classically, the correlator between current and charge is given by the symmetrized correlation function which contains only the real part of S_{IQ}. Imaginary part of S_{IQ}

should be directly related to the linear response coefficient λ of the transistor [13]. Comparison of Eq. (8) to the expression for $\lambda = \delta\langle I\rangle/\delta V_g$ shows that, indeed, the two quantities are related: $(C_g/C_\Sigma)\mathrm{Im}[S_{IQ}] = -\hbar\lambda/4\pi$.

4. RESULTS AND DISCUSSION

In this section, we calculate explicitly the noise spectral densities S_Q, S_I, and S_{IQ} in various limits, and discuss the implications of the obtained relations for the characteristics of transistor as quantum detector. At small bias voltages V, $V \sim T \ll E_j$, the electron excitation energies ω_j can be neglected in comparison to energy barriers E_j, and expressions for the spectral densities obtained in the previous section are simplified:

$$S_Q = \frac{\hbar e^2}{6}\left(\frac{1}{E_1^2} - \frac{1}{E_2^2}\right)^2\left\{g_1g_2[(eV)^2 + (2\pi T)^2]eV \coth\frac{eV}{2T} + (g_1^2 + g_2^2)4\pi^2 T^3\right\},$$

$$S_I = \frac{g_1g_2e^3V}{6\hbar}\left(\frac{1}{E_1} + \frac{1}{E_2}\right)^2[(eV)^2 + (2\pi T)^2]\coth\frac{eV}{2T}, \qquad (9)$$

$$S_{IQ} = i\frac{g_1g_2e^3V}{6}\left(\frac{1}{E_1} - \frac{1}{E_2}\right)\left(\frac{1}{E_1} + \frac{1}{E_2}\right)^2[(eV)^2 + (2\pi T)^2].$$

A fundamental figure-of-merit of a quantum detector is "energy sensitivity" ϵ that is defined in terms of the output noise, back-action noise, and response coefficient of the detector - see, e.g., [13]. Qualitatively, energy sensitivity characterizes the amount of noise introduced by the detector into the measurement process, and is limited from below by $\hbar/2$ due to the quantum mechanical restrictions on the detector dynamics. In the case of SET transistor in the co-tunneling regime (when the current I and charge Q are classically uncorrelated), ϵ can be expressed directly through the three spectral densities S_I (4), S_Q (6), and S_{IQ} (8), as follows:

$$\epsilon = \frac{\hbar}{2}\frac{\sqrt{S_I S_Q}}{|S_{IQ}|}, \qquad (10)$$

Equations (9) show that at small bias voltages the transistor energy sensitivity (10) is:

$$\epsilon = \frac{\hbar}{2}\left[\coth\frac{eV}{2T}\left(\coth\frac{eV}{2T} + \left(\frac{g_1}{g_2} + \frac{g_2}{g_1}\right)\frac{4\pi^2 T^3}{eV[(eV)^2 + (2\pi T)^2]}\right)\right]^{1/2}. \qquad (11)$$

The energy sensitivity (11) approaches the fundamental limit $\hbar/2$ at $eV \gg T$ [14].

At larger bias voltages $eV \sim E_j \gg T$, temperature T can be neglected in equations for the noise spectral densities of the previous section and they give:

$$S_Q = \hbar e^2 g_1 g_2 \left\{ \frac{(eV)^3}{6} \sum_j \frac{1}{(E_j(E_j + eV))^2} + \frac{4eV}{(E_1 + E_2 + eV)^2} - \right.$$
$$\left. 2 \frac{eV(E_1 + E_2) + (eV)^2 + 2E_1 E_2}{(E_1 + E_2 + eV)^3} \sum_j \ln\left(1 + \frac{eV}{E_j}\right) \right\},$$

$$S_I = \frac{e^2 g_1 g_2}{\hbar} \left\{ \left(eV + \frac{2E_1 E_2}{E_1 + E_2 + eV}\right) \sum_j \ln\left(1 + \frac{eV}{E_j}\right) - 2eV \right\}, \qquad (12)$$

$$S_{IQ} = ie^2 g_1 g_2 \left\{ \frac{eV(E_1 - E_2)}{(E_1 + eV)(E_2 + eV)} \left(1 + \frac{eV(E_1 + E_2 + eV)}{2E_1 E_2}\right) - \right.$$
$$\left. \frac{E_1 - E_2}{E_1 + E_2 + eV} \sum_j \ln\left(1 + \frac{eV}{E_j}\right) \right\}.$$

Figure 2 shows the zero-temperature energy sensitivity (10) as a function of the bias voltage calculated from these equations for the SET transistor with equal junction capacitances and several values of the charge q_0 induced by the gate voltage. The energy sensitivity reaches $\hbar/2$ at small bias voltages and diverges when the voltage approaches the Coulomb blockade threshold (see also Eq. (13) below). The non-monotonic behavior of ϵ at small q_0 is caused by the fact that the correlator S_{IQ} vanishes for $E_1 \rightarrow E_2$.

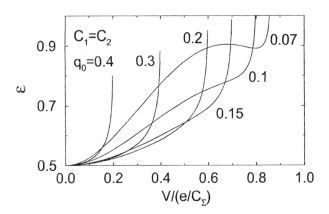

Figure 2: Energy sensitivity, in units of $\hbar$, of the SET transistor in the co-tunneling regime.

The energy barriers E_j decrease with increasing bias voltage V, and one or both of them disappear when V approaches the Coulomb blockade threshold. At the threshold, the spectral densities (12) diverge. For generic values of the gate voltage, only one of the barriers is suppressed at the threshold, e.g., $E_1 \to 0$. In this case,

$$S_Q = \frac{\hbar g_1 g_2 e^3 V}{6 E_1^2}, \quad S_I = \frac{g_1 g_2 e^3 V}{\hbar} \ln\left(\frac{eV}{E_1}\right), \quad S_{IQ} = -i\frac{g_1 g_2 e^3 V}{2 E_1}.$$

The energy sensitivity (10) also slowly diverges:

$$\epsilon = \frac{\hbar}{2} \left[\frac{2}{3} \ln\left(\frac{eV}{E_1}\right)\right]^{1/2}. \tag{13}$$

All these divergencies should be regularized by the finite width of the charge states created by tunneling [11,12]. Qualitatively, this means that for E_1 smaller than $\hbar G_2/C_\Sigma$, the logarithm in Eq. (13) should saturate at $\ln(1/g_2)$. Quantitative treatment of the threshold region is outside the scope of this work.

To summarize, we have studied noise properties of the SET transistor in the co-tunneling regime, and calculated its energy sensitivity as a detector. The energy sensitivity approaches $\hbar/2$ for small bias voltages, and slowly diverges at the Coulomb blockade threshold.

This work was supported by AFOSR.

REFERENCES

[1] D.V. Averin and K.K. Likharev, J. Low Temp. Phys. **62**, 345 (1986).

[2] T.A. Fulton and G.J. Dolan, Phys. Rev. Lett **59**, 109 (1987).

[3] M.A. Kastner, Rev. Mod. Phys. **64**, 849 (1992).

[4] D.V. Averin, *Solid State Commun.* **105**, 659 (1998).

[5] Yu. Makhlin, G. Schön, and A. Shnirman, Nature **398**, 305 (1999).

[6] Y. Nakamura, Yu.A. Pashkin, and J.S. Tsai, Nature **398**, 786 (1999).

[7] M.H. Devoret and R.J. Schoelkopf, Nature **406**, 1039 (2000).

[8] A. Shnirman and G. Schön, Phys. Rev. B **57**, 15400 (1998).

[9] A.N. Korotkov, cond-mat/0008461.

[10] D.V. Averin and Yu.V. Nazarov, in: *"Single Charge Tunneling"*, Ed. by H. Grabert and M. Devoret (Plenum, NY, 1992), p. 217.

[11] A.N. Korotkov, D.V. Averin, K.K. Likharev, and S.A. Vasenko, in: *"Single Electron Tunneling and Mesoscopic devices"*, Ed. by H. Koch and H. Lübbig, (Springer, Berlin, 1992), p. 45.

[12] Yu.V. Nazarov, J. Low Temp. Phys. **90**, 77 (1993).

[13] D.V. Averin, in: *"Exploring the Quantum-Classical Frontier: Recent Advances in Macroscopic and Mesoscopic Quantum Phenomena"*, Eds. J.R. Friedman and S. Han, to be published; cond-mat/0004364.

[14] It should be noted, however, that this fact does not necessarily mean that the regime of relatively small bias voltages, $T \ll eV \ll E_j$, represents optimal operating point of the practical SET transistors. Other noise sources, not included in the model but present in realistic systems, make it important to have large absolute values of the output signal, the condition that is not fulfilled for the SET transistor biased deep inside the Coulomb blockade region.

Charge Fluctuations and Dephasing in Coulomb Coupled Conductors

Markus Büttiker

Département de Physique Théorique, Université de Genève,
CH-1211 Genève 4, Switzerland

Abstract. The dephasing rates in two nearby mesoscopic conductors coupled only via the Coulomb interaction are discussed. The necessity of a non-perturbative, electrically self-consistent treatment is emphasized. Resistances are identified which can serve to characterize the noise power of charge fluctuations.

1. Introduction

Experiments (Buks et. al., 1998; Sprinzak et. al., 1998) in which two mesoscopic conductors are brought in close proximity and in which one of the conductors acts as a controllable dephasor on the other conductor are conceptually very intersting. Fig. 1 shows a mesoscopic cavity which is connected via a single lead to a reservoir and which is in proximity to a quantum point contact (QPC). We can ask: How does the proximity of the QPC affect the dephasing rate of carriers in the cavity? In a number of theoretical discussions such problems have been treated perturbatively taking the charge fluctuations to be those of a non-interacting system (Gurvitz, 1997; Levinson, 1997; Stodolsky, 1999). This leads to dephasing rates proportional to the square of the coupling constant. While it is well accepted that in single electron circuits the charging energy plays an important role, open systems like QPC's are often treated as non-interacting (Makhlin, Schön and Shnirman, 2000; Averin and Korotkov, 2000; Averin, 2000). Here we discuss a self-consistent treatment (Büttiker and Martin, 2000) which takes into account that charging a QPC is also governed by interactions. This approach leads to a dephasing rate that exhibits a more interesting dependence on the coupling strength (and in some cases gives a dephasing rate which is proportional to one over the square of the coupling strength). A self-consistent treatment of charge fluctuations is important not only for dephasing but also in the theory of dynamic conductance of mesoscopic conductors (Büttiker, Thomas and Prêtre, 1993; Büttiker, 1996; Ma, Wang and Guo, 1998; Deblock et. al., 2000) and frequency dependent noise spectra (Büttiker, 1996; Pedersen, van Langen and Büttiker, 1998; Nagaev, 1998).

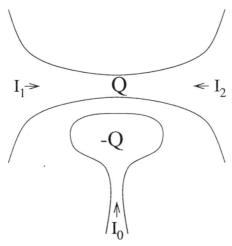

Figure 1. Cavity with a charge deficit $-Q$ in the proximity of a quantum point contact with an excess charge Q. The dipolar nature of the charge distribution ensures the conservation of currents I_0, I_1 and I_2 flowing into this structure. After (Pedersen, van Langen and Büttiker, 1998) .

2. Dephasing rates and potential fluctuations

Consider a scattering state $\Psi_i(\mathbf{r}, E)$ at energy E in conductor i which solves the Schrödinger equation for a fixed potential $U_{eq,i}(\mathbf{r})$. Fluctuations of the potential $U_i(\mathbf{r}, t)$ away from the static (average) equilibrium potential will scatter the carrier out of the eigenstate $\Psi_i(\mathbf{r}, E)$. Below we will for simplicity consider conductors which are so small that the fluctuations of the potential in the interior of the conductor can be taken to be space-independent. Thus the fluctuating potential in the region of interest is a function of time only $U_i(t)$. The effect of the fluctuating potential can then be described with the help of a time-dependent phase $\phi(t)$ which multiplies the scattering state. Thus we consider a solution of the type $\Psi_i(\mathbf{r}, E)\exp(-i\phi(t))$ of the time dependent Schrödinger equation. The equation of motion for the phase is simply, $\hbar d\phi/dt = eU_i(t)$. Let us characterize the potential fluctuations in conductor i by its noise spectrum $S_{U_i U_i}(\omega)$. Using the connection between phase and potential we find that at long times the phase ϕ of the scattering state diffuses with a rate

$$\Gamma_\phi^{(i)} = \langle(\phi(t) - \phi(0))^2\rangle/t = (e^2/2\hbar^2)S_{U_i U_i}(0) \tag{1}$$

determined by the zero-frequency limit of the noise power spectrum of the potential fluctuations.

An evaluation of the dephasing rates requires thus a discussion of the potential fluctuations. In turn, the potential fluctuations can be obtained from the charge fluctuations. Moreover, since we need only the zero-frequency limit of the potential fluctuation spectrum, it is sufficient to find the quasi-static (zero-frequency) charge fluctuations. Considering the conductors of Fig. 1, we assume that each electric field line emanating from the cavity ends up either again on the cavity or on the QPC. There exists a Gauss volume which encloses both conductors which can be chosen so large that the electrical flux through its surface vanishes (Büttiker, 1996). Consequently, the charge on the two conductors is conserved. Charge fluctuations are thus of a dipolar nature: within our Gauss volume, every excess charge in a small volume element is compensated by a charge deficit in an other volume element. To keep the discussion simple, we will consider here only one dipole: We permit the charging of one conductor against the other nearby conductor. The Coulomb interaction is described with help of a single geometrical capacitance C. The fluctuations of the charge Q_1 on the cavity and the charge Q_2 on the QPC (see Fig. 1) are related to the electrostatic potentials U_1 and U_2 on these two conductors via $Q_1 = C(U_1 - U_2)$ and $Q_2 = C(U_2 - U_1)$. The two equations relating charge to potential can be thought of as a poor man's Poisson equation. The potentials at the contacts of all conductors are held fixed. The potential U_i characterizes a deviation away from the equilibrium potential only in the interior of conductor i. The charge on conductor i can also be written in terms of the bare fluctuating charges $e\hat{N}_i$ (calculated by neglecting the Coulomb interaction) counteracted by a screening charge $eD_i e\hat{U}_i$. For Thomas-Fermi screening, D_i is the density of states on conductor i in the volume over which $e\hat{N}_i$ is not screened completely. We arrive thus at the following self-consistent equations relating the true charges Q_i to the bare charges $e\hat{N}_i$ and to the potentials $\hat{U}_i$,

$$\hat{Q}_1 = C(\hat{U}_1 - \hat{U}_2) = e\hat{N}_1 - e^2 D_1 \hat{U}_1, \tag{2}$$

$$\hat{Q}_2 = C(\hat{U}_2 - \hat{U}_1) = e\hat{N}_2 - e^2 D_2 \hat{U}_2. \tag{3}$$

We have written these two equations not for the average quantities but for operators. The fluctuations are determined by the off-diagonal elements of the charge and potential operators. Below we will specify these expressions in detail. Solving these equations for the potential operators, we find $\hat{U}_i = e \sum_j G_{ij} \hat{N}_j$ with an effective interaction G_{ij} given by

$$\mathbf{G} = \frac{C_\mu}{e^2 D_1 e^2 D_2 C} \begin{pmatrix} C + e^2 D_2 & C \\ C & C + e^2 D_1 \end{pmatrix}. \tag{4}$$

Here the electrochemical capacitance $C_\mu^{-1} = C^{-1} + (e^2 D_1)^{-1} + (e^2 D_2)^{-1}$ is the series capacitance of the geometrical contribution and the density of states of the two conductors (Büttiker, Thomas and Prêtre, 1993). Note, that in contrast to perturbation treatments, the effective coupling element G_{12} is not proportional to e^2/C but in general is a complicated function of this energy. Let us now introduce the noise power spectra of the bare charges, $S_{N_i N_i}(\omega)$ for each of the conductors. The bare charge fluctuation spectra on different conductors are uncorrelated, $S_{N_i N_j}(\omega) = 0$ for $i \neq j$. With the help of the effective interaction matrix, we can now relate the potential fluctuation spectra to the fluctuation spectra of the bare charges. In the zero-frequency limit we find,

$$S_{U_i U_i}(0) = e^2 \sum_j G_{ij}^2 S_{N_j N_j}(0). \tag{5}$$

Since the dephasing rate in conductor i is determined directly by $S_{U_i U_i}(0)$, we have found an expression of the dephasing rate in terms of an effective interaction matrix and the bare charge fluctuations. Eq. (5) shows that the potential fluctuations and thus the dephasing rate has two sources: A carrier in conductor i suffers dephasing due to charge fluctuations in conductor $j = i$ itself, and due to charge fluctuations of the additional nearby conductor $j \neq i$. Accordingly, we can also write the dephasing rate in conductor i as a sum of two contributions, $\Gamma_\phi^{(i)} = \sum_j \Gamma_\phi^{(ij)}$ with

$$\Gamma_\phi^{(ij)} = (e^4/2\hbar^2) G_{ij}^2 S_{N_j N_j}(0). \tag{6}$$

At equilibrium, the bare charge fluctuation spectrum, normalized by the density of states D_i of conductor i has the dimension of a resistance. We introduce the charge relaxation resistance

$$2kT R_q^{(j)} \equiv e^2 S_{N_j N_j}(0)/(e^2 D_j)^2. \tag{7}$$

In simple cases, R_q together with an appropriate capacitance determines the RC-time of the mesoscopic structure. The charge relaxation resistance can thus alternatively be determined by investigating the poles of the conductance matrix. The dynamic conductance matrix $G_{\alpha\beta}(\omega) \equiv dI_\alpha(\omega)/dV_\beta(\omega)$ of our mesoscopic structure (QPC and cavity) which relates the currents $dI_\alpha(\omega)$ at a frequency ω at contact α to the voltages $dV_\beta(\omega)$ applied at contact α has at low frequencies a pole determined by $\omega_{RC} = -iC_\mu(R_q^{(1)} + R_q^{(2)})$. Alternatively we could carry out a low frequency expansion of the element $G_{00}(\omega)$ (the element of the conductance matrix which gives the current at the contact of the cavity in response to an oscillating voltage applied to the cavity) to find that $G_{00}(\omega) = -iC_\mu\omega + C_\mu^2 R_q \omega^2 + \dots$ Thus R_q plays a role not only in dephasing but in many other problems in which charge relaxation or charge fluctuations are of crucial importance. The charge relaxation resistance differs from the dc-resistance. For instance a ballistic one-channel quantum wire connecting two reservoirs and capacitively coupled to a gate has for spinless carriers a dc-resistance of $R = h/e^2$ and charge relaxation resistance of $R_q = h/4e^2$. The dc-resistance corresponds to the series addition of resistances along the conductance path, whereas an excess charge on the conductor relaxes via all possible conductance channels to the reservoirs and thus corresponds to the addition of resistances in parallel. This is nicely illustrated for a chaotic cavity connected via contacts with M_1 and M_2 perfectly transmitting channels to reservoirs and capacitively coupled to a gate. Its ensemble averaged dc-resistance is $R = (h/e^2)(M_1^{-1} + M_2^{-1})$, whereas its charge relaxation resistance is $R_q = (h/e^2)(M_1 + M_2)^{-1}$. Thus the dc-resistance is governed by the smaller of the two contacts, whereas the the charge relaxation resistance is determined by the larger contact.

If the conductor is driven out of equilibrium with the help of an applied voltage $e|V|$, the thermal noise described by Eq. (7) can be overpowered by shot noise. For $e|V| \gg kT$ the charge fluctuation spectrum becomes proportional to the applied voltage and defines a resistance

$$2e|V|R_v^{(j)} \equiv e^2 S_{N_j N_j}(0)/(e^2 D_j)^2. \qquad (8)$$

The resistance R_v is thus a measure of the noise power of the charge fluctuations associated with shot noise.

Eq. (7) and Eq. (8) describe the behavior of $e^2 S_{N_j N_j}(0)/(e^2 D_j)^2$ in the limits $kT \gg e|V|$ and $kT \ll e|V|$. For fixed temperature as a function of voltage $e^2 S_{N_j N_j}(0)/(e^2 D_j)^2$ exhibits a smooth crossover from the equilibrium result Eq. (7) to Eq. (8) valid in the presence of shot noise.

From these expressions we learn that a conductor (labeled 2) at equilibrium generates in a nearby conductor (labeled 1) an additional dephasing rate

$$\Gamma_\phi^{(12)} = (e^2/\hbar^2)(e^2 D_2 G_{12})^2 R_q^{(2)} kT \qquad (9)$$

and if conductor 2 is subject to a current generated by a voltage $|V|$ (with $kT << e|V|$) a dephasing rate

$$\Gamma_\phi^{(12)} = (e^2/\hbar^2)(e^2 D_2 G_{12})^2 R_v^{(2)} e|V|. \qquad (10)$$

These dephasing rates are proportional to $(e^2 D_2 G_{12})^2$. Thus $e^2 D_2 G_{12}$ plays the role of an effective coupling constant in our problem. In the limit of a small Coulomb energy ($C \gg e^2 D_1$ and $C \gg e^2 D_2$) we find $e^2 D_2 G_{12} = D_1/(D_1 + D_2)$ *independent* of the geometrical capacitance C. In the limit of a large Coulomb energy ($C \ll e^2 D_1$ and $C \ll e^2 D_2$) the effective coupling constant becomes proportional to the capacitance $e^2 D_2 G_{12} = C/(e^2 D_1)$. The second limit, typically, is the physically relevant limit. Below we focus on this case: We assume that the properties of conductor 2 are varied with the help of a gate voltage (which permits to open or close the QPC). That leaves the effective coupling constant unchanged and the dependence of the dephasing rate on gate voltage is entirely determined by the gate voltage dependence of $R_q^{(2)}$ and $R_v^{(2)}$.

3. Bare Charge Fluctuations and the Scattering Matrix

Let us now determine the charge operator for the bare charges (non-interacting carriers). The operator for the total charge on a mesoscopic conductor can be found from the current operator and by integrating the continuity equation over the total volume of the conductor. This gives a relation between the charge in this volume and the particle currents entering this volume. We obtain for the density operator (Pedersen, van Langen and Büttiker, 1998)

$$\hat{N}(\omega) = \hbar \sum_{\beta\gamma} \sum_{mn} \int dE \, \hat{a}_{\beta m}^{\dagger}(E) \mathcal{D}_{\beta\gamma mn}(E, E+\hbar\omega) \hat{a}_{\gamma n}(E+\hbar\omega), \quad (11)$$

where $\hat{a}_{\beta m}^{\dagger}(E)$ (and $\hat{a}_{\beta m}(E)$) creates (annihilates) an incoming particle with energy E in lead β and channel m. The element $\mathcal{D}_{\beta\gamma mn}(E, E+\hbar\omega)$ is the non-diagonal density of states element generated by carriers incident simultaneously in contact β in quantum channel m and by carriers incident in contact γ in channel n. In particular, in the zero-frequency limit, we find in matrix notation (Pedersen, van Langen and Büttiker, 1998),

$$\mathcal{D}_{\beta\gamma}(E) = \frac{1}{2\pi i} \sum_{\alpha} \mathbf{s}_{\alpha\beta}^{\dagger}(E) \frac{d\mathbf{s}_{\alpha\gamma}(E)}{dE}. \quad (12)$$

Expressions of this type are known from the discussion of quantum mechanical time delay (Smith, 1960). The sum of the diagonal elements of this matrix is the density of states of the conductor

$$D(E) = \sum_{\beta} Tr[\mathcal{D}_{\beta\beta}(E)] = \frac{1}{2\pi i} \sum_{\alpha,\beta} Tr[\mathbf{s}_{\alpha\beta}^{\dagger}(E) \frac{d\mathbf{s}_{\alpha\beta}(E)}{dE}], \quad (13)$$

where the trace is over the quantum channels. In contrast to the discussion of time-delay (density of states), the charge fluctuations are determined by the off-diagonal elements. Proceeding as for the case of current fluctuations (Lesovik, 1989; Büttiker, 1990; Blanter and Büttiker, 2000) we find for the fluctuation spectrum of the total charge

$$S_{NN}(0) = 2h \sum_{\gamma\delta} \int dE \, Tr[\mathcal{D}_{\gamma\delta}^{\dagger} \mathcal{D}_{\delta\gamma}] f_{\gamma}(E)(1 - f_{\delta}(E)). \quad (14)$$

To determine the dephasing rates we have now to find the elements $\mathcal{D}_{\delta\gamma}$.

4. Charge relaxation resistance of a quantum point contact

To illustrate the preceding discussion, we now consider specifically the charge relaxation resistance and subsequently the resistance R_v of a QPC. For simplicity, we consider a symmetric QPC (the asymmetric case (Sprinzak et. al., 1998) is treated in (Büttiker and Martin, 2000; Levinson, 2000)): For a symmetric scattering potential the scattering matrix (in a basis in which the transmission and reflection matrices are diagonal) is for the n-th channel of the form

$$s_n(E) = \begin{pmatrix} -i\sqrt{R_n}\exp(i\phi_n) & \sqrt{T_n}\exp(i\phi_n) \\ \sqrt{T_n}\exp(i\phi_n) & -i\sqrt{R_n}\exp(i\phi_n) \end{pmatrix}, \qquad (15)$$

where T_n and $R_n = 1 - T_n$ are the transmission and reflection probabilities and ϕ_n is the phase accumulated by a carrier in the n-th eigen channel. We find for the elements of the density of states matrix, Eq. (12),

$$\mathcal{D}_{11} = \mathcal{D}_{22} = \frac{1}{2\pi}\frac{d\phi_n}{dE}, \quad \mathcal{D}_{12} = \mathcal{D}_{21} = \frac{1}{4\pi}\frac{1}{\sqrt{R_n T_n}}\frac{dT_n}{dE}. \qquad (16)$$

With these density of states matrix elements, we can determine the particle fluctuation spectrum, Eq. (14) in the white-noise limit, and R_q with the help of Eq. (7)

$$R_q = \frac{h}{4e^2}\frac{\sum_n\left[(\frac{d\phi_n}{dE})^2 + \frac{1}{4T_n R_n}(\frac{dT_n}{dE})^2\right]}{[\sum_n\frac{d\phi_n}{dE}]^2}. \qquad (17)$$

Eq. (17) is still a formal result, applicable to any symmetric (two-terminal) conductor. To proceed we have to adopt a specific model for a QPC. If only a few channels are open the average potential has in the center of the conduction channel the form of a saddle (Büttiker, 1990):

$$V_{eq}(x,y) = V_0 + \frac{1}{2}m\omega_y^2 y^2 - \frac{1}{2}m\omega_x^2 x^2, \qquad (18)$$

where V_0 is the potential at the saddle and the curvatures of the potential are parametrized by ω_x and ω_y. The resulting transmission probabilities have the form of Fermi functions $T(E) = 1/(e^{\beta(E-\mu)} + 1)$ (with a negative temperature $\beta = -2\pi/\hbar\omega_x$ and $\mu = \hbar\omega_y(n + 1/2) + V_0$). As a function of energy (gate voltage) the conductance rises step-like. The energy derivative of the transmission probability $dT_n/dE = (2\pi/\hbar\omega_x)T_n(1 - T_n)$ is itself proportional to the transmission probability times the reflection probability. We note that such a relation holds not only for the saddle point model of a QPC but also for instance for the adiabatic model (Glazman et. al., 1988). As a consequence $(1/4T_n R_n)(dT_n/dE)^2 = (\pi/\hbar\omega_x)^2 T_n R_n$ is proportional

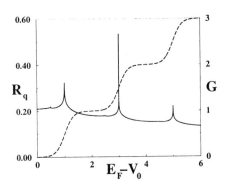

Figure 2. Charge relaxation resistance R_q of a saddle QPC in units of $h/4e^2$ for $\omega_y/\omega_x = 2$ and a screening length of $m\omega_x\lambda^2/\hbar = 2E_\lambda/\hbar\omega_x = 25$ as a function of $E_F - V_0$ in units of $\hbar\omega_x$ (full line). The broken line shows the conductance of the QPC. After (Büttiker and Martin, 2000) .

to $T_n R_n$. Thus the charge relaxation resistance of a saddle QPC is

$$R_q = \frac{h}{4e^2} \frac{\sum_n \left[\left(\frac{d\phi_n}{dE}\right)^2 + \left(\frac{\pi}{\hbar\omega_x}\right)^2 T_n R_n \right]}{[\sum_n \frac{d\phi_n}{dE}]^2}. \tag{19}$$

To find the density of states of the n-th eigen channel, we use the relation between density and phase (action) and $D_n = \sum_i \mathcal{D}_{n,ii} = (1/\pi)d\phi_n/dE$. We evaluate the phase semi-classically. The spatial region of interest for which we have to find the density of states is the region over which the electron density in the contact is not screened completely. We denote this length by λ and the associated energy by $E_\lambda = (1/2)m\omega_x^2\lambda^2$. The density of states is then found from $D_n = 1/h \int_{-\lambda}^{\lambda} \frac{dp_n}{dE} dx$ where p_n is the classically allowed momentum. A calculation gives a density of states (Pedersen, van Langen and Büttiker, 1998) $D_n(E) = (4/(h\omega_x))$ asinh$[E_\lambda/(E - E_n)]^{1/2}$, for energies E exceeding the channel threshold E_n and gives a density of states $D_n(E) = (4/(h\omega_x))$ acosh$[E_\lambda/(E_n - E)]^{1/2}$ for energies in the interval $E_n - E_\lambda \leq E < E_n$ below the channel threshold. Electrons with energies less than $E_n - E_\lambda$ are reflected before reaching the region of interest, and thus do not contribute to the DOS. The resulting density of states has a logarithmic singularity at the threshold $E_n = \hbar\omega_y(n + \frac{1}{2}) + U_0$ of the n-th quantum channel. (A fully quantum mechanical calculation gives a density of states which exhibits also a peak at the threshold but which is not singular). The charge relaxation resistance for a saddle QPC is shown in Fig. 2 for a set of parameters given in the figure caption. The charge relaxation resistance exhibits a sharp spike at each opening of a quantum channel. Physically this implies that the relaxation of charge, determined by the RC-time is very rapid at the opening of a quantum channel.

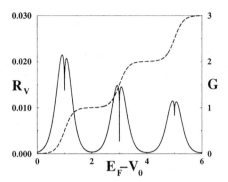

Figure 3. R_v (solid line) for a saddle QPC in units of h/e^2 and G (dashed line) in units of e^2/h as a function of $E_F - V_0$ in units of $\hbar\omega_x$ with $\omega_y/\omega_x = 2$ and a screening length $m\omega_x\lambda^2/\hbar = 25$. R_v and G are for spinless electrons. After (Büttiker and Martin, 2000).

Similarly, we can find the resistance R_v for a QPC subject to a voltage $e|V| >> kT$. Using the density matrix elements for a symmetric QPC given by Eqs. (16), we find (Pedersen, van Langen and Büttiker, 1998)

$$R_v = \frac{h}{e^2} \frac{\sum_n \frac{1}{4R_n T_n} \left(\frac{dT_n}{dE}\right)^2}{[\sum_n (d\phi_n/dE)]^2} = \frac{h}{e^2} \left(\frac{\pi}{\hbar\omega_x}\right)^2 \frac{\sum_n T_n R_n}{[\sum_n (d\phi_n/dE)]^2}. \tag{20}$$

The resistance R_v is shown in Fig. 3.

As emphasized earlier, according to Eq. (9) and Eq. (10), in an experiment in which the QPC is opened with the help of a gate, the dephasing rate follows, just R_q at equilibrium or R_v in the non-equilibrium case. Note that the resistance R_v is proportional to the shot noise power $S_{II}(0) = 2e(e^2/h)|V|\sum_n T_n R_n$. This statement is in contrast to arguments (Buks et. al., 1998; Sprinzak et. al., 1998; Aleiner, Wingreen, and Meir, 1997) which lead to a dephasing rate proportional to $(\Delta T)^2|V|^2/S_{II}$, where ΔT is the change in the transmission probability due to the variation of the charge on the QPC and S_{II} the shot noise power. That the identification of $(TR)^{-1}$ with shot noise is paradoxical can be seen from the fact that $(TR)^{-1}$ appears already in the equilibrium charge relaxation resistance R_q.

Without screening R_v would exhibit a bell shaped behavior as a function of energy, i. e. it would be proportional to $T_n(1 - T_n)$ in the energy range in which the n-th transmission channel is partially open. Screening, which in R_v is inversely proportional to the density of states squared, generates the dip at the threshold of the new quantum channel at the energy which corresponds to $T_n = 1/2$. It is interesting to note that the experiment (Buks et. al., 1998) does indeed show a double hump behavior of the dephasing rate.

5. Discussion

We have presented a self-consistent discussion of dephasing rates of Coulomb coupled conductors. This approach emphasizes that also in open conductors, like QPC's, charge fluctuations are associated with a Coulomb energy. In such a self-consistent treatment, the dephasing rates are typically not simply proportional to a coupling constant. We have attributed only a single potential to each conductor, but the theory is not in fact limited to such a simplification and permits the treatment of an arbitrary potential landscape (Büttiker and Martin, 2000; Martin and Büttiker, 2000).

Acknowledgements

This work was supported by the Swiss National Science Foundation and the TMR network.

References

I. L. Aleiner, N. S. Wingreen, and Y. Meir, Phys. Rev. Lett. **79**, 3740 (1997).
D. V. Averin. In *Exploring the Quantum-Classical Frontier: Recent Advances in Macroscopic and Mesoscopic Quantum Phenomena*, Eds. J.R. Friedman and S. Han, (unpublished). cond-mat/0004364
Ya. M. Blanter and M. Büttiker, Physics Reports, (unpublished). cond-mat/9910158
E. Buks, R. Schuster, M. Heiblum, D. Mahalu and V. Umansky, Nature **391**, 871 (1998).
M. Büttiker, Phys. Rev. Lett. **65**, 2901 (1990); Phys. Rev. B **46**, 12485 (1992).
M. Büttiker, Phys. Rev. B **41**, 7906 (1990).
M. Büttiker, H. Thomas, and A. Prêtre, Phys. Lett. A **180**, 364 (1993).
M. Büttiker, J. Math. Phys., **37**, 4793 (1996).
M. Büttiker and A. M. Martin, Phys. Rev. B**61**, 2737 (2000).
R. Deblock, Y. Noat, H. Bouchiat, B. Reulet, and D. Mailly, Phys. Rev. Lett. **84**, 5379 (2000).
L. I. Glazman, G. B. Lesovik, D. E. Khmel'nitskii, and R. I. Shekhter, JETP Lett. **48**, 238 (1988).
S. A. Gurvitz, Phys. Rev. B **56**, 15215 (1997).
A. N. Korotkov, D. V. Averin, (unpublished). cond-mat/0002203
G. B. Lesovik, JETP Lett. **49**, 592 (1989).
Y. B. Levinson, Europhys. Lett. **39**, 299 (1997).
Y. B. Levinson, Phys. Rev. B **61**, 4748 (2000).
Z.-s. Ma, J. Wang and H. Guo, Phys. Rev. B **57**, 9108 (1988).
Y. Makhlin, G. Schön, A. Shnirman, (unpublished). cond-mat/0001423
A. M. Martin and M. Buttiker, Phys. Rev. Lett. **84**, 3386 (2000).
K. E. Nagaev, Phys. Rev. B **57**, 1838 (1998).
M. H. Pedersen, S. A. van Langen and M. Büttiker, Phys. Rev. B **57**, 1838 (1998).
F. T. Smith, Phys. Rev. **118** 349 (1960).
D. Sprinzak, E. Buks, M. Heiblum and H. Shtrikman, Phys. Rev. Lett. (unpublished). cond-mat/9907162
L. Stodolsky, Phys. Lett. B **459**, 193 (1999).

THE MEASUREMENT OF PHOTON
NUMBER STATES USING CAVITY QED

Simon Brattke[a,b]
Benjamin Varcoe[a]
and Herbert Walther[a,b]

[a] *Max Planck Insitute for Quantum Optics*
[b] *Section Physik Univiersität München*

Abstract   In two recent experiments it was demonstrated that number or Fock states can be generated. Two types of experiments have been performed. In the first one the number states are achieved in steady state via the so-called trapping states of the micromaser field. In the second experiment the number states were generated in a dynamical experiment by state reduction. In the latter case the generated field was afterwards probed by observing the dynamics of a probe atom in the field. This is the first unambiguous measurement of the purity of number states in cavity quantum electrodynamics.

Keywords: Fock states, number states, sub-Poissonian Statistics, One-Atom Maser

1.   INTRODUCTION

The one atom maser or micromaser represents the most basic system in cavity quantum electrodynamics (QED). In this maser a single atom interacts with a single mode of the radiation field [1]. It thus represents the system treated in the Jaynes Cummings model in the early days of masers and lasers [2]. In the micromaser highly excited Rydberg atoms interact with a single mode of a superconducting cavity with a high quality factor that leads to very long photon lifetimes. The steady-state field generated in the cavity has already been the object of detailed studies of the sub-Poissonian statistical distribution of the field [3], the quantum dynamics of the atom-field photon exchange represented in the collapse and revivals of the Rabi nutation [4], atomic interference

[5], bistability and quantum jumps of the field [6], atom-field and atom-atom entanglement [7].

The quantum treatment of the radiation field uses the number of photons in a particular mode to characterise the quantum states whereby in the ideal case the modes are defined by the boundary conditions of a cavity giving a discrete set of eigen-frequencies. The ground state of the quantum field is represented by the vacuum state consisting of field fluctuations with no residual energy. The states with fixed photon number are usually called Fock or number states. They are generally used as the basis for the quantum representation of radiation fields. number states thus represent the most basic quantum states and are maximally distant from what one would call a classical field. Although number states of vibrational motion are routinely observed in ion traps [8], number states of the radiation field are very fragile and very difficult to produce and maintain. They are perfectly number-squeezed, extreme sub-Poissonian states in which intensity fluctuations vanish completely. In order to generate these states it is necessary that the mode considered has minimal losses and the thermal field, always present at finite temperatures, has to be eliminated since it causes photon number fluctuations.

In this paper we present two methods of creating number states in the micromaser. The first one uses the well known trapping states, which are generated under steady-state conditions and represent number states at certain experimental parameters. We also present a second method using state reduction for preparation of the number states. The latter method allows to determine the purity of the number states produced.

The micromaser setup used for the experiments is shown in Fig. 1 and has been described in detail previously [9]. Briefly, in this experiment, a $^3He-^4He$ dilution refrigerator houses the microwave cavity which is a closed superconducting niobium cavity. A rubidium oven provides two collimated atomic beams: a central one passing directly into the cryostat and a second one directed to an additional excitation region. The second beam was used as a frequency reference. A frequency doubled dye laser ($\lambda = 297$ nm) was used to excite rubidium (^{85}Rb) atoms to the Rydberg 63 $P_{3/2}$ state from the $5S_{1/2}$(F = 3) ground state. The cavity is tuned to a 21.456 GHz transition from the $63P_{3/2}$ state to the $61D_{5/2}$ state, which is the lower or ground state of the maser transition. In a recent experiment a cavity with a Q factor of 4×10^{10} was produced with a field decay time of 0.6s which corresponds to a photon lifetime of 0.3s. The decay was measured using ring down spectroscopy the result of which is presented in Ref. [10]. This is the largest cavity Q factor ever obtained for this type of experiment and is more than two orders of magnitude larger than has been seen in related experiments [12].

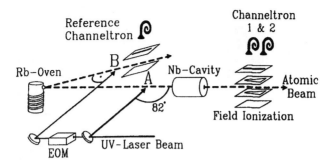

Figure 1.1 The experimental setup. The atoms leaving the rubidium oven are excited into the $63P_{3/2}$ Rydberg state using a single step UV excitation at an angle of 11°. After interacting with the cavity the atoms are detected using state selective field ionisation. Tuning of the cavity is performed using two piezo translators. The reference beam stabilises the laser to a Stark shifted atomic resonance allowing the selected velocity subgroup to be changed continuously within the range of the velocity distribution of the atoms.

2.   TRAPPING STATES: STEADY STATE NUMBER STATES

The trapping states are a steady-state feature of the maser field peaked in a single photon number, they occur in the micromaser as a direct consequence of field quantisation. A prerequisite is that the number of blackbody photons in the cavity mode is reduced [13, 9]. They occur when the atom field coupling, Ω, and the interaction time, t_{int}, are chosen such that in a cavity field with n photons each atom undergoes an integer number, k, of Rabi cycles. This is summarised by the condition,

$$\Omega t_{int} \sqrt{n+1} = k\pi. \tag{1.1}$$

When Eq. 1.1 is fulfilled the cavity photon number is left unchanged after the interaction of an atom and hence the photon number is "trapped". This will occur over a large range of the atomic pump rate N_{ex}. (N_{ex} stands for the number of atoms per decay time of the cavity). The trapping state is therefore characterised by the photon number n and the number of integer multiples of full Rabi cycles k.

The trapping states are detected by a reduced emission probability of the pumping atoms leading to a dip in the atomic inversion. Once in a trapping state the maser will remain there regardless of the pump rate. There is, in general, a slight deviation from a pure number state resulting from photon losses. A lost photon will be replaced by the next

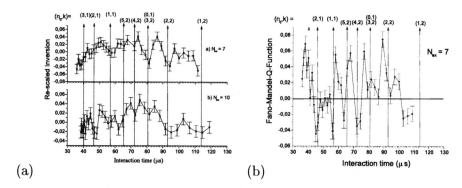

(a)   (b)

Figure 1.2 The first experimental measurement of trapping states in the maser field (Weidinger et al.[10]). (a) Rescaled inversion vs interaction time. The rescaling was necessary owing to a differential loss of atoms resulting from the finite lifetime of the two maser levels (488 μs and 244 μs for the excited and ground states respectively). Trapping states appear as a reduced probability of finding a ground state atom. (b) Mandel Q - parameter showing the deviation from Poissonian statistics. The presence of number states is indicated by sub-Poissonian statistics. The theoretical positions of trapping states are marked with the label (n,k). The coincidence of theoretical indicators and experimental positions of both sub-Poissonian statistics and lower ground state probabilities indicates that trapping states are produced.

incoming atom, however, there is a small time interval until this actually happens depending on the availability of an atom. This leads to a small probability of finding lower photon numbers in the field (see for example Fig. 1.4c).

Figure 1.2 shows experimental results in which trapping states of up to $n=5$ were found in the micromaser field [9]. Simulations of the trapping state photon number distribution was performed for the experimental parameters of Fig. 1.2 indicate that there is a high fidelity number state created even under those non ideal conditions [11]. Trapping states are in fact a very strong feature of maser dynamics and trapping state scans over the $(0,1)$, $(1,1)$ and $(1,2)$ trapping states are routinely used to calibrate the apparatus.

3.   DYNAMICAL NUMBER STATE PREPARATION

There is also a dynamical way to generate number states with a pulsed atomic beam. When the atoms leave the cavity of the micromaser they are in an entangled state with the cavity field [7]. A method of state reduction was suggested by Krause et al.[14] to observe the build up of the cavity field to a known number state. By state reduction of the outgoing atom also the field part of the entangled atom-field state is projected out and the photon number in the field either increases or decreases depending on the state of the observed outgoing atom. If the field is initially in a state $|n\rangle$ then an interaction of an atom with the cavity leaves the cavity field in a superposition of the states $|n\rangle$ and $|n+1\rangle$ and the atom in a superposition of the internal atomic states $|e\rangle$ and $|g\rangle$.

$$\Psi = \cos(\phi)|e\rangle|n\rangle - i\sin(\phi)|g\rangle|n+1\rangle \qquad (1.2)$$

where ϕ is an arbitrary phase. The state selective field ionisation measurement of the internal atomic states also reduces the field to one of the states $|n\rangle$ or $|n+1\rangle$. State reduction is independent of interaction time, hence a ground state atom always projects the field onto the $|n+1\rangle$ state independent of the time spent in the cavity. This results in an *a priori* probability of the maser field being in a specific but unknown number state [14]. If the initial state is the vacuum, $|0\rangle$, then a number state created is equal to the number of ground state atoms that were collected within a suitably small fraction of the cavity decay time. The purity of the number state can then be measured using the Rabi-dynamics of the following atom.

In the presence of a cavity photon number, $|n\rangle$, the relative populations of the excited and ground states of an atom will oscillate at a frequency $\Omega\sqrt{n+1}$. Experimentally the atomic inversion, given by $I = P_{\rm g} - P_{\rm e}$, is measured; where $P_{\rm g(e)}$ is the probability of finding a ground (excited) state atom. If the field is not in a number state the inversion is given by

$$I(n, t_{\rm int}) = -C\sum_{\rm n} P_{\rm n}\cos(2\Omega\sqrt{n+1}t_{\rm int}) \qquad (1.3)$$

where $P_{\rm n}$ is the probability of finding n photons in the mode, $t_{\rm int}$ is the interaction time of the atoms with the cavity field. Rabi cycling of a probe atom can therefore be used as an unambiguous test of the photon distribution in the cavity by analysing the dynamics by means of Eq. 1.3. The factor C considers the reduction of the signal amplitude due to dark counts.

An experiment to observe these states was performed recently (see ref. [16]). To create and detect an n photon number state in the cavity, $N = n+1$ atoms are required. That is n atoms to create the number state and the final atom as a probe of the state. However owing to the non-perfect detector efficiency and atomic decay there are missed counts. By using a laser pulse of short duration the number of excited-state atoms entering the cavity per pulse is rather low. Hence we know that when N atoms

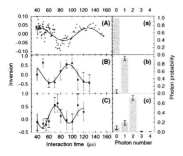

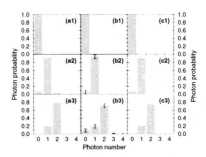

Figure 1.3 (A),(B) and (C): Three Rabi oscillations are presented, for the number state n=0,1 and 2. (a),(b) and (c): plots presenting the coefficients P_n resulting from fits to Eq. 1.3. In each fit there is a clear maximum photon number which corresponds to the target number state.

Figure 1.4 Comparison between theory and experimental results on the purity of number states. Shown is (a) a theoretical simulation of the current experiment; (b) the current experimental results; and (c) the steady state trapping states.

per pulse were detected the probability of having $N + 1$ atoms per pulse was negligibly small. With 40 % detector efficiency and the assumption that the probability of missing a count is statistically independent, there is a probability of about 1 % of the state preparation being incorrect because an atom escapes detection [15]. For the measurement of an n-photon number state, the detection of the probe atom is triggered by the detection of n ground state atoms within the length of the laser pulse.

To ensure that the cavity is in the vacuum state at the start of a measurement, there is a delay of 1.5 cavity decay times between the laser pulses. Hence the compromise that the Q-value be lower than ultimately possible in our setup, since a higher Q would lead to an increase of the data collection time. Even with the reduced cavity life time of 25 ms that was used in this experiment and large delay times between the laser pulses a cyclically steady state maser field can build up in the cavity [17]. The time delay between pulses was selected as a compromise between limiting the growth of the maser field and the length of the data collection time.

Figure 1.3(A-C) displays the Rabi cycles obtained by measuring the inversion of a probe atom that followed the detection of n=0, 1 or 2 ground state atoms respectively. Figure 1.3(a-c) displays normalised values of the coefficients obtained from these curve fits.

The fact that we do not measure pure number states is the result of dissipation in the time interval between production and analysis of the cavity field. Incorperating dissipation from the cavity indicates that number states with a purity of 99 % for the $n=1$ state and 95 % for the $n=2$ state were produced in the moment of creation.

The limited resolution of the $n=1$ and $n=2$ Rabi oscillations is due to the cumulative effect of velocity and detuning fluctuations, a 3-5 % miscount rate in the state selective detectors and detector dark counts. The presence of number states, however, can be determined independently of the resolution because the only important factor is the accumulated Rabi phase of the probe atom as it passes through the cavity. A detailed calculation has been performed to model the dissipation of the field before detection and is described in detail in [17]. The results of these calculations are compared to the experimental results in Fig. 1.4a and Fig. 1.4b respectively.

As dissipation is the most essential loss mechanism, it is interesting to compare the purity of the number states generated by the current method with that expected for trapping states (Fig. 1.4c). The agreement between these results is striking. The trapping state photon distribution is generated in the steady-state, which means that whenever the loss of a photon occurs the next incoming atom will restore the old field with a high probability after a small delay. The atom rate used in these calculations was 25 atoms per cavity decay time, or an average delay of 1 ms. This can be compared to the delay between the preparation and probe atoms in the present experiment. In the steady state simulation loss due to cavity decay determines the purity of the number state. It can therefore be concluded that dissipative loss due to cavity decay in the delay to a probe atom, largely determines the measured deviation from a pure number state. The steady state simulation was performed for a temperature of 100 mK, which makes the influence of the thermal field in the steady-state correspondingly more closely to that of the dynamical experiment.

4.   SUMMARY

Here we have reported on the generation and unambiguous measurement of number states in the micromaser field. In the future it would be interesting to consider methods of creating number states using some combination of the methods presented above that would allow one to create number states in a more efficient manner. However the fact that number states can be readily created opens many opportunities, especially in connection with the investigation of quantum information and the investigation of macroscopic entanglement of particles such as GHZ states. The decoherence of Schrödinger cat states can be largely attributed to dissipation, the very low influence of dissipation in this experiment therefore provides an excellent environment in which to investigate decoherence and nonlocal quantum phenomena such as entanglement of atoms.

References

[1] D. Meschede, H. Walther, and G. Müller, Phys. Rev. Lett. **54**, 551 (1985).

[2] E.T. Jaynes and F.W. Cummings, Proc. IEEE. **51**, 89 (1963).

[3] G. Rempe, F. Schmidt-Kaler, and H. Walther, Phys. Rev. Lett. **64**, 2783 (1990).

[4] G. Rempe, H. Walther, and N. Klein, Phys. Rev. Lett. **58**, 353 (1987).

[5] G. Raithel, O. Benson, and H. Walther, Phys. Rev. Lett. **75**, 3446 (1995).

[6] O. Benson, G. Raithel, and H. Walther, Phys. Rev. Lett. **72**, 3506 (1994).

[7] B.-G. Englert, M. Löffler, O. Benson, B. Varcoe, M. Weidinger, and H. Walther, Fortschr. Phys. **46**, 897 (1998).

[8] D. Leibfried, D. M. Meekhof, B. E. King, C. Monroe, W. M. Itano, and D. J. Wineland, Phys. Rev. Lett. **77**, 4281–4285 (1996).

[9] M. Weidinger, B.T.H. Varcoe, R. Heerlein, and H. Walther, Phys. Rev. Lett. **82**, 3795–3798 (1999)

[10] B. T. H. Varcoe, S. Brattke, B.-G. Englert, and H. Walther, Fortschr. Phys. **48**, 679 (2000).

[11] B. T. H. Varcoe, S. Brattke, and H. Walther, J. Opt. B. **2** 154 (2000).

[12] G. Nogues, A. Rauschenbeutel, S. Osnaghi, M. Brune, J. M. Raimond and S. Haroche, Nature (London) **400**, 239 (1999).

[13] P. Meystre, G. Rempe, and H. Walther, Opt. Lett. **13**, 1078–1080 (1988).

[14] J. Krause, M.O. Scully, and H. Walther, Phys. Rev. A. **36**, 4547–4550 (1987)

[15] E. Wehner, R. Seno, N. Sterpi, B.-G. Englert and H. Walther, Opt. Commun. **110**, 655 (1994).

[16] B. T. H. Varcoe, S. Brattke, M. Weidinger, and H. Walther, Nature (London) **403**, 743 (2000).

[17] S. Brattke, B.-G. Englert, B. T. H. Varcoe, and H. Walther, J. Mod. Opt. (in print) (2000).

ABRUPT MAGNETORESISTANCE JUMPS IN Ni-WIRE SYSTEMS AND COULOMB BLOCKADE UNDER ELASTIC ENVIRONMENT IN SINGLE JUNCTION/CARBON NANOTUBE SYSTEM

J. Haruyama, I. Takesue, S. Kato, K. Takazawa, and Y. Sato
Aoyama Gkuin University, Dept. Electrical Engineering and Electronics
6-16-1 Chitosedai, Setagaya, Tokyo 157-8572 Japan

Keywords: Ferromagnetic nanowire, magenetoresistance jump, macroscopic quantum tunneling, localization, Coulomb blockage, singe junction, external environment, energy transfer, Carbon nanotube

ABSTRACT

We report on: **1.** An abrupt magnetoresistance (MR) jump in a Ni-nanowire system and on; **2.** Coulomb blockade (CB) associated with localization effect in its external environment in a multi-walled Carbon Nanotube (MWNT)/single tunnel junction system. The observed MR hysteresis, wire diameter, and temperature dependences indicate a possibility that the MR jump originates from macroscopic quantum tunneling of a magnetic domain wall in the Ni-wire. Its correlation with phase coherence of electron waves, weak localization, is also discussed. Since one can pick up MQT as a large MR jump with keeping phase coherence in the system, it may be attractive for magnetic quantum computation. CB in single tunnel junctions is also observable even under non-dissipative high impedance external environment, localization, in MWNT system. Since discussion about it is associated with the spirit of Caldeira and Leggett, it may be meaningful for MQC. The CB is also very sensitive to phase modulation in its external environment.

1. INTRODUCTION

Macroscopic quantum coherence (MQC), coherent oscillation of macroscopic particle among multi-levels by quantum tunneling, is one of the key factors to realize quantum computation. It is observable in some material systems. The most promising one is superconductor systems, in which phase of electron wave is macroscopically conserved, with Josephon junction. There are so many works

427

reporting MQC and computation (MQC^2) related phenomena there. The other typical one is ferromagnetic material systems, in which phase of electron spin is macroscopically conserved. Although some works have reported macroscopic quantum tunneling (MQT) in magnetic material systems, the number of reports are relatively smaller compared with those in superconductor systems.

Here, in Section 3, we report on abrupt magnetoresistance (MR) jumps, which have a possibility of MQT of magnetic domain walls (DWs), in Ni-nanowires, and the correlation with phase coherence effect of electron waves in diffusive regime (i.e., localization effect). A typical measurement method for electron spin moment was a magnetization measurement by SQUID, whereas MR measurement has recently attracted much attention, because MR is more sensitive to spin flipping and, hence, one can pick up change of a small spin moment as a large electrical signal. There are many interesting phenomena observed using MR measurement and MR jump is a typical one. An interpretation for MR jump is MQT of magnetic DWs, spin transition layers including a large number of electrons, across a pinning potential barrier. If the MQT is magnetically controllable, it can be a candidate for quantum computation. It is also interestingly discussed whether the MQT of magnetic DW causes dephasing of electron waves in Ni-nanowires.

The other up coming candidate for MQC^2 system may be organic material systems. Carbon nanotubes, conducting molecular wires, are such typical examples. Single-walled Carbon nanotubes have exhibited a mean free path as large as micron order providing very unique phenomena. In contrast, multi-walled Carbon nanotubes (MWNTs), multiple graphene sheets coaxially rolled into seamless hollow cylinders, have been interpreted only by large scale phase interference effects of electron waves in diffusive regime (e.g., weak localization, Altshuler-Aronov-Spivak (AAS) effect, and universal conductance fluctuation). In Section 4 we report on Coulomb blockade (CB), a typical single electron tunneling phenomenon, associated with localization effects in MWNTs directly connected to single tunnel junctions.

It is of core importance for single junctions sytem to have an impedance of external electromagnetic environment higher than quantum resistance, h/e^2, to yield CB. Phase correlation theory, explaining a possibility of emergence of CB in single junction systems, implies that CB is caused by that tunneling electrons transfer the energy to such a high impedance environment. It basically replaced the external environment to a series of LC circuits (harmonic oscillators) with each energy quantum $\hbar \omega$ and, hence, the energy transfer is carried out by exciting this LC mode.This is just the spirit of Caldeira-Leggett, the important concept for MQC with energy dissipation in external enivronment. Although single electron tunneling itself is not MQT, discussion about this environmental effect will be meaningful also for MQC. Here, localization, yielding a high impedance environment and CB in our system, in MWNT is elastic without any energy dissipation. Whether CB is observable under such a condition and how it is associated with phase correlation theory are discussed. It is also shown that the CB is modulatable by AAS effect of the MWNT.

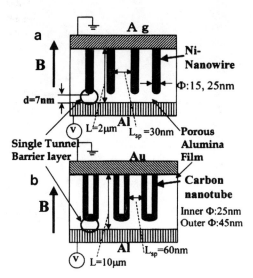

Figure 1. (a) Schematic cross section of Al/Al$_2$O$_3$/Ni-nanowires array. **(b)** An array of Al/Al$_2$O$_3$/MWNTs.

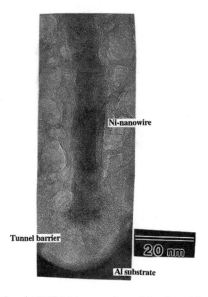

Figure 1. (c) Cross sectional HRTEM image of one Ni-wire with a tunnel barrier around the bottom.

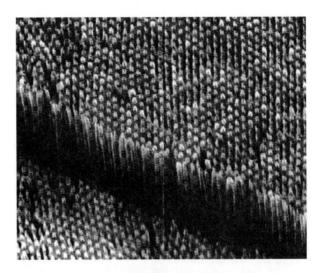

Figure 1. (d) SEM image of the exposed MWNT array, after the Al₂O₃ layer near sample surface was etched out.

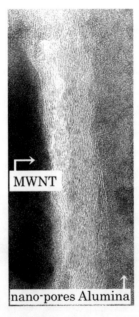

Figure 1. (e) Cross sectional HRTEM image of one-side shells of MWNT with about 26 shells and the mean radius of 17.5nm.

2. SAMPLE STRUCTURES: NANO-POROUS ALUMINA TEMPLATE

We have utilized a unique material, nano-porous Alumina film template, to fabricate nano-structures and reported some interesting mesoscopic phenomena.[1] Figure 1 shows the schematic cross sectional views of the samples. Nano-porous Alumina film has a nano-sized diameter pores located like honey comb on the top view. Since the nano-pores has a straight-line shape in perpendicular to Al substrate as shown in Figure 1 and any materials can be electrochemically deposited into the pores, one can easily attain a variety of nano-structures. Ni was deposited for Section 3 and Carbon nanotube was deposited by chemical vapor phase deposition for Section 4, leading to the arrays of Ni-wire/Al_2O_3/Al (Fig.1 (a)) and MWNT/Al_2O_3/Al (Fig. 1(b)). TEM and SEM images of the actual samples are shown in Fig. 1(c–e).

3. ABRUPT MAGNETORESISTANCE JUMPS AND WEAK LOCALIZATION IN Ni-NANOWIRE SYSTEMS

3.1. Background

Abrupt magnetoresistance (MR) jumps have been observed in some magnetic systems and some interpretations have been proposed for them. MQT of magnetic domain walls (DWs) is one of the attractive interpretations. Hong *et al.* reported on a MR jump and a stochastic behavior of critical magnetic fields for it in a Ni-wire.[2,3] The results were interpreted by macroscopic quantum tunneling (MQT) of a magnetic domain walls (DW), spin transition layers between two magnetic domains including huge number of electrons, through pinning potential barriers. Since DW abruptly escapes from the potential valley at a critical field applied, the total magnetization abruptly changes, leading to an abrupt MR jump. In principle, such a MQT (or MQC) has been theoretically and experimentally reported in superconductor systems with Josephson junction, leading to some successful works.[4,5] In contrast, the identification of MQT, however, is not yet clear in ferromagnetic materials, particularly in ferromagnetic nanowires.[6] Even Hong's experiments did not give direct evidence, leaving some ambiguous points.

In addition, the other interpretation for MR jump has been recently presented. It is a destruction of phase interference of electron waves in diffusive regime, weak localization, by motion of a DW proposed by Tatara and Fukuyama.[7] Weak localization is strongly coupled with electron spin in ferromagnetic materials. Hence, they calculated quantum correction of Boltzmann equation for self-energy type Cooperon and argued that spin flip scattering by DW lead to the destruction of weak localization and that the fluctuation caused by the motion of DW resulted in further destruction, yielding a MR jump. In this work we report on an abrupt MR jump and discuss its correlation with MQT of magnetic DWs and weak localization.

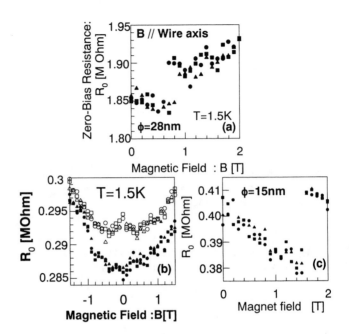

Figure 2. (a) Typical magenetoresistance (MR) jump observable in our Ni-nanowire array. (b) Hysteresis loop in the MR. (c) MR jump in a smaller wire diameter sample.

3.2. Abrupt MR Jump and MQT of Magnetic Domain Walls

Figure 2(a) shows a typical MR jump observable in our sample. It clearly exhibits a MR jump around B = 0.8 T. The MR also exhibits a hysteresis loop as shown in Figure 2 b. Since the presence of a hysteresis loop in ferromagnetic materials means that spin flippin can not follow the magnetic field reversed in some parts, it indicates the presence of geometry partrs with different spin orientations (i.e., magnetic domains). Hence, it also implies the presence of magnetic DWs, spin transition layers among the magnetic domains. Although multi-domains generally become difficult to exist as the wire diameter is reduced, this result suggests that at least two magnetic domains exist in one wire.

Figure 2c shows the MR jump in the sample with a smaller wire diameter. The MR jump shifts to a higher magnetic field with the larger jump amplitude. In contrast it disappears in the sample with the large wire diameter. These wire diameter dependences imply a possibility that the MR jump is strongly associated with MQT of a DW, in accordance with Hong's report. Based on it, a DW is pinned around the narrowest position of the wire, which has the lowest potential energy and can be a potential valley. In a decrease of wire diameter, the depth of potential valley can qualitatively become deeper. Consequently, higher critical magnetic field is required for a DW to escape from the valley byMQT. On the contrary the pinning is difficult to take pleace in the larger wire diameter wire beause of shallower potential valley and, hence, MQT dissappears.

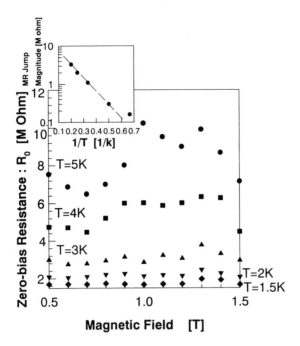

Figure 3. Temperature dependence of MR. **Inset:** Temperature dependence of the magnitude of MR jump, ΔR.

Figure 3 shows the temperature dependence of MR. The critical magnetic field around B=1.3T is mostly independent on the temperature up to 3K, whereas it is smeared by the other MR peak near 0.9T at 4K and 5K. The inset shows the temperature dependence of the magnitude of MR jump, ΔR. Above T=2 K it clearly exhibits a thermal activation type characteristic, $\Delta R \propto exp(-E/kT)$, which qualitatively agrees with classical regime of Clarke's result, (*i.e.,* $\Gamma=(\omega/2\pi)exp(-U/kT)$, where ω and U are the plasma frequency of a particle in the potential well and the depth of potential well, respectively), reporting MQT in Josephson junction system.[4] It explained Γ as the thermal escape rate of a particle from the quantum potential well. In our case this characteristic implies Γ of a magnetic DW from the pinning potential. U can be estimated to be about 0.6 meV from the slope of the inset here, considering the number of measured wires. In contrast, the linear relation is collapsed below about T=2 K, at which the characteristics of MR jump shown in Figure 2 was measured. These results support the possibility that the MR jump at T=1.5 K originates from quantum tunneling process of the DW though the pinning potential barrier, i.e., MQT of the DW.

3.3. Correlation of MQT with Weak Localization

Figure 4 shows the temperature dependence of magnetoconductance at

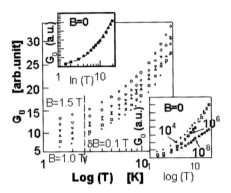

Figure 4. Logarithmic temperature dependence of magnetoconductance at each magnetic field in Figure 3 sample. **Inset:** Logarithmic magnetic field dependence of magnetoconductance in Figure 2(a) sample.

each magnetic field. They are qualitatively following the formula of two-dimensional (2D) weak localization with magnetic impurities and without spin-orbit interaction (*i.e.,* logarithmic behavior at high temperatures and its saturation at low temperatures), although the absolute value is changed depending on the magnetic field applied. The negative MR around B=0T shown in Fig. 2 also exhibits a logarithmic dependence as magnetoconductance on an increase of magnetic field as shown in the inset and it is also qualitatively consistent with weak localization. In addition, our recent analysis has revealed that the experimental result is in good agreement with the theory of 2D weak localization with magnetic impurities. This implies that phase coherence of electron waves is still survived, although it is slightly affected by spin flip scattering in Ni-wire itself.

Here, it is the most important to notice that this temperature dependence is independent of the abrupt MR jump. The trend does not basically change even before and after the MR jump. It implies that MR jump does not destroy weak localization. In the sense, this abrupt MR jump can not be interpreted by Tatara's theory. In contrast, logarithmic dependence on magnetic field looks to be destroyed by MR jump. Since, however, MR transits to the other regime (*i.e.,* decrease of MR by Lorentz force) around the MR jump, the correlation can not be confirmed only from it.

In conclusion, we reported on the abrupt MR jumps in Ni-nanowires. The observed MR hysteresis loop, the wire-diameter and temperature dependences of MR jump implied a possibility that the jump was attributed to MQT of magnetic DWs. We also found that the MQT did not cause dephasing of electron waves, weak localization, in the Ni-wire. These results imply if the MQT of magnetic

DWs is controllable by magnetic field in time domain, it will lead to a kind of quantum computer, because one can pick up the MQT as a large MR change with two coherent electrical levels. Our recent experiments (e.g., the influence of nano-tunnel barrier attached to the end of Ni-wire to the MR jump and detailed temperature dependence of jump amplitude), are reconfirming the presence of this MQT. On the contrary, there still remained some unsolved problems. Much further study is required in order to fully understand the phenomena.

4. COULOMB BLOCKADE ASSOCIATED WITH LOCALIZATION IN A SINGLE JUNCTION/CARBON NANOTUBE SYSTEM

4.1. Background

Coulomb blockade (CB), a typical phenomenon associated with single electron tunneling, has been successfully observed in a variety of mesoscopic systems. Although the correlation with phase coherence of electron waves and spins, in the external electromagnetic environment (*EME*), has recently attracted much attention,[8] no work has reported it in single-junction systems. It is because correlation between CB and its external environment in single junction systems has characteristics quite different from those in multi-junctions systems. It has been well known as phase correlation (PC) theory.[9,10]

In general, phase correlation theory implies that the external electromagnetic environment plays the following two key roles, **1.** The tunneling electron transfers its energy to the external environment by exciting environmental modes, yielding Coulomb blockade. **2.** Fluctuation of the external environment fluctuates the junction surface charges (e.g., through a commutation relation between phase fluctuation $\tilde{\varphi}$ and charge fluctuation $\tilde{Q}$ on junction surface (i.e., $\tilde{\varphi}$, $\tilde{Q}$]=ie)), smearing out Coulomb blockade. To realize the first role and avoid the second, the real part of the total impedance of the external environment ($\mathrm{Re}[Z_t(\omega)]$) must be much larger than resistance quantum ($R_Q = h/e^2 \sim 25.8\mathrm{K}\Omega$). This is the key factor in phase correlation theory. A part of $Z_t(\omega)$ also must be closely located to a single junction to avoid the second role as a high impedance transmission line R_L ($\gg R_Q$). A part of phase correlation theory has been also experimentally confirmed.[11-13]

Here, we investigate the correlation of Coulomb blockade in a single junction with localization effects, phase interference effects of electron waves in diffusive regime, in its external environment, based on phase correlation theory. We ask the question, **"Can localization play two such roles as a high $\mathrm{Re}[Z_t(\omega)]$ in PC theory and yield Coulomb blockade?"** In PC theory, external environment is basically replaced by circuit model like a set of *LC* circuits (i.e., a set of harmonic oscillators with energy quanta $\hbar\omega$) or *RC* circuit. Energy transfer is performed by exciting this *LC* (*RC*) mode. In particular, when a single junction is coupled only with a resistive wire (an Ohmic resistor) described by the frequency independent impedance $Z(\omega)=R$, the system is represented by RC circuit. In this case, charging energy is transferred by exciting RC mode, which may directly correspond to electron-phonon scattering in the Ohmic resistance in realistic systems because RC mode basically has no fluctuation factor, unlike LC mode. In contrast, localization can also yield high impedance ($> R_Q$) at low temperature.[14] Since, however, localization effects are elastic processes any energy transfer can not occur there, unlike electron-phonon scattering in RC mode. Therefore, if one follows PC theory (at least from the viewpoint of the role of energy transfer), no Coulomb blockade may be observed under a high impedance external environment caused by localization. We try to examine this hypothesis by directly connecting multi-walled Carbons (MWNT) with localization to single tunnel junctions.

Of course, Landauer formula assumes the energy dissipation in electrodes connected to mesoscopic regime, whereas high impedance environment as described by PC theory does not generally exist in such Ohmic metal electrodes. This is animportant difference for the interpretation of the results. Although Coulomb oscillation also has been already reported in single-walled Carbon nanotube systems with multi-tunnel junctions,[15] it should be emphasized that this work has a quite different meaning from those.

4.2. Coulomb Blockade in Single Junction System

Figure 5(a) clearly exhibits a zero-bias conductance (G_0) anomaly. The shape of G_0 anomaly is drastically varied near T=5K and the shape at 2K is quite different from the inset.[1] It is evident that the nano-materials connected to single tunnel junctions strongly contribute to the G_0 anomaly. The G_0 anomaly in Fig. 5(a) can be well fit by phase correlation (PC) theory as shown in Fig. 5(b) and as explained in the next paragraph, whereas that in the inset of Fig. 5(a) can not be directly fit by PC theory. Its first derivative (*i.e.*, dG/dV vs. V curve) is fit by Nazarov's theory introducing a mutual Coulomb interaction to the external environment of Coulomb blockade[1] as shown in the inset of Fig. 5(b). This large difference is also evidence that Coulomb blockade in the single junction system can be much influenced by mesoscopic phenomena in the nanowires.

Figure 5(c) is distinguished by the following three temperature regions, **1. Above 10K**: linear G_0 vs. log(T) relation, **2. 5K-10K**: its saturation region, **3. Below 5K**: linear G_0 vs. temperature relation (see the upper inset). This linear G_0 vs. temperature relation is also observable in the lower inset. We have argued that it provides strong qualitative and quantitative evidence of Coulomb blockade in an array of single junctions located in parallel (*i.e.*, as a temperature dependence of averaged G_0).[1] The temperature of 5K also agrees with that at which the shape of the G_0 anomaly starts to change in Fig. 5(a), suggesting that the G_0 anomaly below 5K originates from Coulomb blockade.

In order to confirm Coulomb blockade in Al/Al$_2$O$_3$/MWNTs array and clarify its connection with PC theory, we first numerically calculate a G vs. V curve, normalized by tunneling resistance (R_t) and the number of junctions, using eqs.(1–3) from PC theory. We then fit the G vs. V curve measured at 2K in the Coulomb blockade regime in Fig. 5(a) by the calculation result. Here, we know the R_t of 300K Ω ($\gg R_Q$) from the measurement and also employed a lumped *RC* coupling mode as a simplest case, because the inductance (*L*) related characteristics of Carbon nanotube is not yet clarified. Hence, the fitting parameters are the resistance of external environment (R_{ext}) and junction capacitance (*C*) as R_Q/R_{ext} and $\hbar \omega_{RC}/kT$, where $\omega_{RC} = 1/(R_{ext} C)$.

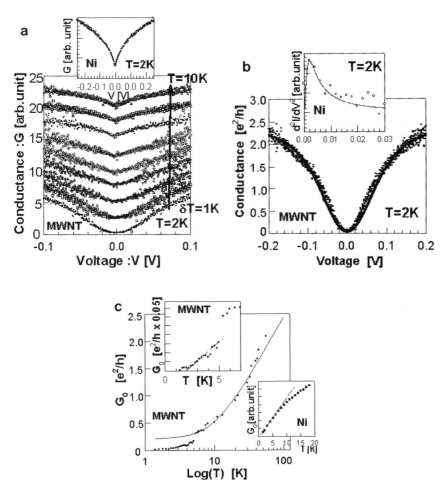

Figure 5. **(a) Al/Al$_2$O$_3$/MWNT array;** Temperature dependence of a typical G_0 anomaly. **Inset: Al/Al$_2$O$_3$/Ni-nanowires array.** **(b) MWNT system;** Data fitting to G vs. V curve of (a) by phase correlation theory. The solid line was numerically calculated from eqs.(1–3). **Inset: Ni-nanowire system;** Data fitting to dG/dV vs. V curve of the inset of (a) by Nazarov's theory. **(c) MWNT system;** Temperature dependence of G_0 shown in Fig.5(a). The solid line is the result calculated by 2D weak localization formula of MWNT. **Upper inset:** Linear G_0 vs. temperature relation below 5K. **Lower inset: Ni-nanowire system.**

$$I(V) = \frac{1 - e^{-\beta eV}}{eR_t} \int_{-\infty}^{+\infty} dE \frac{E}{1 - e^{-\beta E}} P(eV - E) \quad (1)$$

$$P(E) = \frac{1}{2\pi\hbar} \int_{-\infty}^{\infty} dt e^{J(t) + i\frac{E}{\hbar}t} \quad (2)$$

$$J(t) = 2 \int_{-\infty}^{\infty} \frac{d\omega}{\omega} \frac{Re[Z_t(\omega)]}{R_Q} \frac{e^{-i\omega t}}{1 - e^{-\beta\hbar\omega}} \quad (3)$$

where β is the $1/kT$ and $Z_t(\omega)=1/[i\omega C+Z(\omega)^{-1}]$ is the total EME impedance consisting of junction capacitance C in parallel with an external environment impedance $Z(\omega)=R_{ext}$ in a *RC* circuit model. $J(t)$, $P(E)$, and $I(V)$ are phase correlation function, Fourier transform of $J(t)$, and the tunnel current, respectively. These three equations well represent the general argument of PC theory mentioned in the introduction part. Since here is no space, it is explained other somewhere.[1,9]

As shown by the solid line in Fig. 5(b), the measurement and calculation results are in excellent agreement in our weak tunneling case (*i.e.*, R_t of 300KΩ>R_Q). The best fitting gives the R_{ext} of 450KΩ and the $Re[Z_t(\omega)]$ with the same order value as R_{ext}. This R_{ext} of 450KΩ mainly corresponds to the resistance of MWNT (R_{NT}) in our system, because only MWNT was directly connected to the single junction and the resistance of the Gold contact layer with the Gold/MWNT interface, which may provide energy dissipation based on Landauer formula, was at most on the order of 100Ω. In general, the interface between Carbon nanotube and metal contact layer is highly resistive, whereas our low interface resistance originates from high diffusion between Gold particles and Carbon by high temperature annealing. Consequently, we can not find the resistance of 450KΩ in any parts of our system other than the MWNT. The value of 450KΩ as the resistance of MWNT is also in good agreement with that in a previous report.[16] In addition, note that this 450KΩ actually yields $Re[Z_t(\omega)]$ larger than R_Q. Therefore, we conclude that our Coulomb blockade is consistent with phase correlation theory, and that the MWNT acts as a high impedance external electromagnetic environment.

4.3. Localization Effect in Carbon Nanotube

The origin of this high impedance of MWNT is the key point for this work. As we expected, it can be qualitatively understood as a result of weak localization (WL) from the curve fitting shown in Fig.5(c). As shown by the solid line, the G_0 vs. temperature characteristic is in nice agreement with the following formula of two dimensional (2D) WL of MWNT,[16] except for the Coulomb blockade temperature region.

$$G(T) = G(0) + \frac{e^2}{2\pi^2\hbar} \frac{n\pi d}{L} \ln[1 + (\frac{T}{T_c(B,\tau_s)})^p] \quad (4)$$

where n, d, L, and τ_s are the number of shells, the diameter of the inner shell of the MWNT, the length of the MWNT, and the relaxation time of spin flip scattering, respectively. The contribution of the number of MWNTs and R_t were taken into account by the term of G(0), whereas they are not considered in the term of temperature dependence, because we simply assumed that the measurement result was the superposition of the characteristics of 10^6 MNWTs which independently work due to the enough spacing. The best fitting gives n=18, p=2.1, and T_c=10K. Here, a main difference from the past reports[16] is the T_c as high as 10K. This high T_c is understandable by the presence of magnetic impurities (Cobalt), which was deposited into the nano-pores as a reactant for MWNT growth, in the MWNT.

It can not be confirmed whether this localization effect in the MWNT is weak localization or anti-localization (i.e., influence of spin-orbit interaction) because CB prevents us from the observation at temperatures below T=5K. Figure 6, however, as shown in the latter section exhibits emergence of negative magnetoconductance at T=2K, in a increase of the magnetic field (B) from B=0T. This implies the presence of anti-localization, destructive phase interference of electron waves. Our recent experiments reveal that this id due to diffusion of Gold atoms, deposited on the all top of MWNTs as a contact layer, into the MWNT by high temperature annealing.[18] Since Gold is an atom much heavier than Carbon its diffusion into the MWNT can lead to spin flipping in electron trajectories for phase interference by strong spin-orbit interaction and, hence, anti-localization.

Based on these confirmations (*i.e.*, **1.** Coulomb blockade following PC theory with near 450KΩ as the external impedance, **2.** The external resistance of 450KΩ can exist only in the MWNT in our system, **3.** The highly resistive MWNT is strongly associated with localization), we finally conclude the possibility that the localization in the multi-walled Carbon nanotube yields the real part of the total impedance (Re[$Z_t(\omega)$]) higher than R_Q by coupling a junction capacitance C as a RC mode and contributes to Coulomb blockade.

This conclusion appear to contradict our hypothesis mentioned in the introduction and strange. However, it may be understood from the following interpretation.

4.4. Interpretation

Unless the tunneling electrons can not transfer its energy to the environment, tunneling event is allowed and, hence, Coulomb blockade can not be caused in any cases, if one follows PC theory (*i.e.*, eqs. (1–3)). This is of core importance also for MQC (MQT) following the spirit of Caldeira and Leggett. MQT is much smeared by energy dissipation in its external environment. There may exist, however, the following two exceptions:[9] **1.** Debye-Waller factor *exp(-ρ)* in Mößbauer effect in *LC* mode, **2.** Infrared divergence in *RC* mode at finite temperature.

The latter comes from the analogy between the motion of free Brownian particle and the *RC* environmental effect with the impedance $Z_t(\omega)=R$ and is relevant for the *RC* environment with the impedance much larger than R_Q. With $Z_t(\omega)=R$, $Re[Z_t(\omega)]$ is given by $R/(1+(\omega RC)^2)$ and then $(\pi/C)\delta(\omega)$ for very large *R* at finite temperature. Hence, energy transfer is carried out only around $\omega=0$ in accordance with $P(E)=\delta(E - E_c)$. This is so called infrared divergence. In this case, only thermal fluctuation yields a time evolution of phase J(t) (i.e., diffuse the free Brownian particles), leading to $J(t) \propto t$ for long time, because *RC* mode at very low frequency has basically no time-fluctuated electromagnetic factor. Therefore, the tunneling electrons can have a chance to transfer the charging energy E_c by exciting *RC* mode through thermal fluctuation even in the non-dissipative electromagnetic environment, particularly in the environment sensitive to electron phase fluctuation like localization. This new interpretation means a possibility that thermal environments is taken into account as P(E) in eq.(1), if it is coupled with the electromagnetic environment. Our result may be qualitatively interpreted by this case at least from the following three points: **(1)** $Re[Z_t(\omega)]$ of about $450K\Omega$ is much larger than R_Q; **(2)** Resistance of MWNT is basically frequency independent; and **(3)** the measurement temperature is 2K.

Otherwise, P(E) in equations 1 and 2 may have to be reinterpreted as "the other probability", not associated with the energy transfer probability of the tunneling electrons to the EME. Here P(E) is Fourier transform of $J(T)=<[\tilde{\varphi}(t)-\tilde{\varphi}(0)]\tilde{\varphi}(t)>$, a time evolution of phase fluctuation $\tilde{\varphi}$ in the EME, and the origin of the phase was defined as $\tilde{\varphi}(t)=e/\hbar \int dt\mathbf{V(t)}$, where $V(t)=Q(t)/C_j$ is the voltage across the tunnel junction. Phase interference effect in localization also originates from this definition. Hence, J(t) is a time evolution of $\tilde{\varphi}$ but should be attached to localization effect so as not to destroy phase coherence in the MWNT. In the sense, P(E) may be reinterpreted as a transmission probability of electrons, associated with J(t) in the localization regime, in the MWNT. This

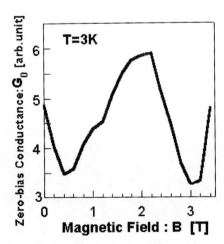

Figure 6. G_0 vs. magnetic field at Coulomb blockade temperature regime. Magnetic field was aligned in parallel to the tube axis.

does not deny the energy dissipation in external environment. In this case, energy transfer may be taken place in the contact metal electrodes with lower impedances.

In addition, we may have to perform more careful data fitting from the following points. 1. Junction capacitance C: We used C obtained in our past report.[1] Since the C was estimated from data fitting by Nazarov's theory, it is not yet experimentally confirmed. 2. Parasitic capacitance C_p: We did not take into consideration the influence of the C_p of MWNT. When we define L=$\tau \times c$ (where $\tau \sim h/eV$, c is the velocity of light in vacuum) as the geometry for an effective C_p based on Horizontal model and included the C_p in the data fitting, it does not exhibit perfect agreement. C_p for better agreement should be smaller than that calculated from L=$\tau \times c$. To explain this difference, a smaller velocity instead of "c" may have to be employed, because our MWNT has very disordered surface. 3. LCR model: We also have employed RC mode as a lumped circuit model in PC theory here. Since MWT has distributed L, R, and C including this C_p in the actual system, *LCR* transmission line model will have to be introduced. However, even if apart from these data fitting problems, the linear G_0 vs. temperature dependence, the high impedance EME only in the MWNT directly connected to the junction, and its dependence on the localization effect will support our conclusion.

4.5. Modulation of Coulomb Blockade by External Phase Interference: AAS Effect in MWNT

Here, we try to control this CB by modulating phase interference of MWNTs. As shown in Figure 6, the G_0 vs. magnetic field (B) relation exhibits periodic oscillation. Such oscillation in MWNT has been understood as Altshuler-Aronov-Spivak (AAS) effect in a graphite cylinder,[17] which originates from phase interference of the electron waves encircling the cylinder in opposite directions and modulated by magnetic flux enclosed, with an oscillation period $\Delta B=(h/2e)/(\pi r^2)$ where r is the radius of cylinder. Here, the radius can be estimated (B) relation to be 17.3 nm from the first oscillation peak. This radius is in excellent agreement with the mean radius of our actual MWNT of 17.5 nm. Our recent experiment[18] is further reconfirming the relevance of this diameter dependence. It is a straightforward indication that this conductance oscillation is strongly associated with AAS effect. In addition, it also clarified whether the phase interference is constructive or destructive (i.e., weak localization anti-localization) in AAS oscillation strongly depends on the diffusion of the atoms of the top contact materials. This result supports that our Coulomb blockade is very sensitive to phase interference in the MWNT, because Coulomb blockade is basically independent of applied magnetic field. Phase coherence in the system is the key factor for MQC. It should be conserved in a long time range. This result may indicate that MQT is controllable by modulating the phase in its external environment at high frequency. It may be useful for quantum computation.

5. CONCLUSION

We reported on (1) the abrupt MR jump, related to MQT of DWs, and its correlation with weak localization, in the Ni-nanowire system and on (2) Coulomb blockade depending on localization, elastic phenomenon, in its external environment and its association with the spirit of Caldeira and Leggett, in the single junction Carbon nanotube system. Both phenomena provided possibilities for magnetic and phase modulatable quantum computation.

Acknowledgment

We sincerely thank M. Buttiker, M. Ueda, B. L. Altshuler, Y. Imry, X.H. Wang, Y. Yamamoto, and W. Oliver for very useful discussions and suggestions, J.M. Xu group for sample preparation and SEM image. This work was financially supported by the MST, and the scientific research project both on the basic study B and on the priority area of Japanese Ministry of Education, Science, Sport, and Culture.

REFERENCES

1)   J. Haruyama, K. Hijioka, *et al.*, Phys.Rev.B, 62, 8420 (2000-II); D. Davydov, J. Haruyama, et al., Phys.Rev.B 57, 13550 (1998); J. Haruyama, Y. Sato, *et al.*, Appl.Phys.Lett. 76, 1698 (2000); J. Haruyama, I. Takesue, Y. Sato, *et al.*, Phys.Rev.B 63 (2001) and Appl.Phys.Lett. 77, 2891 (2000); J. Haruyama, I. Takesue, *et al.*, in "Quantum Mesoscopic Phenomena and Mesoscopic Devices in Microelectronics" edited by I. Kulik, 145 NATO ASI series (Plenum 2000).
2)   K. Hong and N.Giordano, J.Mag.Mag.Mat., 151, 396 (1995).
3)   K.Hong and N.Giordano, in "Quantum Tunneling of Magnetization" edited by L. Gunther *et al.*, NATO ASI E-301, 257 (1995).
4)   J. Clarke, A.N. Cleland, M.H. Devoret, D. Estive *et al.*, Science 239, 992 (1988); Y. Nakamura, J.S. Tsai, *et al.*, Nature 398, 786 (1999).
5)   A.J. Leggett *et al.*, Rev.Mod.Phys. 59, 1 (1987); A.O. Caldeira and A.J. Leggett, Phys.Rev.Lett. 46, 211 (1981).
6)   A. Garg, Phys.Rev.Lett. 74, 1458 (1995); "Quantum Tunneling of Magnetization" edited by L. Gunther *et al.*, NATO ASI E-301 (1995).
7)   G. Tatara and H. Fukuyama, Phys.Rev.Lett. 78, 3773 (1997).
8)   A. Yacoby, M. Heiblum, D. Mahalu, and H. Shtrikman, Phys.Rev.Lett.74, 4047 (1995).
9)   G.-L. Ingold, Y.V. Nazarov, in "Single Charge Tunneling" edited by H. Grabert and M.H. Devoret, NATO ASI Series B-294, 21 (1991).
10)   M.H. Devoret, D. Estive, *et al.*, Phys.Rev.Lett. 64, 1824 (1990).
11)   A.N. Cleland, J.M. Schmidt, *et al.*, Phys.Rev.Lett. 64, 1565 (1990).
12)   S.H. Farhangfar, J.P. Pekola, *et al.*, Europhys.Lett. 43, 59 (1998); T. Holst, D. Esteve, C. Urbina and M.H. Devoret *et al.*, Phys.Rev.Lett. 73, 3455 (1994).
13)   X.H. Wang and K.A. Chao, Phys.Rev.B 56, 12404 (1997-I) and 59, 13094 (1999-II); P. Joyez, D. Esteve, and M.H. Devoret, Phys.Rev.Lett. 80, 1956 (1998).
14)   Y. Imry in "Introduction to Mesoscopic Physics", Oxford University Press (1997).
15)   S.J. Tans, M.H. Devoret, C. Dekker *et al.*, Nature 386, 474 (1997) and 394, 761 (1998).
16)   L. Langer, V. Bayot, *et al.*, Phys.Rev.Lett. 76, 479 (1996).
17)   A. Bachtold, C. Strunk, *et al.*, Nature 397, 673 (1999).
18)   J. Haruyama, I. Takesue, and T. Hasegawa, submitted to Nature.

Spontaneous generation of flux in a Josephson junction loop

Raz Carmi, Emil Polturak, and Gad Koren

Physics Department, Technion-Israel Institute of Technology,

Haifa 32000, Israel

Abstract

We report experimental measurements of flux generated spontaneously inside a $YBa_2Cu_3O_{7-\delta}$ superconducting loop made of 214 Josephson junctions connected in series. The flux is generated spontaneously during cooldown from the normal into the superconducting state. Stable random values of the flux measured at the end of each cooldown follow a normal distribution. The data are consistent with the total phase difference around the loop being a sum of the instantaneous values of the fluctuating phase differences across each junction. This agreement is conditional upon the assumption that the phase can be unambiguously defined relative to a phase standard. Our result indicates that topologically equivalent circuits will always contain random flux in their initial state.

The experiment described here is part of an ongoing search for formation of topological defects in a superconductor as it is cooled through T_c [1, 2]. Such effects are expected in condensed matter systems on general grounds, due to a departure from thermal equilibrium near T_c, where the relaxation time diverges [3, 4, 5]. The experiment was designed to test the very basis of the hypothesis, that of the connection of superconducting segments having random phases. In our setup we use a loop made of 214 Josephson junctions in series. As the loop is cooled through T_c , separate superconducting sections are created while the loop as a whole has no phase coherence. At lower temperatures, the junctions become phase locked and phase coherence around the loop is established. The remarkable result is that a spontaneously established non-zero phase difference can exist around the loop in its final state, in contrast to what one would expect from the ground state of the system in zero field. Our result is relevant to the way one defines the phase of an isolated superconductor [6]. Additionally, we expect this work to apply to any circuit made of Josephson junctions with a similar topology. Details of the experiment are given in Ref. [1]. Here, we focus the discussion on the meaning of the phase of the order parameter of an isolated system.

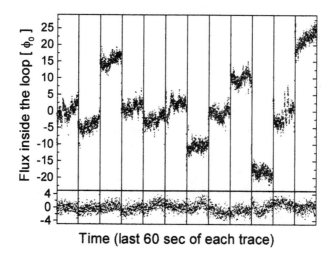

Figure 1: Top part shows a typical sequence of stable values of the spontaneous flux in the loop measured by the SQUID during several consecutive cooling cycles. The bottom part shows identical reference measurements with a blank substrate.

In the experiment we used a $YBa_2Cu_3O_{7-\delta}$ loop, interrupted by 214 grain-boundary Josephson junctions. The rectangular loop is patterned from a 70 nm epitaxial c-axis oriented YBCO film, grown on a symmetric $24°$-$SrTiO_3$ bi-crystal substrate [7, 8]. The pattern is a 20 μm wide meander line. Each time this line crosses the grain-boundary, another Josephson junction is added. We use a high-T_c SQUID adjacent to the loop to perform a continuous measurement of the flux through the loop while cooling the sample from 100 K to 77 K. At $T_c(= 90$ K$)$, segments of the film separating the junctions become superconducting. However, the junctions are still normal and so the superconducting segments are effectively separate. As the temperature decreases, the critical currents of the junctions increase, and eventually the loop becomes coherent. At the lowest temperature during each cooldown, we record some constant value of the flux, which however changes randomly from one cooldown to the next. A series of such measurements is shown in Fig.1. In Fig.2 we show the flux measured by the SQUID during two complete heating-cooling cycles. At $t \sim 5$ sec. the sample goes through T_c, and reaches 77 K after about 70 sec. The flux level becomes stable after some large oscillations. We believe that these random flux jumps, seen down to a few K below T_c, are due to large thermal fluctuations of the phase difference across the Josephson junctions. We found that the distribution of the stable flux level values, obtained during 166 cooldowns follows a normal distribution around zero with a standard deviation of $(7.4 \pm 0.7)\phi_0$.

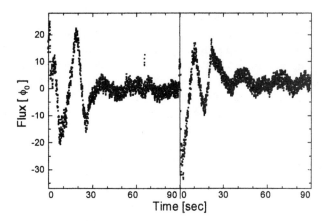

Figure 2: An example of flux measurements in the Josephson junction loop, during two heating-cooling cycles. The heating takes about 2 sec at the beginning of each measurement. The flux in the loop reaches a stable level a few K below T_c.

In Ref. [1] we show that the mechanism by which the loop becomes coherent is a non equilibrium process, and it can be considered as instantaneous. In addition, we show that the spontaneous flux can not be created by thermal fluctuations in equilibrium. The time scale for a flux quantum to pass through a junction, in our case, is $\tau_J \approx 10^{-11}$ sec at the moment the loop becomes finally closed. The time takes the order parameter to adjust across the whole loop in this range of temperature is $\tau_{op} = (1 \text{ cm}) \times \tau/\xi \approx 10^{-5}$ sec. The adjustment of the order parameter around the loop is the dominant delaying process for the loop to reach equilibrium. Comparing the two relevant time scales, we get $\tau_J \ll \tau_{op}$, meaning that the final connection of the superconducting segments into a coherent loop can be considered as an instantaneous nonequilibrium process[3].

In the experiment, we see trapped flux up to $\sim 30\phi_0$. The energy needed to support such a flux in the loop is of the order of $90k_BT$, which means that this large value of the flux would be unlikely to appear as a result of fluctuations at thermal equilibrium. Two other possible scenarios have even smaller probabilities: (a) the loop becomes coherent via sequential locking of more and more junctions; and (b) thermally activated flux trapping inside the loop via single flux quanta passing through individual junctions during an equilibrium cooling process. Both these processes should leave no flux at the end of the cooldown since the loop will decay into the ground state. The spontaneous flux may be created only via a non equilibrium mechanism, while the required energy comes from the coupling energy of the junctions during the closure of the loop.

We propose that flux reflects the total phase difference around the loop resulting from the instantaneous thermal fluctuations of the junctions at the time the loop becomes coherent. In our experiment the physical size of the superconducting segments is constant ($\simeq 60 \ \mu$m) and each segment reaches internal equilibrium long before the junctions. During cooldown, the segments between adjacent junctions become superconducting at T_c while the maximum Josephson current I_c, is still small. At this stage, the coupling energy of the junctions, $E_J \equiv I_c \phi_0 / \pi$ is much smaller than $k_B T$. The various segments are effectively uncoupled and the loop is incoherent. Thermal fluctuations of individual junctions are described by a probability to have a random phase difference δ_i given by $P(\delta_i) \sim \exp[-I_c \phi_0 / 2\pi k_B T \times (1 - \cos(\delta_i))]$ where i refers to the i-th junction [7, 9]. Large fluctuations changing δ_i by $\geq 2\pi$ are dominant down to a few kelvin degrees below T_c. Any time such fluctuation occurs, the whole loop becomes incoherent, meaning that the topological charge has changed. As the temperature decreases, I_c increases and with it the coupling energy. Individual fluctuations still exist, but δ_i is less likely to jump by 2π. At this stage, the loop becomes coherent and a supercurrent can flow around the loop. The instantaneous flux Φ is given by the sum of all the phase differences around the loop $\Phi \simeq \phi_0 \sum_i \delta_i / 2\pi$. While the loop is closed, small thermal fluctuations still exist but now the topological charge of the loop is fixed. It is important to notice that the topological charge can change only if at least one junction becomes open (via a 2π fluctuation). In Fig.3 we show an illustration of the phase winding around the loop. The black line represents the phase of the order parameter. Different points around the solid part of the torus (the gray ring in the figure) represent different values of the phase. For example, if the black line is wrapped around the torus only once, it is represents a cumulative phase of 2π. We show three different cases: (a) In a state of strong fluctuations (near T_c) , the phase is "discontinuous" in some places and the winding number is not well defined. (b) In the true ground state, which can be reached by a cooldown process while in equilibrium, the final winding number is zero and no flux is present. (c) The loop is closed via a non-equilibrium process (instantaneous connection) and is trapped with some non-zero fixed winding number, and hence, spontaneous flux. At this stage the winding number can not be changed by small fluctuations, which can "shake" the black line without unwinding it.

In general, the assignment of a phase to the order parameter in each segment is problematic because an absolute phase in an isolated system is not a well defined quantity [6]. Our analysis deals only with phase differences in the system. However, even in this case there is an additional degree of freedom which allows an arbitrary value of the flux in the loop due to the local gauge symmetry, as discussed below. This issue is especially important in our case as the phase differences are not constrained by a minimization of the energy of the junction, since $E \propto \cos \delta$. The problem is resolved if we adopt an approach similar to the "phase standard" [6, 11] which proposes a consistent method to assign a phase to the order parameter. In a Josephson junction with a well defined energy and supercurrent the phase difference in a gauge invariant form is written as: $\delta = \delta^o - 2\pi/\phi_0 \times \int_a^b \mathbf{A} \cdot d\mathbf{l}$ where δ^o is the phase difference of the

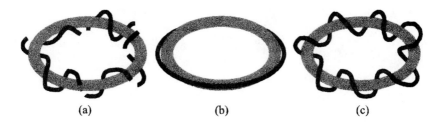

(a)   (b)   (c)

Figure 3: Illustration of the phase winding of the order parameter (black line). (a) Strong thermal fluctuations. (b) A global equilibrium ground state. (c) Final state containg topological defect having a large winding number, after a non-equilibrium connection.

order parameter across the junction. The value of $\delta^o \equiv \varphi_a - \varphi_b$ depends on the gauge choice of $\mathbf{A}$, and the integral path is taken directly through the junction. Summing around a closed loop gives the relation $\sum_{loop} \delta_i = 2\pi q + 2\pi\Phi/\phi_0$ [10]. The arbitrary integer q comes from the condition $\oint \nabla\varphi \cdot d\mathbf{l} = 2\pi q$, even if we keep δ_i within the interval -2π to 2π. In its coherent state, the maximal phase difference δ across the junction is $\leq |2\pi|$ because otherwise a flux quantum is added or taken out of the loop. Thus, in this case the flux in the loop is not completely fixed by the initial phase differences. However, if we assign a priori a definite phase φ_i to each segment relative to a single phase standard and assume $\delta^o_{ab} = \delta^o_{as} - \delta^o_{bs}$ (s is the standard, a and b are two segments) we *always* get $\oint \nabla\varphi \cdot d\mathbf{l} = \sum_{loop}(\delta^o)_i = \delta^o_{as} - \delta^o_{bs} + \delta^o_{bs} - \delta^o_{cs} + \delta^o_{cs} \cdots - \delta^o_{as} = 0$. The flux is now determined unambiguously by $\sum_{loop} \delta_i = 2\pi\Phi/\phi_0$. The fact that our experimental results are consistent with the last equation, without an additional arbitrary phase winding (meaning that $q = 0$) indicates that the phase standard is a meaningful and necessary concept. It is argued [6, 11] that the phase standard approach breaks down if one has completely separate segments which however are in thermal equilibrium with the environment. This is not a crucial problem in our case because some degree of coupling between the junctions exists as soon as E_J become finite.

   In conclusion, the present experiment tests the basic idea of "phase connection" leading to a spontaneous flux. Experimental results support this idea, and at the same time show that the concept of a "phase standard" is essential in order to understand our result. The fact that a random flux appears in any topologically equivalent circuit containing Josephson junctions (no matter how slowly it is cooled) may be relevant to the way the initial state of superconducting logic circuits is defined.

References

[1] R. Carmi, E. Polturak, and G. Koren, Phys. Rev. Lett. **84**, 4966 (2000).

[2] R. Carmi and E. Polturak, Phys. Rev. B **60**, 7595 (1999).

[3] W.H. Zurek, Nature (London) **317**, 505 (1985); W.H. Zurek, Phys. Rep. **276**, 177 (1996).

[4] T.W.B. Kibble, J. Phys. A **9**, 1387 (1976); T.W.B. Kibble and A. Vilenkin, Phys. Rev. D. **52**, 679 (1995).

[5] A. J. Leggett, Physica Scripta **T76**, 199 (1998).

[6] A. J. Leggett and F. Sols, Found. Phys. **21**, 353 (1991); A. J. Leggett, Found. Phys. **25**, 113 (1995).

[7] R. Gross *et al.*, Phys. Rev. Lett. **64**, 228 (1990).

[8] D. Dimos *et al.*, Phys. Rev. B. **41**, 4038 (1990).

[9] S. Schuster *et al.*, Phys. Rev. B. **R48**, 16172 (1993).

[10] A. Barone and G. Paterno, *Physics and applications of the Josephson Effect.*, John Wiley & Sons, Inc (1982).

[11] J. A. Dunningham and K. Burnett, Phys. Rev. Lett. **82**, 3729 (1999).

SOLITON QUANTUM BIT

Noriyuki Hatakenaka and Hideaki Takayanagi
NTT Basic Research Laboratories

Keywords: soliton, qubit

1.   INTRODUCTION

Quantum computers promise extremely fast computation through massive parallelism based on the quantum-mechanical superposition principle. Only two quantum logic gates, a quantum controlled-NOT gate and a rotation gate, are sufficient for any arbitrary quantum computation [1]. The elementary unit of quantum logic gates is a quantum bit, or qubit, which is produced by the quantum-mechanical superposition of two states.

So far qubits have been implemented in several quantum systems, such as cavity quantum electrodynamic systems [2], ion traps [3], and nuclear spins of large numbers of identical molecules [4]. In these systems, quantum coherence is high, but it seems difficult to realize a large number of interacting qubits. In contrast, solid state circuits are adequate for large-scale integration by recent development in nanofabrication. Along this line, future possible implementations of qubits with semiconductors include the spins of individual donor atoms in silicon [5] and spin states in quantum dots [6]. Since all the states of electrons are the same in superconducting materials, a superconducting element can maintain the superposition state for long periods. Thus, superconducting systems seem to have a great advantage for quantum computers. It has been suggested that qubits for quantum computing based on superconducting elements could be created by using 0-π states in d-wave superconductors [9], charge states in a Cooper-pair box [7, 8] and flux states in superconducting quantum interferometer devices [10]. Recently, coherent charge oscillations in a Cooper-pair box have been observed [11]. However, these existing qubits strongly depends on nanometer-scale fabrication and control techniques.

In this paper, we propose a novel scheme in which a qubit, *a self-formed qubit*, is created by using the quantum nature of a particle captured within the interacting soliton potential. Our qubit does not rely on any nanofabrication techniques. By this scheme, simpler and easier creations of qubit is expected.

2.   SOLITON

A solitary wave is a kind of nonlinear waves formed by balancing nonlinearity and dispersion in nonlinear dispersive media. Since a solitary wave behaves like a particle in media it is often called a "soliton". A typical soliton equation is the Korteweg-de Vries (KdV) equation

$$u_t - 6uu_x + u_{xxx} = 0, \tag{1.1}$$

where u is the normalized amplitude of the wave. The one-soliton solution is known to have the form

$$u = -2\kappa^2 sech^2[\kappa x - 4\kappa^3 t - \delta], \tag{1.2}$$

where κ is a parameter corresponding to the wave number. Since $v = \omega/\kappa = 4\kappa^2$, soliton velocity is propotional to the soliton amplitude; The larger soliton amplitude, the faster the soliton propagation.

A typical example of the particle nature of a soliton can be seen in soliton interactions. A soliton conserves its profile after colliding with another soliton as shown in Fig. 1. There are two collision types depending on the ratio ($\lambda = \kappa_1/\kappa_2$) of the parameters ($\kappa_1 > \kappa_2$) of interacting solitons[12]: at an interaction center, solitons fuse when $\lambda > \sqrt{3}$, while they separate when $\lambda < \sqrt{3}$. Two-soliton interaction can be expressed as

$$u(x,t) = 8[\kappa_1^2 e^{2\eta_1} + \kappa_2^2 e^{2\eta_2}$$
$$+ \{(\kappa_1 + \kappa_2)^2 \left(\frac{\kappa_1 - \kappa_2}{\kappa_1 + \kappa_2}\right)^2 + (\kappa_2 - \kappa_1)^2\} e^{2(\eta_1+\eta_2)}$$
$$+ \kappa_2^2 \left(\frac{\kappa_1 - \kappa_2}{\kappa_1 + \kappa_2}\right)^2 e^{2\eta_1} e^{2(\eta_1+\eta_2)}$$
$$+ \kappa_1^2 \left(\frac{\kappa_1 - \kappa_2}{\kappa_1 + \kappa_2}\right)^2 e^{2\eta_2} e^{2(\eta_1+\eta_2)}]/\phi, \tag{1.3}$$

where

$$\phi = 1 + e^{\eta_1} + e^{\eta_2} + \left(\frac{\kappa_1 - \kappa_2}{\kappa_1 + \kappa_2}\right)^2 e^{\eta_1+\eta_2}, \tag{1.4}$$

$$\eta_n = \kappa_n x - \kappa_n^3 t. \tag{1.5}$$

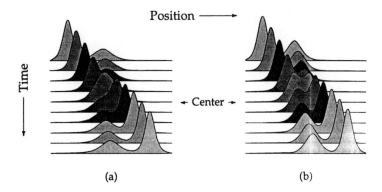

Figure 1.1 Time evolution of two interacting solitons: (a) single-peak formation and (b) double-peak formation.

3. SOLITON QUANTUM BIT

The indivisible unit of classical information is a bit, which takes one of two possible values, 0 or 1. Quantum information can be reduced to elementary units, called quantum bits or qubits. A qubit is a two-state quantum system. The soliton qubit we propose here is also not exceptional. The bases of our soliton-qubit creation scheme are two key ideas that make full use of soliton properties. First, a soliton can be regarded as a potential for a particle. A hole-shaped soliton may act as an attractive potential and capture a particle. This captured particle is then transported by soliton propagation. Second, the soliton with the captured particle can be made to interact with an empty soliton under the $\lambda < \sqrt{3}$ condition. For $\lambda < \sqrt{3}$, two interacting solitons are distinguishable from each other at all times, as shown in Fig. 1-(b). Therefore, two states for each soliton can be defined, i.e., the left state for $|L\rangle$ and the right state for $|R\rangle$. Near the interaction center, the captured particle can tunnel between the two states quantum-mechanically, *i.e.*, the particle shares these states. Therefore, the particle is in a quantum-mechanical superposition state, which makes it a qubit. The soliton qubit is a kind of flying qubits and has an EPR nature since it consists of two solitons.

4.   QUANTUM LOGIC GATES

4.1   THE ROTAION GATE

The rotation gate, that is, basic one-bit operation, can be achieved by operating the two-dimensional unitary transformation on the qubit:

$$U(\theta) = \begin{pmatrix} \cos(\theta/2) & \sin(\theta/2) \\ -\sin(\theta/2) & \cos(\theta/2) \end{pmatrix}, \tag{1.6}$$

where θ is the rotation angle. In general, this can be accomplished by irra-
diating microwave onto a quantum well for a short-time. Consider a particle
in a double-well potential as depicted in Fig. 2, where the bonding and the
antibonding states with the energy separation ΔE are formed by quantum tunneling. The particle occupies the right quantum well at an initial time. When
a mircowave of a specific frequency ($\omega = \Delta E/\hbar$) is irradiated onto the quantum well for irradiation time τ, the probability of finding the particle in each
well changes. During irradiation, the particle will go back and forth between
the wells. The rotation angle is defined as $\theta = \Delta E \tau/\hbar$. Thus, an inversion
gate, which changes from '0' to '1', and a $\pi/2$ rotation gate, which makes
superposition of '0' and '1', should be realized by adjusting the irradiation.

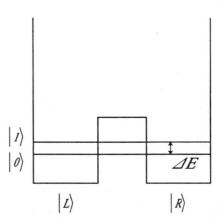

Figure 1.2   Quantum-mechnical superposition of $|L\rangle$ and $|R\rangle$ states in a double-well potential.

A distinctive feature of our scheme is that the soliton qubit itself works
in part as the natural rotation gate required for quantum computing without
irradiaton because of the probability that a particle exists in the soliton potential
changes in accordance with the interacting soliton evolutions, as shown in Fig.

2. The ground (ζ_2) and excited (ζ_1) state eigenfunctions are expressed as

$$\zeta_1 = \frac{\sqrt{2\kappa_1}e^{-\eta_1}}{\phi}(1+\gamma_2), \tag{1.7}$$

$$\zeta_2 = \frac{\sqrt{2\kappa_2}e^{-\eta_2}}{\phi}(1-\gamma_1), \tag{1.8}$$

where γ_n is the mixing parameter of the eigenfunctions given by

$$\gamma_n = e^{-2\eta_n}\frac{\kappa_1-\kappa_2}{\kappa_1+\kappa_2}. \tag{1.9}$$

In general, the eigenfunctions for bound states in the potential are related to the KdV soliton such that

$$u(x,t) = 2\frac{d}{dx}\left(\sum_{n=1}^{N}c_n\zeta_n(x)e^{-\kappa_n x}\right). \tag{1.10}$$

The n-th bound state eigenfunction $\zeta_n(x)$ is obtained by solving

$$\zeta_n(x) + \sum_{m=1}^{N}\frac{c_n c_m e^{-(\kappa_n+\kappa_m)x}}{\kappa_n+\kappa_m}\zeta_m(x) - c_n e^{-k_n x} = 0. \tag{1.11}$$

4.2   THE QUANTUM CONTROLLED-NOT GATE

The quantum controlled-NOT logic gate, which is indispensable for quantum computing, performs the following action on the two qubits: with the first (controller) qubit in the state $|x\rangle$ and the second (target qubit) in the state $|y\rangle$ the operation shall leave the target qubit unchanged if $x=0$, while flipping the target bit between 0 and 1 when $x=1$. In matrix notation, this is expressed as

$$U_{CNOT} = \begin{pmatrix} 1 & 0 \\ 0 & \sigma_x \end{pmatrix}. \tag{1.12}$$

In soliton qubits, the quantum controlled-NOT logic gate can also be designed when there are four solitons and mutual interactions between soliton qubits (one for the control bit and another for the target bit) are introduced to acquire quantum-mechanical entanglements. An energy diagram for non-interacting qubits is illustrated in Fig. 4-(a). For the controlled-NOT operation, quantum transition only between $|1\rangle|0\rangle$ and $|1\rangle|1\rangle$ must be induced. To do so, let us introduce Coulomb interactions between two soliton qubits, i.e., the captured particle in the interacting soliton potential has a charge. Suppose that the particle is an electron and two soliton qubits, four solitons, interact at a certain point for simplicity. In this case, four solitons merge into a hole-shaped

Position ⟶

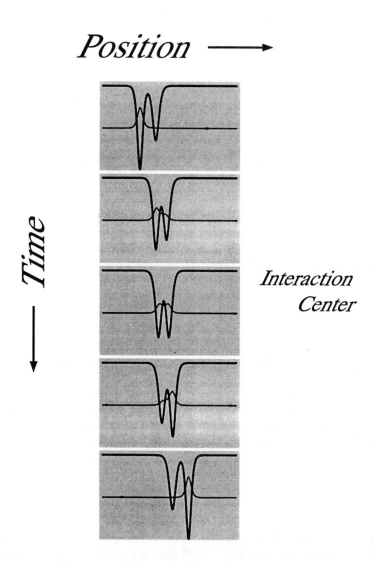

Time

Interaction
Center

Figure 1.3 Soliton qubit. Thick solid lines show soliton potential to a particle and thin solid lines show the probability that the particle exists.

potential with two elcteons. The situation resembles a helium atom which binds two electrons. In this case, the combined energy state for different obitals separates into spin-singlet and spin-triplet states due to an exchange interaction. The energy level is shifted as

$$\Delta E = C \pm A, \qquad (1.13)$$

where

$$C = \int \psi_T^2(1)\psi_C^2(2)\frac{e^2}{r}d\tau_1 d\tau_2, \qquad (1.14)$$

$$A = \int \psi_T(1)\psi_C(2)\psi_T(2)\psi_C(1)\frac{e^2}{r}d\tau_1 d\tau_2. \qquad (1.15)$$

The energy diagram is changed to Fig. 4-(b). Therefore, quantum transition between $|1\rangle|0\rangle$ and $|1\rangle|1\rangle$ can be caused selectively.

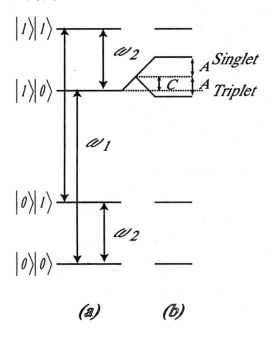

Figure 1.4   Energy levels of two soliton qubits without and with the coupling induced by the Coulomb interaction between captured charged particles.

5.   PROMISING SYSTEMS

The most promising candidate for our scheme is a surface acoustic soliton in a thin film deposited on a semi-infinite substrate [13]. Since the surface acoustic waves propagate along the solid surface concentrating their energy within about one wavelength from the surface, the lattice anharmonicity is enhanced. This anharmonicity could be balanced with the dispersion effects due to the intrinsic or extrinsic origins. Sakuma et al. predicted the formation of surface acoustic solitons. In particular, they found that KdV-type surface acoustic solitons are formed in a thin film deposited on the substrate which can support the Love wave [13]. Later, they developed a theory of electron transport by soliton propagation. Under certain circumstances, the suface acoustic solitons act as an attractive potential for two-dimensional conduction electrons in a thin film to trap and transport them. Finally they proposed a new mechanism of two-dimensional electron transport caused by a surface acoustic soliton of the KdV type in thin film. This may be called two-dimensional "quasi-superconductivity" owing to the stability of the charge-carrying suface acoustic soliton [14]. Surface acoustic solitons have not been observed yet, but expriments are now in progress at some institutes.

Other possible candidates are solitonic charge transport in a quasi-one-dimensional organic system [15] and polaritons trapped by a soliton near excitonic resonance [16].

6.   SUMMARY

We have proposed a novel scheme in which a *self-formed qubit* is created by using a quantum particle captured within a interacting soliton potential. The discussion analogized quantum logic gate for quantum computing with semiconductor quantum well systems.

The greatest technical advantage of this scheme is that states can be prepared using the feature that solitons are spontaneously formed from the initially given shapes. Unlike previous schemes for creating qubits, ours require no special techniques, such as nanofabrication, for initial state preparation. Furthermore, any quantum operation can be designed by controlling the initial conditions.

In this paper, we have developed only a general soliton qubit theory. So we cannot discuss the decoherence time of qubits because it strongly depends on the specific system. However, we believe that it may be possible to maintain a superposition state for a long time due to soliton stability.

Although the soliton is a classical object and does not seem to be related to quantum mechanics, soliton physics was developed based on the principles of quantum mechanics. This can be seen in the inverse scattering method where soliton evolutions are governed by a differential equation similar to the Schrödinger equation in quantum mechanics. The proposed qubit should thus also be of considerable interest for fundamental studies of the intimate relationship between quantum mechanics and solitons, in addition to its quantum computing potential.

Acknowledgments

We would like to thank Drs. Hirotaka Tanaka, Yoshiaki Sekine, and Shiro Saito, Ken -i. Matsuda for their stimulating discussions. We also would like to thank Ms. Akiko Nishiyama for her kind assistance.

References

[1] A. Barenco, C. H. Bennett, R. Cleve, D. P. DiVincenzo, N. Margolus, P. Shor, T. Sleator, J. A. Smolin, and H. Weinfurter, Phys. Rev. A52, 3457 (1995).

[2] Q. A. Turchette, C. J. Hood, W. Lange, H. Mabuchi, and H. J. Kimble, Phys. Rev. Lett. 75 4710 (1995).

[3] C. Monroe, D. M. Meekhof, B. E. King, W. M. Itano, and D. J. Wineland, Phys. Rev. Lett. 75, 4714 (1995).

[4] N. Gelshenfeld and I. Chuang, Science 275 350 (1997).

[5] B. Kane, Nature, 393, 133 (1998).

[6] D. Loss and D. DiVincenzo, Phys. Rev. A 57, 120 (1998).

[7] A. Shnirman, G. Schön, and Z. Hermon, Phys. Rev. Lett. 79, 2371 (1997).

[8] D. V. Averin, Solid State Commun. 105, 659 (1998).

[9] L. B. Ioffe, V. B. Geshkenbein, M. V. Feigel'man, A. L. Fauchere, and G. Blatter, Nature 398 679 (1999).

[10] J. E. Mooij, T. P. Orlando, L. Levitov, L. Tian, C. H. van der Wal, S. Lloyd, Sicence 285 1036 (1999).

[11] Y. Nakamura, Yu. A. Pashkin, J. S. Tsai, Nature 398, 786 (1999).

[12] N. Hatakenaka, Phys. Rev. E 48 4033 (1993).

[13] T. Sakuma and Y. Kawanami, Phys. Rev. B 29, 869 (1984).

[14] T. Sakuma and N. Nishiguchi, Jpn. J. Appl. Phys. 30 Supplement 30-1 137-139 (1991).

[15] J. S. Zmuidzinas, Phys. Rev. B17 3919-3925 (1978).

[16] I. B. Talanina and M. A. Collins, Phys. Rev. B49, 1517 (1994).

INDEX